MARITIME-PORT TECHNOLOGY AND DEVELOPMENT

PROCEEDINGS OF THE CONFERENCE ON MARITIME-PORT TECHNOLOGY (MTEC 2014), TRONDHEIM, NORWAY, 27–29 OCTOBER 2014

# Maritime-Port Technology and Development

*Editors*

Sören Ehlers & Bjørn Egil Asbjørnslett
*Norwegian University of Science and Technology (NTNU), Trondheim, Norway*

Ørnulf Jan Rødseth & Tor Einar Berg
*Norwegian Marine Technology Research Institute (MARINTEK), Trondheim, Norway*

CRC Press is an imprint of the
Taylor & Francis Group, an **informa** business
A BALKEMA BOOK

*CRC Press/Balkema is an imprint of the Taylor & Francis Group, an informa business*



Typeset by V Publishing Solutions Pvt Ltd., Chennai, India
Printed and bound in Great Britain by CPI Group (UK) Ltd, Croydon, CR0 4YY



Published by: CRC Press/Balkema
P.O. Box 11320, 2301 EH Leiden, The Netherlands
e-mail: Pub.NL@taylorandfrancis.com
www.crcpress.com – www.taylorandfrancis.com

ISBN: 978-1-138-02726-8 (Hbk + CD-ROM)
ISBN: 978-1-315-73162-9 (eBook PDF)

*Maritime-Port Technology and Development – Ehlers et al. (Eds)*
*© 2015 Taylor & Francis Group, London, ISBN 978-1-138-02726-8*

# Table of contents

*Maritime-Port Technology and Development – Ehlers et al. (Eds)*
*© 2015 Taylor & Francis Group, London, ISBN 978-1-138-02726-8*

# Foreword

We are pleased to host the International Maritime and Port Technology and Development Conference in Trondheim, Norway, in 2014. This conference has served in the past as an important and internationally recognised platform to disseminate the latest research results in the field of Maritime and Port Technology and Development with strong contributions in the field of efficiency and reliability of seaborne transport and operations.

The preparations of this conference and proceedings would not have been possible without the support of the numerous reviewers and the efforts of Martin Bergsröm, Jitapriya Das, Boris Erceg, Sandro Erceg, Aleksandar-Sasa Milakovic and Drazen Polic in formatting the manuscripts where needed. Furthermore, we would like to thank the steering committee for promoting and supporting the conference. The financial support of our sponsors is greatly acknowledged. Finally, we are wishing all participants a fruitful, stimulating and professionally rewarding stat and the Marine Technology Center in Trondheim, Norway.

*Sören Ehlers*
*Ørnulf Jan Rødseth*
*Tor Einar Berg*
*Bjørn Egil Asbjørnslett*

*Maritime-Port Technology and Development – Ehlers et al. (Eds)*
*© 2015 Taylor & Francis Group, London, ISBN 978-1-138-02726-8*

# Congestion and truck service time minimization in a container terminal

D. Ambrosino
*DIEC-Department of Economics and Business Studies, University of Genova, Italy*

C. Caballini
*DIBRIS-Department of Informatics, BioEngineering, Robotics and Systems Engineering, University of Genova, Italy*

ABSTRACT: The objective of this paper is to address the minimization of truck service times at container terminals while respecting a certain level of congestion. Truck congestion at terminal gates is a major concern for container terminals, especially considering the increasing volumes of goods they have to manage. Truck arrivals, if not properly managed, can result in long queues of trucks, decreasing their service level and can lower the terminal productivity, affecting all the other areas of the terminal. The terminal road cycle is described through some mathematical relations implemented in a spreadsheet; these relations are the basis of a decision support system that, for each truck having executed the check-in, decides if it should be allowed to enter the terminal and, if yes, which service level it will be given. Based on a big container terminal located in Northern Mediterranean, the proposed approach has been successfully tested on different scenarios, with different level of terminal congestions, yard filling and trucks arrivals. The results obtained have shown that the tool is able to effectively reduce congestion inside the terminal by indicating if a truck can enter the gate in order to be served within a predetermined service time, or if it has to stop outside due to the high congestion level inside the terminal. Moreover, a series of KPIs have been defined and analyzed to assess various scenarios which differ in the number of truck arrivals and in the initial state of the terminal; this evaluation also enables to size the number of equipment needed by the container terminal in order to properly carried out all its operations.

## 1 INTRODUCTION

The increasing volume of goods passing through ports, and more general logistic nodes, imposes to carefully manage trucks at port gates and inside terminal areas in order to guarantee short and certain service times to road transport and, at the same time, to not affect the productivity of all the other activities inside the terminal (ships, trains, etc.).

Big container terminals, managing up to and beyond one million TEUs per year, can have to handle 1500–2000 trucks per day. Unluckily, this flow of trucks, is not uniformly distributed along check-in hours unless particular systems are implemented. During a day, high flow peaks usually occur, risking to cause high congestion and delays both to trucks and to the other transport means of the terminal. Truck appointment systems allow terminals to know in advance the trucks that will arrive at the terminal gate in each time window of the day. However, these functionalities may cause other organizational issues as stressed in (Sharif et al., 2011).

Hence, if not properly managed, the truck cycle in a container terminal can strongly affect overall terminal performances. For this reason, advanced methodologies should be found in order to guarantee a certain level of service to trucks while both not decreasing terminal throughput and productivity, and not generating negative externalities (environmental pollution, social congestion, noise, etc.).

Congestion issues at container terminals have been addressed by several authors in the past few years. Chena et al. (2013) tried to reduce gate congestion by applying the Vessel Dependent Time Windows (VDTWs) method to control truck arrivals. Han et al. (2008) developed a model and a tabu search heuristic to decrease traffic congestion of prime movers inside a container terminal. Zehendner & Feillet (2014) proposed to reduce truck turnaround and, consequently, terminal congestion, by developing a mixed integer linear programming model devoted to size a Truck Appointment System (TAS). They simultaneously determined the number of truck appointments to offer and allocates straddle carriers to different transport modes. Also Zhang et al. (2013) provided an optimization model for truck appointments, by developing a BCMP queuing network to describe the queuing process of trucks in the terminal.

To solve the model, a method based on Genetic Algorithm (GA) and Point wise Stationary Fluid Flow Approximation (PSFFA) was designed.

Zhaoa W. & Goodchildb (2010) utilized truck arrival information to reduce terminal rehandles so improving terminal operations, while Chena et al., 2011 proposed to optimizing truck arrivals at ports by utilizing time-varying tolls.

The goal of the present paper is to analyze import service times of trucks in a container terminal with the constraint of respecting a maximum level of congestion both at the gate area and inside the terminal. More specifically, the main aim is to control the import road cycle without modifying the Terminal Operating System (TOS) and the Gate Operating System (GOS), its yard layout and without introducing a truck appointment system.

The paper is organized as follows. Section 2 describes the problem under study, while Section 3 presents the mathematical relations used for describing the system. The case study is depicted in Section 4 where different scenarios are discussed. Finally, Section 5 outlines some conclusions and future research.

## 2 DESCRIPTION OF THE PROBLEM

Let us describe more in detail the problem under consideration. The terminal layout considered in this study presents an import area organized in blocks which are operated by Rubber Tired Gantry-RTG cranes. Moreover, adjacent to each block area, there is a lane, called "truck lane" in which trucks wait in order to receive their containers by RTG cranes.

A generic truck import process in a container terminal is presented in Figure 1. The process outlined follows these steps:

1. The truck arrives at the terminal check-in and executes the required activities. Usually the check-in process is performed outside the terminal area in order to allow truck hauliers to park their means of transport in a dedicated zone and carry out the documentary process. At this stage, the truck is informed about the position in the yard of the import container it has to pick up (block and exact position in the corresponding truck lane).
2. The truck moves to the gate-in area.
3. Usually, the truck may have to queue in order to perform the gate-in process.
4. The gate-in procedure is carried out.
5. Having passed the gate, the truck travels to the indicated block area in which it is stored the container it has to picked up.
6. When arriving to the truck lane of the indicated block, the truck halts in the specific truck lane slot in correspondence to its import container waiting for the RTG crane to deliver it.
7. The RTG crane picks up the specific container (eventual re-handles are made) and delivers it on the back of the truck.
8. The truck travels from the truck lane to the gate-out zone.
9. The truck may have to queue for a while in the gate-out area.
10. Gate-out is performed and the truck is free to leave the terminal towards its final destination.

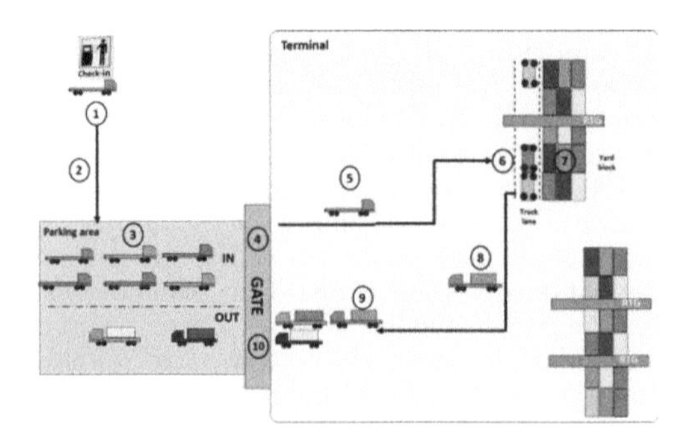

Figure 1. Process framework of the road cycle in a generic container terminal.

The goal of this process framework is to monitor and control critical points of the terminal where possible queues can be formed, by controlling trucks accesses to the terminal and/or changing terminal equipment productivity.

In particular, three activities can be considered crucial, or rather may cause queues and congestion: referring to Figure 1 they are points 4 (the gate in), 6–7 (the truck lane) and 10 (the gate out).

A truck, after having executed the check-in, has to execute the gate-in process for entering the terminal (4). A truck may pass a certain period of time in the queue before carrying out the gate-in activities. The time depends on the number of gate-in lanes activated and their productivity.

A truck reaching the truck lane (6) may spend time dependent on a set of factors, for instance the level of occupancy of the yard, the number and productivity of the RTGs serving the specific yard block. Each truck, in order to be served by the RTG, must park in the truck lane slot adjacent to the slot of the yard block where the container is stacked. If the truck lane slot is occupied, the truck should wait outside the truck lane, which means increasing the terminal congestion. Generally, the specific truck lane slot has to be free when the truck arrives, otherwise it will have to park outside the truck lane so contributing to create viability problems in the terminal.

The waiting time of the truck in the truck lane and its service time, strongly depend on the number and productivity of the RTGs serving a specific truck lane, and on the occupancy level of each yard block related to the specific truck lane.

Besides, it must be noted that in this paper RTG cranes are considered as yard handling equipment but the approach can be easily applied to terminals where other types of handling means, such as straddle carriers or reach stackers, are used.

The last crucial point is the gate-out (10). Before a truck performs the gate-out activities it might spend same time in queue. As for the gate-in queue, this time depends on both the number of gate-out lanes activated and their productivity.

In general, the average speed of a truck inside the terminal depends on the congestion level.

Finally, it has to be specified that, in the present work, each truck is performing a single operation in the terminal, that is the picking up of an import container. Other typologies of movements (two import containers, export cycle and double cycle) are not dealt in this paper, even if the proposed approach can be extended to cover such situations.

## 3 MATHEMATICAL RELATIONS

In this section the activities of the truck import cycle of a container terminal are described in more details through some mathematical relations. More specifically, these relations are the basis of a decision support system that, for each truck having executed the check-in, decides if it should be allowed to enter the terminal and, if yes, which service level it will be given. The decision is based on analysis of the degree of congestion (or occupancy) of the terminal. The crucial activities (e.g. 4, 6–7, 10) are analyzed in the following.

### 3.1 *Input data*

Input data that has to be considered are the following:

- Layout and structure of the terminal under analysis that affect the following parameters;
  - time needed by a truck to cover the distance from the check-in to the gate-in area,
  - time needed by a truck to cover the distance from the gate-in to any terminal yard block,
  - time needed by a truck to cover the distance from any terminal yard block to the gate-out area.
- Number of terminal equipment, for each typology, and their productivity;
  - gate-in lanes,
  - gate-in operators,
  - RTG,
  - gate-out lanes,
  - gate-out operators,
  - technological level (i.e. different levels of automation).
- Arrival pattern of trucks at the terminal check-in;
  - exact time of arrival,
  - yard slot in which the import container for each truck is stored.
- Service Level Agreement (SLA) of the terminal, which corresponds to the service time agreed by the terminal with the trucking companies. More specifically this time is the acceptable service time that the terminal must guarantee to a generic truck to perform the required operations of the road terminal cycle.

In this first attempt to model the road cycle in a maritime terminal, deterministic truck arrival times are considered.

The following notation is used:

- $T = \{1,2, \ldots, t\}$ is the set of time periods composing the considered time horizon;
- $V = \{1,2, \ldots, v\}$ is the set of trucks approaching the terminal during the considered time horizon;
- $L = \{1,2, \ldots, l\}$ is the set of truck lanes adjacent to the import area blocks in which trucks wait for the RTG cranes to deliver their containers;
- $S_l$ is the set of slots composing the truck lane $l$, $\forall l \in L$;
- $s_v$ is the slot corresponding to the block in the yard where the import container to be loaded by truck $v$ is stored (see Fig. 2), $\forall v \in V$;
- $r_s$ is the average degree of occupancy of the yard block adjacent to slot $s$ of truck lane $l$, $\forall s \in S_l$; $\forall l \in L$;

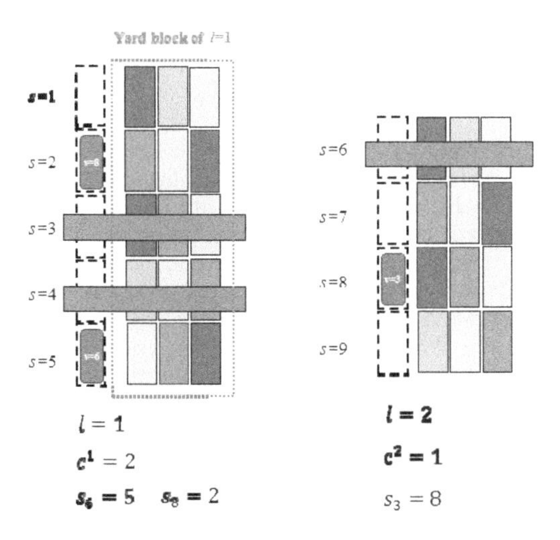


Figure 2. Truck lanes and yard slots framework.

- $d_s^I$ is the distance of slot $s$ from the gate-in, $\forall s \in S_l; \forall l \in L$;
- $d_s^O$ is the distance of slot $s$ from the gate-out, $\forall s \in S_l; \forall l \in L$.

Figure 2 helps clarifying the notation used. It can be noticed that the yard block on the left side is served by two RTG cranes ($c^1 = 2$) and is related to a truck lane ($l = 1$) composed of 5 slots. Moreover, at the time period considered, two trucks are occupying the truck lane: truck number 8 ($v = 8$) is parked in slot 2—corresponding to the slot position nearest to its container storage position in the yard block of truck lane $l$—while truck number 6 is in slot 5.

### 3.2 *Relation 1: Gate-in queue*

The queue at the gate–in is an important factor to take into account and it is modelled by equation (1). The variable, $q_t^I$, represents the number of trucks queuing at gate-in at time period $t$, and it is function of:

- $q_{t-1}^I$, the number of trucks queuing at gate-in at time period $t$–1;
- $n_t^{QI}$, the number of trucks which have arrived at the gate-in queue at time period $t$ and have executed the check-in at time period $t - \tau^{CI}$, where $\tau^{CI}$ is the time needed by a truck to cover the distance between the check-in building and the gate;
- $n_t^{EI}$, the number of trucks that leaves the gate-in queue at time period $t$ and are ready to execute the gate-in. It is given by the product of:
  - $g_t^I$, the number of gate-in lanes activated at time period $t$;
  - $\rho^I$, the productivity of a generic gate-in lane, which is supposed to be constant.

$$q_t^I = q_{t-1}^I + n_t^{QI} - n_t^{EI} \tag{1}$$

$$q_t^I \geq 0 \tag{2}$$

Note that the number of trucks exiting the gate-in queue has to be not greater than the number of vehicles already present in the queue plus the ones entering it. This is assured by equation (2).

Having spent the time needed in the queue, the truck carries out the gate-in activity which requires a time equal to $1/\rho^I$ where $\rho^I$ is the gate-in productivity. When the gate-in has been performed, the truck can enter the terminal and moves to its truck lane adjacent to its destination yard block.

### 3.3 *Relation 2: Truck lane queue*

The queue at a generic truck lane $l$ is modelled by equation (3). The truck lane queue, $q_t^l$ represents the number of trucks queuing at truck lane $l$, and is a function of:

- $q_{t-1}^l$, the number of trucks queuing at truck lane $l$ at time period $t$–1;
- $n_t^{Ql}$, the number of trucks entering the truck lane queue at time period $t$ and that has finished the gate-in at time period $t - \tau^{Il}$, where $\tau^{Il}$ represents the time periods needed by the truck to cover the distance between the gate-in and truck lane $l$;
- $n_t^{El}$, the number of trucks exiting the truck lane queue at time $t$ and is a function of $\rho_t^l$, i.e. the productivity of lane $l$ at time $t$ that depends on:
  - $C_t^l$, the number of RTG cranes serving truck lane $l$ at time period $t$;
  - $\rho_c^l$, the productivity of crane $c$ working at lane $l$;
  - $r_t^l$, the degree of occupancy of the yard block adjacent to truck lane $l$ at time period $t$.

$$q_t^l = q_{t-1}^l + n_t^{Ql} - n_t^{El} \quad \forall l \in L \tag{3}$$

$$q_t^l \geq 0 \quad \forall l \in L \tag{4}$$

$$q_t^l \leq |S_l| \quad \forall l \in L \tag{5}$$

Analogously to Relation (2), it is highlighted that the number of trucks exiting truck lane $l$ queue cannot be greater than the number of vehicles already present in the queue plus the ones entering the queue. This is assured by equation (4). Equation (5) guarantees that no trucks wait outside the truck lanes' slots, avoiding to create circulation problems inside the terminal.

Having spent the time needed in the truck lane queue, trucks are served in a service time that is function of $\rho_c^l$ and $r_t^l$. Then the truck leaves the truck lane and moves to the gate-out.

### 3.4 *Relation 3: Gate-out queue*

The gate-out queue, $q_t^O$ is the number of trucks queuing at gate-out at time period $t$; it is modeled by equation (6) and is function of:

- $q_{t-1}^O$, the number of trucks queuing at gate-out at time period $t$–1;
- $n_t^{QO}$, the number of trucks which have left their truck lane and enter the gate-out queue at time period $t$, i.e. $\Sigma_l n_{t-\tau^{lO}}$ where $\tau^{lO}$ represents the time periods needed by a truck to cover the distance between the truck lane $l$ and the gate building;
- $n_t^{EO}$, the number of trucks that leaves the gate-out queue at time period $t$ and are ready to execute the gate-out. This is a function of:
  - $g_t^O$, the number of gate-out lanes activated at time period $t$;
  - $\rho^O$, the productivity of a generic gate-out lane.

$$q_t^O = q_{t-1}^O + n_t^{QO} - n_t^{EO} \quad (6)$$

$$q_t^O \geq 0 \quad (7)$$

Having spent the time needed in the queue, the truck carries out the gate-out activity which requires a time equal to $1/\rho^O$ where $\rho^O$ is the productivity of a generic gate-out lane.

When the gate-out has been performed, the truck can leave the terminal.

### 3.5 *Timing of a generic truck from check-in to gate-out*

For a specific vehicle (truck) $\bar{v}$, starting from the check-in up to the gate-out, a certain number of times (ten) can be calculated, as shown in Figure 3.

1. $t_{\bar{v}}^C$ is the period of time of check-in acceptance for vehicle $\bar{v}$.
2. $t_{\bar{v}}^{QI}$ is the period of time of arrival of vehicle $\bar{v}$ at the gate-in queue (Equation 8).

$$t_{\bar{v}}^{QI} = t_{\bar{v}}^C + \tau^{CI} \quad (8)$$

3. $t_{\bar{v}}^{EI}$ is the period of time of entering the gate-in by vehicle $\bar{v}$ (Equation 9).

$$t_{\bar{v}}^{EI} = t_{\bar{v}}^{QI} + \tau^{QI} \quad (9)$$

where $\tau^{QI}$ represents the time periods spent by a truck in the gate-in queue (Equation 10).

$$\tau^{QI} = \frac{q_{t_{\bar{v}}^{QI}}^I}{g_{t_{\bar{v}}^{QI}}^I \rho^I} \quad (10)$$

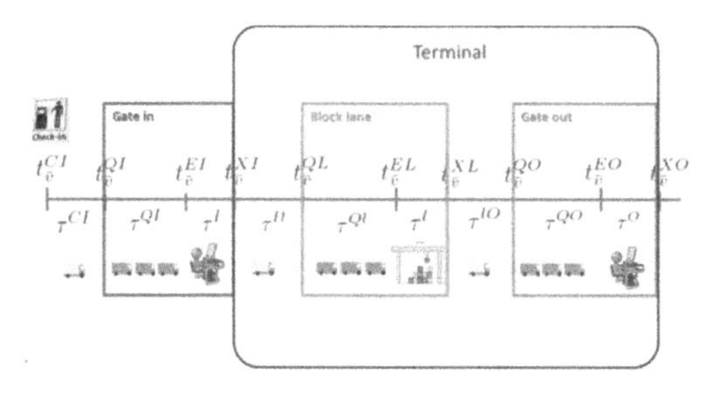


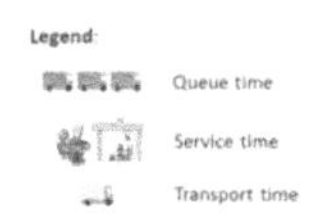


Figure 3. Timing of a truck $\bar{v}$ from check-in to gate-out.

Note that the product $g_{t_{\bar{v}}^{QI}}^I \rho^I$ represents the number of vehicles exiting the gate-in queue at time $t_{\bar{v}}^{QI}$ (i.e. $n_{t_{\bar{v}}^{QI}}^{EI}$).

4. $t_{\bar{v}}^{XI}$ is the period of time of exiting the gate-in by vehicle $\bar{v}$ (Equation 11).

$$t_{\bar{v}}^{XI} = t_{\bar{v}}^{EI} + \tau^I \quad (11)$$

where $\tau^I$ represents the time of periods needed by a truck to pass the gate-in and it is given by $1/\rho^I$.

5. $t_{\bar{v}}^{Ql}$ is the period of time of arrival at the truck lane by vehicle $\bar{v}$ that will start its queue in order to be served (Equation 12).

$$t_{\bar{v}}^{Ql} = t_{\bar{v}}^{XI} + \tau^{Il} \quad (12)$$

where $\tau^{Il}$ represents the time periods spent by a truck to get to the truck lane queue (Equation 13).

$$\tau^{Il} = \frac{d_{s_{\bar{v}}}^I}{\mu_{t_{\bar{v}}^{XI}}^{IL}} \quad (13)$$

where:

- $d_{s_{\bar{v}}}^I$ is the distance of slot assigned to vehicle $\bar{v}$ from the gate;
- $\mu_{t_{\bar{v}}^{XI}}^{Il}$ is the speed assumed by the trucks that move from the terminal gate-in to the import yard at time period $t_{\bar{v}}^{XI}$. This speed is calculated taking into account the maximum speed allowed in the terminal corrected by a proper terminal congestion factor.

6. $t_{\bar{v}}^{El}$ is the time period corresponding to the starting of vehicle $\bar{v}$ service at truck lane $l$;

$$t_{\bar{v}}^{El} = t_{\bar{v}}^{Ql} + \tau^{Ql} \quad (14)$$

where $\tau^{Ql}$ represents the periods of time spent by a truck in the truck lane queue (Equation 15).

$$\tau^{Ql} = \frac{q_{t_{\bar{v}}^{Ql}}^l}{n_{t_{\bar{v}}^{Ql}}^{El}} \quad (15)$$

7. $t_{\bar{v}}^{Xl}$ is the time period of exiting the truck lane $l$ by vehicle $\bar{v}$, once it has been served (Equation 16).

$$t_{\bar{v}}^{Xl} = t_{\bar{v}}^{El} + \tau^l \quad (16)$$

where $\tau^l$ is the time spent by a truck to be served by the RTG crane in order to receive the import container (Equation 17).

$$\tau^l = \beta^l \frac{1}{\rho_c^l} \quad (17)$$

where $\beta^l$ is a correction factor that takes into account the level of occupancy of the yard block corresponding to truck lane $l$.

8. $t_{\bar{v}}^{QO}$ is the time of entering the gate-out queue by vehicle $\bar{v}$ (Equation 18).

$$t_{\bar{v}}^{QO} = t_{\bar{v}}^{XI} + \tau^{lO} \tag{18}$$

where $\tau^{lO}$ represents the periods of time needed by a truck, once picked up its container, to cover the distance between the truck lane $l$ and the gate-out queue (Equation 19).

$$\tau^{lO} = \frac{d_{s_{\bar{v}}}^{O}}{\mu_{t_{\bar{v}}^{XO}}^{XI}} \tag{19}$$

where:

- $d_{s_{\bar{v}}}^{O}$ is the distance of slot assigned to vehicle $\bar{v}$ from the gate;
- $\mu_{t_{\bar{v}}^{XI}}^{XI}$ is the speed assumed by the trucks that move from the import yard block at time period $t_{\bar{v}}^{XI}$ to the terminal gate for the gate-out. This speed is calculated taking into account the maximum speed allowed in the terminal corrected by a proper terminal congestion factor.

9. $t_{\bar{v}}^{EO}$ is the time of entering the gate-out by vehicle $\bar{v}$ (Equation 20).

$$t_{\bar{v}}^{EO} = t_{\bar{v}}^{QO} + \tau^{QO} \tag{20}$$

where $\tau^{QO}$ represents the periods of time spent by a truck in the gate-out queue (Equation 21).

$$\tau^{QO} = \frac{q_{t_{\bar{v}}^{O}}^{QO}}{g_{t_{\bar{v}}^{QO}}^{O} \rho^{O}} \tag{21}$$

where $g_{t_{\bar{v}}^{QO}}^{O} \rho^{O}$ is the number of vehicles exiting the gate-out at time $t_{\bar{v}}^{QO}$ (i.e. $n_{t_{\bar{v}}^{QO}}^{EO}$).

10. $t_{\bar{v}}^{XO}$ is the time of exiting the gate-out by vehicle $\bar{v}$ (Equation 22).

$$t_{\bar{v}}^{XO} = t_{\bar{v}}^{EO} + \tau^{O} \tag{22}$$

where $\tau^{O}$ represents the periods of time needed by a truck to pass the gate-out stage and it is defined by $1/\rho^{O}$ in which $\rho^{O}$ is the productivity of a generic gate-out lane.

## 4 THE DSS AND SCENARIOS ANALYSIS

The mathematical relations depicted in Section 3 have been implemented in an Excel spreadsheet, used both as a DDS and as a simulation tool for analyzing a set of scenarios. Based upon one hour time horizon split into 60 time periods of one minute and given different terminal initial conditions, various trucks arriving processes are considered.

As initial conditions, we consider different queue lengths and different yard storage situations. Moreover, truck arrivals can assume values equal to "low", "medium" and "high" for each time horizon analyzed (i.e. per one hour).

Table 1 provides the features of the six scenarios. As can be seen, the arrival process can assume the following three states:

- low, which means 40 trucks arriving per hour;
- medium, i.e. 100 trucks arriving per hour;
- high, i.e. 180 trucks arriving per hour.

These different arrival processes usually alternate during a typical day of a terminal (opening hours of check-in usually range from early in the morning till evening). Moreover, we consider the flow of vehicle arrivals to be uniformly distributed in the considered time horizon. Different sequences of arrival affect the terminal queues and the service time of trucks in a different way.

The initial state of the terminal can assume two conditions, low or high, as better explained in Table 2.

A certain number of assumptions has been made:

- the number of operative lanes at the gate-in and at the gate-out are 5 and 3, respectively, corresponding to a productivity of 2.5 and 3 vehicles per minute;
- the capacity of each truck lane is 5 trucks: each truck occupies four 20′ slots and between two adjacent trucks a slot must be set free for maneuvering;
- each RTG crane has a productivity of 20 moves per hour when the yard has a medium degree

Table 1. Scenarios framework.

| Scenario | Arrival process | | |
|---|---|---|---|
| Initial state | Low (L) | Medium (M) | High (H) |
| Low (L) | LL | LM | LH |
| High (H) | HL | HM | HH |

Table 2. Initial states of the terminal.

| | Queue length (number of trucks) | | | |
|---|---|---|---|---|
| | Gate-in | Truck lane | Gate-out | Yard block |
| Low | 3 | 2 | 2 | 40% occupancy |
| High | 10 | 5 | 6 | 70% occupancy |

of occupancy, while when considering a high occupancy level of the yard this productivity is reduced to 12 moves per hour since same reshuffles must been executed;

- at most 2 RTG cranes can be used for each truck lane.
- each truck approaching the check-in in a certain time-slot cannot be influenced by trucks arriving in the future.

Last assumption is valid because we deal with single-import cycle, so all trucks leaving the check-in stage has to move directly to import areas. The one moving first of course will arrive before those that will move later. This is generally not true in case of mixed export-import cycles, where truck A that move to export first, and then to import, could be influenced by presence of truck B, arrived later, but directly moving to import slots.

### 4.1 *Description of the Decision Support System (DSS)*

In this sub-section, the DSS implemented by using the Excel spreadsheet for computing the service time of each truck entering the terminal is described. In Table 3, a screenshot of this Excel tool is provided.

The goal of this instrument is to suggest to a generic truck if to enter the terminal immediately ("GO!") or stop outside ("STOP!") due to a congested situation in terms of high queues and service time. The results "STOP!" or "GO!" depend on the Service Level Agreement.

Table 3. Timing of a truck.

| Row # | **INPUT DATA** | |
|---|---|---|
| 1 | Arrival time (t.p.) | 60 |
| 2 | Import truck lane number | 1 |
| | **TRUCK TIMING** | |
| 3 | arrive at the check-in (t.p.) | 60 |
| 4 | arrive at the gate-in queue (t.p.) | 62 |
| 5 | ***time spent at the gate-in queue (t.p.#)*** | ***0*** |
| 6 | entering the gate-in (t.p.) | 62 |
| 7 | exiting the gate-in (t.p.) | 63 |
| 8 | arrive at the truck lane (t.p.) | 65 |
| 9 | ***time spent at the truck lane (t. p.#)*** | ***20*** |
| 10 | exiting the truck lane (t.p.) | 85 |
| 11 | arrive at the gate-out queue (t.p.) | 87 |
| 12 | ***time spent at the gate-out queue (t.p.#)*** | ***0*** |
| 13 | entering the gate out (t.p.) | 87 |
| 14 | exiting the gate-out (t.p.) | 88 |
| | **RESULTS** | |
| 15 | Service Time (t.p. #) | 28 |
| 16 | Stop or Go? | STOP! |

More specifically, when a truck approaches the check-in, its arrival time and its container slot triggered the truck service time calculation. Starting from the terminal picture concerning all its queues and all the vehicles carrying out import cycle operations, the timing of the truck (Table 3) is determined by applying the mathematical equations (1)–(22).

The difference between the "exiting gate out" time and "arrive at check-in" time provides the service time of the truck under consideration.

In the same window it is possible to visualize if the truck can enter or it has to stop outside the terminal. In the case shown in Table 3, the truck service time is equal to 28 minutes, so, being the SLA equal to 25 minutes, the result is "STOP". Note that, in Table 3, "t.p." represents the time period, while "t.p.#" indicates the number of time periods.

Times reported in rows 5, 9 and 12 of Table 3 derive from the calculations obtained by the Excel Spreadsheets presented in Figures 4, 5 and 6, which represent the core of the support system, and permit to know for each period of time, how many trucks are present in each queue (gate-in, truck lanes and gate-out).

At each time period, the three queues are updated. Referring to Figure 4, the number of vehicles arriving at the check-in are registered and, taking into account the time needed to perform the check-in operation, they are inserted in the gate-in queue. The last column of Figure 4 indicates the truck waiting time in the gate-in queue and is computed by considering the number of vehicles queuing and the productivity at the gate-in (equation (10)).

The reasoning is analogous both for the truck lane and gate-out queues.

**GATE-IN QUEUE**

| time period | # vehicles arrived at check-in | # vehicles arriving at gate-in queue | # vehicles queueing at gate-in | waiting time at gate-in queue |
|---|---|---|---|---|
| 1 | 2 | 0 | 10 | 4 |
| 2 | 2 | 0 | 8 | 3 |
| 3 | 2 | 2 | 7 | 3 |
| 4 | 2 | 2 | 7 | 3 |
| 5 | 2 | 2 | 6 | 3 |
| 6 | 2 | 2 | 6 | 3 |
| 7 | 2 | 2 | 5 | 2 |
| 8 | 2 | 2 | 5 | 2 |
| 9 | 2 | 2 | 4 | 2 |
| 10 | 2 | 2 | 4 | 2 |
| 11 | 2 | 2 | 3 | 2 |
| 12 | 2 | 2 | 3 | 0 |
| 13 | 2 | 2 | 2 | 0 |
| 14 | 2 | 2 | 2 | 0 |

Figure 4. The gate-in queue in the scenario HM.

**TRUCK LANE 1 QUEUE**

| time period | # vehicles arriving at truck lane 1 | # vehicles queueing at truck lane 1 | waiting time at truck lane 1 queue |
|---|---|---|---|
| 1 | 0 | 4 | 13 |
| 2 | 0 | 4 | 12 |
| 3 | 0 | 3 | 11 |
| 4 | 0 | 3 | 10 |
| 5 | 0 | 3 | 9 |
| 6 | 1 | 3 | 11 |
| 7 | 0 | 3 | 10 |
| 8 | 1 | 4 | 12 |
| 9 | 1 | 4 | 14 |
| 10 | 1 | 5 | 16 |
| 11 | 0 | 5 | 15 |
| 12 | 1 | 5 | 17 |
| 13 | 0 | 5 | 16 |
| 14 | 0 | 5 | 15 |

Figure 5. The truck lane queue in the scenario HM.

**GATE-OUT QUEUE**

| time period | # vehicles arriving at gate-out queue | # vehicles queueing at gate-out queue | waiting time at gate-out queue |
|---|---|---|---|
| 15 | 4 | 4 | 2 |
| 16 | 2 | 3 | 0 |
| 17 | 2 | 2 | 0 |
| 18 | 0 | 0 | 0 |
| 19 | 2 | 2 | 0 |
| 20 | 1 | 1 | 0 |
| 21 | 2 | 2 | 0 |
| 22 | 1 | 1 | 0 |
| 23 | 2 | 2 | 0 |
| 24 | 0 | 0 | 0 |
| 25 | 4 | 4 | 2 |
| 26 | 0 | 1 | 0 |
| 27 | 0 | 0 | 0 |
| 28 | 4 | 4 | 2 |
| 29 | 4 | 5 | 2 |
| 30 | 2 | 4 | 2 |

Figure 6. The gate-out queue in the scenario HM.

Note that the vehicles approaching the gate-out queue are given by the sum of all the vehicles arriving from the different terminal truck lanes.

For each tuck lane, an Excel Spreadsheet similar to the one shown in Figure 7 is utilized in order to compute the number of trucks waiting in the truck lane queue in each time period. In particular, it can be noticed that at time period 7, two trucks approach the check-in, and, having the

**TRUCKS FOR TRUCK LANE 1**

| | | | | | | | | | |
|---|---|---|---|---|---|---|---|---|---|
| arrive at the check-in | 1 | 2 | 3 | 4 | 5 | 6 | 7 | 7 | 8 |
| arrive at the gate-in queue | 3 | | 5 | 6 | 7 | | 9 | 9 | 10 |
| *time spent at the gate-in queue* | 0 | | 0 | 0 | 0 | | 0 | 0 | 0 |
| entering the gate-in | 3 | | 5 | 6 | 7 | | 9 | 9 | 10 |
| *exiting the gate-in* | 4 | | 6 | 7 | 8 | | 10 | 10 | 11 |
| arrive at the truck lane | 6 | | 8 | 9 | 10 | | 12 | 12 | 13 |
| *time spent at the truck lane (queue* | 8 | | 6 | 11 | 10 | | 11 | 14 | 19 |
| exiting the truck lane | 13 | | 13 | 19 | 19 | | 22 | 25 | 31 |
| arrive at the gate-out queue | 15 | 20 | 15 | 21 | 21 | 24 | 24 | 27 | 33 |
| *time spent at the gate-out queue* | 0 | | 0 | 0 | 0 | | 0 | 0 | 0 |
| entering the gate out | 15 | | 15 | 21 | 21 | | 24 | 27 | 33 |
| exiting the gate-out | 16 | | 16 | 22 | 22 | | 25 | 28 | 34 |
| **RESULTS** | | | | | | | | | |
| Service Time | 15 | | 13 | 18 | 17 | | 18 | 21 | 26 |
| $\sigma^P$ | 12 | | 10 | 15 | 14 | | 15 | 18 | 23 |

Figure 7. Analysis of trucks for truck lane 1 in scenario HM.

same truck lane as destination, they contribute to increase the service times. Note that at time periods 2 and 6, no truck approaching the check-in is directed to pick-up a container at truck lane 1.

## 4.2 *Analysis of the results obtained*

The idea is to evaluate the whole import truck process in the terminal, analyzing the elements of major interest, as already mentioned. Some Key Performance Indicators (KPIs) has been defined and analyzed:

- $\sigma$, the average time, expressed in minutes, spent in the terminal by the trucks in the considered time horizon from their check-in to their exit from the terminal;
- $\sigma^P$, the average time, expressed in minutes, spent in the terminal by trucks in the considered time horizon from their check-in to their exit from their truck lane;
- $\Omega^I$, $\Omega^l$ and $\Omega^O$ the average number of trucks in the gate-in, truck lane and gate-out queues, respectively, in the considered time horizon.

Results related to the different scenarios of Table1 are reported in Table 4.

By looking at Table 4, it can be noted that, starting from the same initial state of the terminal (L or H), an increase of the truck arrival number (L, M or H) increases truck service time significantly (compare LL, LM and LH, versus HL, HM and HH).

By fixing truck arrivals and varying the initial state of the terminal, no increase in the service time is obtained when the truck arrival number is Low (compare LL with HL); in fact $\sigma$ passes from 11.5 to 11.6 minutes and $\sigma^P$ from 8.9 to 9 minutes. A sensible increase is perceived when truck arrivals is M or H; more specifically $\sigma$ passes from 15,2 to 22.3 minutes in case of truck arrival equal to M and from 25.7 to 33.3 minutes in case of truck arrival of H type. An analogous trend is obtained for $\sigma^P$. By looking at the last three columns of Table 4 it is possible to identify as bottleneck of

Table 4. KPIs values by varying the scenario under analysis.

| KPI / Scenario | $\sigma$ (minutes) | $\sigma^p$ (minutes) | $\Omega^I$ (# trucks) | $\Omega^l$ (# trucks) | $\Omega^O$ (# trucks) |
|---|---|---|---|---|---|
| **LL** | 11.5 | 8.9 | 0 | 2 | 0 |
| **LM** | 15.2 | 12.0 | 0 | 3.5 | 0 |
| **LH** | 25.7 | 23 | 0 | 19 | 0 |
| **HL** | 11.6 | 9.0 | 0.2 | 5 | 0 |
| **HM** | 22.3 | 19.0 | 2.1 | 5.8 | 0.4 |
| **HH** | 33.3 | 30.2 | 1.4 | 24.8 | 0.3 |

the system the truck lanes. Thus, the terminal can decide which actions should take in order to either increase productivity of critical areas or reduce truck service time. The assignment of the various terminal equipment to the different areas involved must be one of the main aim of each terminal which wants to guarantee certain service levels.

For example, the results for the last scenario (i.e. see last row of Table 4, HH) suggest to use two RTG cranes in order to reduce truck service time. In this case, only by adding a RTG crane, the average queue at the truck lane can be reduced from 24.8 to 7.2 minutes, thus permitting to obtain a reasonable service time.

When the truck lane presents a queue and a certain number of trucks are approaching the truck lane, the only way for not reducing too much the productivity of the terminal is to add a crane in the yard block.

Just to give an idea, the graph presented in Figure 8 shows the behavior of the truck lane queue when one and two RTG cranes are used.

Moreover, another important factor to be analyzed is how the service time is partitioned among terminal activities.

In Figures 9 and 10, two different partitions of the time spent by a truck in the terminal are provided.

This analysis is used to investigate the benefits of the proposed "STOP/GO" DDS for the terminal. Figure 8 refers to the case of free accesses, while Figure 9 refers to the case of a "STOP/GO" SSD used by the terminal. However, in the first case (Fig. 8—free accesses), the truck spends more time inside the terminal and, more specifically, in truck lane queue (80% of its total time), while in the second case (Fig. 9—controlled accesses), it is possible to make the truck wait outside the terminal. In this second case, even if the total time spent is the same, the truck remains inside the terminal for a shorter time, thus avoiding to increase congestion, pollution and to incur in safety and security risks.

Hence, compared with the case of "free accesses", the "controlled accesses" solution presents higher benefits. First of all, being obliged to wait outside the terminal, truck drivers has the opportunity to have access to a series of services (such as bars, toilets, banks, gas stations, etc.) which are not

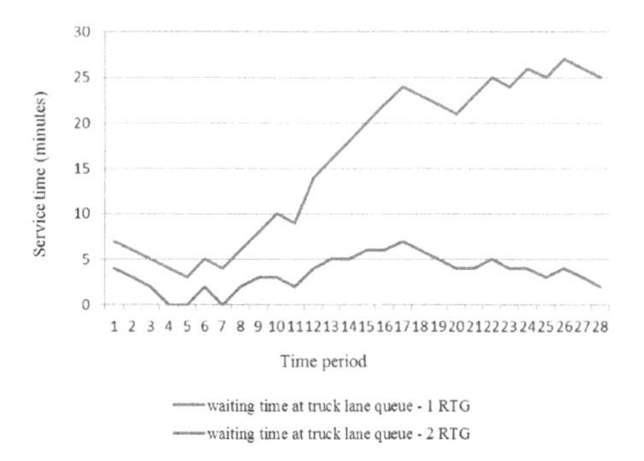


Figure 8. Behavior of the waiting time at the truck lane queue by varying the number of RTGs used.

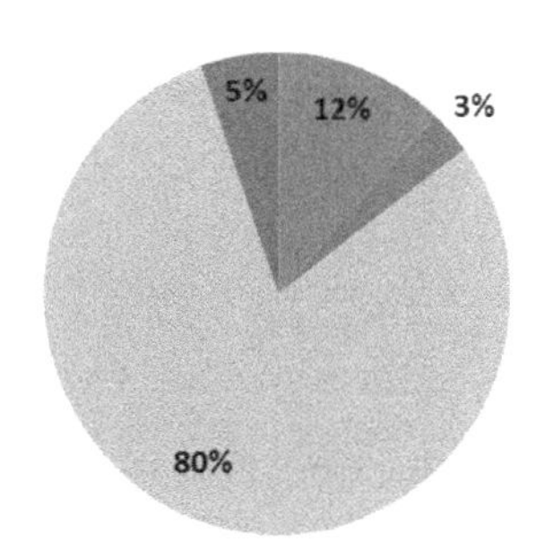


Figure 9. Partition of service time in case of trucks' free accesses.

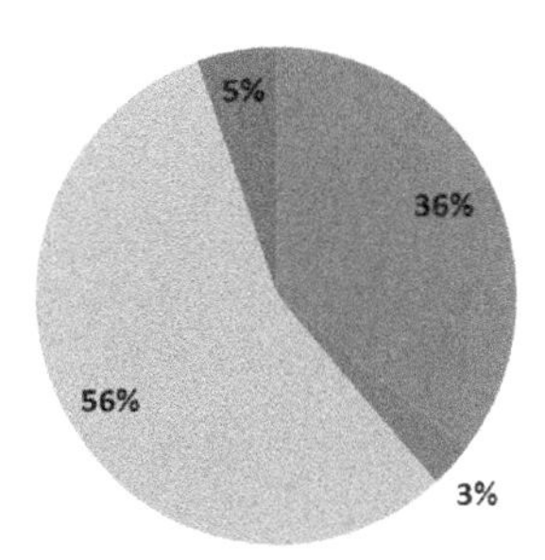


Figure 10. Partition of service time in case of trucks' controlled accesses.

present inside the terminal. Moreover, terminal resources can be dedicated to more priority activities (such as ship loading/unloading), without having too many trucks creating congestion and encumbering operations. This positively affects the service level given to all the terminal customers, both direct (such as shipping companies) and indirect (such as truck carriers); an increased service level improves terminal attractiveness with the consequence of fostering the choice of that terminal by shipping companies for loading/unloading cargo. This is even more important if considering the augmented role played by shipping companies in land transportation (*carrier haulage* transport).

### 4.3 *Discussion of the proposed approach*

The approach here proposed provides a tool for analyzing and addressing congestion issues at container terminal gates. The points of strength of this first attempt are represented by the fact that a simple tool is defined and implemented and it is able to provide important indications to trucks if they can or cannot enter the terminal, with the final goal of reducing terminal congestion and guaranteeing predefined truck service levels. Moreover, this tool enables to properly size terminal equipment (e.g. RTG cranes) on the basis of the workload related to the import road cycle.

However, being a starting base, this methodology is affected by a certain number of assumptions and limitations. First of all, future arrival of trucks (forecasting) are not considered and no stochastic elements are embedded in the mathematical equations. In real scenarios, the check-in can be valid up to the closing of the gate within a specific day. Then, it can happen that, after the check-in execution, the truck driver does not go directly to the gate-in area but arrives there with a certain delay. This can, for instance, occur during lunch time in which drivers prefers to go for lunch in order to avoid possible gate-in queues. This case is not modeled in the present paper.

Secondly, only the road import cycle in relation to the picking up of one container is considered (export cycle and double cycle are excluded from the analysis).

Finally, when the results of the DDS is "STOP" the truck cannot enter and no indication is given about the moment in which it will be allowed to enter.

All these issues will be addressed in future research.

## 5 CONCLUSION

In the present paper, the import road cycle in a container terminal has been described through a series of mathematical relations. The timing of each truck has been properly monitored and presented. In order to test the mathematical equations formulated, an Excel spreadsheet has been used and different scenarios have been simulated by varying both trucks arrivals and the initial occupancy level of the whole terminal. The results obtained have shown that a variation of these parameters strongly affect vehicles' service time.

Providing an indication to a truck about the opportunity for it to enter the terminal is definitely an important indication because it allows to respect the SLA and to reduce vehicles service time, with the consequence of offering a higher service level to truck drivers. An improved service level, in turn, can help fostering the competitiveness of the container terminal and, consequently, lead to better gross throughput.

Moreover, the results obtained have pointed out that a critical point of the import road cycle is represented by the truck lane; this issue can be improved by increasing its productivity (for instance doubling the RTGs cranes) or managing truck arrivals in a proper way.

Future research will be devoted to insert, in the mathematical relations, some stochastic elements in order to simulate more realistic scenarios, taking into account that, in real cases, some input data are not known a priori.

Later on, a proper optimization model will be defined with the goal of minimizing trucks' service time by scheduling trucks entrance at the terminal gate-in.

## REFERENCES


Chena G. & Govindanb K. & Yangc Z. 2013. Managing truck arrivals with time windows to alleviate gate congestion at container terminals. *International Journal of Production Economics*. Volume 141(1): 179–188.

Chena X. & Zhoua X. & Listb G. 2011. Using time-varying tolls to optimize truck arrivals at ports. *Transportation Research Part E: Logistics and Transportation Review*. 47(6): 965–982.

Han, Y. & Lee, L. & Chew, E. & Tan, K. 2008. A yard storage strategy for minimizing traffic congestion in a marine container transshipment hub. *OR Spectrum*. 30(4): 697–720.

Hea J. & Zhanga W. & Huangb Y. & Yanb W. 2013. A simulation optimization method for internal trucks sharing assignment among multiple container terminals. *Advanced Engineering Informatics*. 27(4): 598–614.

Sharif, O. & Huynh, N. & Vidal, J.M. 2011. Application of El Farol model for managing marine terminal gate congestion. Research in Transportation Economics. 32: 81–89.

Zehendner, E. & Feillet D. 2014. Benefits of a truck appointment system on the service quality of inland transport modes at a multimodal container terminal. *European Journal of Operational Research* 235: 461–469.

Zhang X. & Zeng Q. & Chen W. 2013 Optimization Model for Truck Appointment in Container Terminals. *Intelligent and Integrated Sustainable Multimodal Transportation Systems Proceedings*. 96: 1938–1947.

Zhaoa W. & Goodchildb A. 2010. The impact of truck arrival information on container terminal rehandling. *Transportation Research Part E: Logistics and Transportation Review*. 46(3): 327–343.

*Maritime-Port Technology and Development – Ehlers et al. (Eds)*
 *ISBN 978-1-138-02726-8*

# A dynamic discrete berth allocation problem for container terminals

Masoud Moharami Gargari
*Department of TPR, Antwerp University, Antwerp, Belgium*

Mohammad Saeid Fallah Niasar
*Department of Applied Economics, Antwerp University, Antwerp, Belgium*

ABSTRACT: In recent years, improving the process of loading and unloading ships has been the subject of much attention from the terminals operators. The importance of this issue encouraged engineers to design the best simulation for a particular model to reduce time and, subsequently, the costs of the process. The study investigates the Dynamic Berth Allocation Problem (DBAP), in which vessels are assigned to discrete positions based on their length in order to minimize vessels waiting time and berth idle time. We analysed the model by employing two highly effective meta-heuristics in the form of Local Search and Population based methods to obtain optimal solutions subject to the Variable Neighbourhood Search concept (VNS) and Genetic Algorithm (GA), respectively, and compared to each other. The results obtained through an extensive numerical analysis showed that the GA performs better than VNS to this kind of problem. In contrast, with respect to runtime in the same number of iteration, VNS shows the increase in runtime is much lower when problem size rise compared to GA.

*Keywords*: Container transportation, Port operations, GA, VNS

## 1 INTRODUCTION

During the last few decades, the total number of container transports at seaport container terminals has dramatically increased (UNCTAD 2011). Accordingly, terminal operators or port authorities are constantly attempting to boost terminal throughput capacity by considering terminal design, handling equipment as well as technical and tactical applications. Shipping time and cost are fundamental aspects in the efficiency of sea container transportation. A container terminal, as an intersection of shipping routes, performs as an interchange of the different modes involved in the whole transportation network; therefore, an improvement of the efficiency and productivity in terminal operations is essential in decreasing the total voyage time and costs (Imai et al., 2007). As a consequence, marine container terminals are some of the most important and challenging links in the global supply chain. The main critical role of container terminals authorities or operators is to smoothen the progress of the transition between transport modes within the supply chain and present short term storage of boxes in transhipment, export and import.

The economic efficiency and effectiveness of a port relies on making a trade off between the extremely costly facilities (e.g., building quays, container storage yard, cranes, straddle carriers, stacking cranes, etc.) and service time which should be defined by the demand of the market. As the maritime transport industry has developed, it persistently focuses on route improvement, computerization and decision support to achieve good organization and productivity.

Optimization is a strategic way for achieving better solutions in the container handling process in order to increase port efficiency and reduce delays during the process as a whole. In the current study, a mathematical model is proposed to address the economic aspects of improving and updating a port's managerial and infrastructural programs for large container ships to serve faster with more reliable equipment, logistics and efficient management. The berth allocation problem and terminals operations are an interesting problem that still needs a lot more research. Reaching an optimal handling time to enhance the rate of throughput in a port is considered as important factor of port effectiveness. Beneficial handling times originated from following factors: effective factor in choosing quay and number of available facilities (which is referred to as the Berth Allocation Problem), effective factors in allocating equipment (which states Quay Crane Allocation problem and Truck

Allocation Problem) and effective factors concerning the storage of commodity from yard (which defines the Yard Allocation Problem).

The contribution of this study is mainly related to the minimization of the opportunity costs that are caused by the loss of the allocated time devoted to anchored ships. Moreover, it aims at increasing terminals efficiency and productivity, in addition to customer satisfaction by reducing the total service time from arrival in the port to the completion of cargo handling. This will aid in minimizing operating costs and vessels turnaround time as well as increasing port throughput. In turn this will lead to higher revenues and increased competitiveness of the port and a reduction in environmental impact' which are the main purposes of this research through the use of modelling tools.

There are many related studies that cover the different operational aspects of container terminal operation and handling. The efficiency of operation in container ships, container ports and their terminals is an important issue in annual port throughput and customer satisfaction. Employing numerical modelling and using previous knowledge in a practical manner can be useful in generating appropriate planning. Conventionally, berth allocation formulations focus on formulating extra physical constraints and goals of minimizing service time (Xu et al. 2012). Generally, there are different ways of formulating to solve BAP. The most frequently observed cases are Discrete versus Continuous berthing space, (Imai et al., 2001, 2005, Han et al., 2013, Seyedalizadeh, et al., 2010). The important issue needs to study for the continuous berth allocation problem is optimal allocation of ships to quay length (Javanshir and S.R. Seyed 2011). Another kind of BAP, is Static versus dynamic ship arrival time.

Programming modeled on this research mainly falls into two categories: one is to assign the arrival of a ship to a berth through decision variables and to consider the berthing sequence and quay crane for the current ship. Monaco and Sammarra (2007) considered a heuristic algorithm to solve the discrete berth allocation problem. This model is carried out on a small instance in large sample cases. The genetic algorithm to study the Continuous Berth Allocation Problem (CBAP) was developed by Seyedalizadeh et al., (2010) and Javanshir and Seyedalizadeh (2011) and Der-Horng Lee et al., (2010) which was simulated by a greedy randomized adaptive search solution based local search. Further study aiming at optimizing the integrating problem was provided by Salido et al., (2011). Giallombardo and Moccia (2010) investigated the two kinds of tactical level problem in a container terminal. In this paper, they proposed an algorithm for solving the berth allocation problem and the container stacking problem independently. Yang et al., (2012) and Birger et al., (2011) studied the efficient models for the integrated berth allocation and quay crane assignment problem. According to Beirwirth and Meisel (2010), the ship entrance process would be considered static or dynamic. In the former case, it can be assumed that all of the ships are at port at the time of scheduling and berth immediately. Dynamic ships arrival time is scheduled in advance and it cannot berth before expected arrival time (Monaco and Sumaro 2007). On the other hand, static case appears when the vessel intends to serve at time zero (Imai et al., 2001).

The structure of this paper is as follows. The next section presents the model formulation while the solution representation and meta-heuristics are presented in the section 4. A number of comparison results and validity of both of the meta-heuristics are presented in the penultimate section, and the last section concludes the article.

## 2 DEFINITIONS OF MODEL PARAMETERS AND VARIABLES

This section gives a description of the problem we approach in this research. First, the aim of berth allocation problem from two points of view is discussed and then we will propose a mathematical formulation which is considered in this study. While the ship comes to the berth, if there is no available berth to recalling ships, these ships have to wait and have a delay in departure time. Hence, from a ship owner point of view, minimization of the ship waiting time influences the aim of the BAP. On the other hand, if the port is available and no ship comes, the berth has to wait for ship. In this case, berth idle time is formed. Hence from a port owner point of view, minimization of berth idle time follows the aim of BAP. According to the discussion proposed above, this problem consists of two types of cost: minimization of spending time from port owner and ship owner point of view. This is the exactly the aim of this paper.

In this paper we consider the formulation of the Dynamic Berth Allocation Problem as following:

$$Min \sum_{i\in W}\sum_{j\in B}\sum_{k\in U}(T-k+1)(CT_{ij})+S_j-A_i)X_{ijk} + \sum_{i\in W}\sum_{j\in B}\sum_{k\in U}(T-k+1)y_{ijk} \quad (1)$$

Subject to

$$\sum_{j\in B}\sum_{k\in U}X_{ijk}=1, \quad \forall i \quad (2)$$

$$\sum_{i\in W}X_{ijk}\le 1, \quad \forall j,k \quad (3)$$

Table 1. Discusses the symbols used in this model.

Symbols used in model

Indices:
$i\ (1,......,T) \in W$ Set of vessels,
$j\ (1,......,I) \in B$ Set of berths,
$k\ (1,......,T) \in U$ Set of service orders,

Parameters:
$QC_j$: Number of QCs available at berth $j$,
$C_{ij}^{min}$: Minimum number of QCs that can be allocated to ship $i$ at berth $j$
$C_{ij}^{max}$: Maximum number of QCs than should be allocated to ship $i$ at berth $j$
$C_{ij}$: $(C_{ij}^{min}\ ...............\ C_{ij}^{max})$ range of QCs that should be allocated to ship $i$ served at berth $j$,
$S_j$: time when berth $j$ becomes available,
$A_i$: Vessel $i$ arrival time at berth,
$HT_{ij}$: Handling time spent of vessel $i$ served at berth $j$
$t_{tub}$: Turnaround time of each QC for loading and unloading of a single container
$CT_{ij}$: Completion time of the ship $i$ at berth $j$
$M_i$: The average number of cranes allocated to vessel $i-th$,
$CE_{ib}$: Number of export containers to be stored at block $b$ from vessel $i$,
$CI_{ib}$: Number of import containers to be stored at block $b$ from vessel $i$,
$Z_j$: Total number of cranes at berth $j$,

Decision Variables

$X_{ijk}$: 1 if vessel $i$ is served at berth $j$ as the $k$-th vessel
$W_{ijk}$: waiting time of vessel at berth $j$ between the $(i-1)$ th ship and the $i$-th ship,
$Y_{ijk}$: Idle time of berth $i$ between start of service of vessel $j$, and the departure of its immediate predecessor

$$S_j \geq A_i, \forall i, j \tag{4}$$

$$\sum_{i \neq m} \sum_{h<k \in U} \left((HT_{mj})X_{mjh} + y_{mjh}\right) + y_{ijk} - (A_i - S_j)X_{ijk} \geq 0, \forall i, j, k \tag{5}$$

$$WT_{ij} \geq \sum_{i \neq m} \sum_{h<k} (HT_{mj} X_{mjh}) - A_i + S_j - M(1 - X_{ijk}), \quad \forall i,j,k \tag{6}$$

$$\sum_{t \in W} f_i X_{ijk} \leq Z_j, \forall i, j \text{ and } k \tag{7}$$

$$X_{ijk}, \in \{0,1\}, y_{ijk} \geq 0, WT_{ij} \geq 0 \tag{8}$$

The objective function (1) minimizes the total service time in the whole process of container handling. Constraint (2) guarantees that each vessel is served in one position at berth $j$, while constraint (3) presents each berth serve no more than one ship at a time. Constraint (4) ensures that the berthing time is greater than the arrival time. Constraints (5) and (6) show the berth idle time and ship waiting time respectively. Constraint (7) ensures that a sufficient number of cranes is assigned at a quay for handling a ship. Constraint (8) presents the variables. Note that variable $WT_{ij}$ and $y_{ijk}$ are defined as while $(A_i \geq CT_{mi})$ (i.e the arrival time of next ship is later than the completion time of previous ship) means the idle time due to berth $(y_{ijk})$. In contrast, while $(CT_{mi} \geq A_i)$ (i.e the arrival time of next ship is earlier than the completion time of previous ship), a waiting time to vessel $(w_{ijk})$ occurs. In the first case, the berth must wait for the incoming ship and for later, the vessel must wait for which berth to become available. A handling time $CT_{ij}$ is weighted by $(T-k+1)$. This results from the observation that the handling time $CT_{ij}$ of a specific ship serviced at berth $i$ contributes to waiting time to the ships to be serviced at the same berth after it. In other words, the waiting time of a particular vessel is represented by the cumulative handling time of its predecessors.

Moreover, in this model the $CT_{ij}$ is handling time and defined as the time of container handling for vessel $j$:

$$CT_{ij} = \frac{t_{tub} * (CI_{ib} + CE_{ib})}{M_i}$$

$t_{tub}$: The average variable service time of a single crane per container. It is considered as 1.7 min/container.

The value of the completion time is calculated from the sum of the starting berthing time and the service time.

$$CT_{ij} = CT_{ij} + S_j$$

## 3 PROBLEM INSTANCE GENERATION

The random parameters in this model consist of the vessel arrival time, the vessel length the number of import and export goods and the number of quay cranes. All of these variables are involved in the computation in order to obtain an optimized solution. The number of quay cranes that should be allocated to the ship is dependent on the length of the ship. This value for a specified berth is calculated based on the length of the calling ship divided by 50. The capacity of the cranes assumed in the model is considered 25, 30 and 35 move container lifts per hour for berth area.

## 4 SOLUTION REPRESENATION

Two sub strings (layers) of chromosomal representation are incorporated to solve the proposed model. These strings are marked as X1 and X2. Layer 1 is constructed by calling vessels that are ordered by their arrival time. Layer 2 is formed by the number of possible berths. For 2 berths and 7 vessels, the principle of chromosomal representation is shown in table 2. As seen in this table the vessels 3, 5 and 6 are assigned at berth 1 and vessels 7, 1 and 2 are assigned at berth 2. Considering our previous explanations, the main aim of this study is to have a feasible allocation, minimizing total waiting time (ship waiting time and berth idle time). Before allocating the vessels to suitable berth and creating layer X2, we need to consider the following strategy. The process starts by creating a list A that contains all vessels, sorted in ascending order based on the number of allowable berths. Then, the process begins with the first vessel in the list. Corresponding with the allowable number of the berth for the first vessel in the list A, one berth is selected randomly. After allocation of the first vessel, this procedure goes on for the next vessels until the end of the list A, from top to bottom, sequentially in a similar situation. A scheme of the procedure can be found in table 3. As seen from the created list A, for vessel 4, there is possibility to assign to two berths (berth 1and 2). Therefore, one berth is selected, randomly.

### 4.1 *Meta-heuristics*

Broadly speaking, one of the main challenges in meta-heuristic methods is becoming trapped in a local optimum. The consequence of this trap is a deviation of this optimum result from the global optimum solution. A number of strategies can be considered to avoid the danger of becoming trapped in the local optimal solution. These different efficient strategies are common methods which synchronize simple heuristics and rules to obtain optimal accurate solutions to computationally hard integer optimization problems.

In this study, two kinds of meta-heuristic will be discussed namely GA and VNS. The former is based on population based and while the other is constructed based on local search. Contrary to the literature on the BAP, the motivation to use GA is that the majority of previous studies employed it. It implies that GA is a powerful meta-heuristic to handle this kind of problem. On the other hand, only a limited amount of research has considered the BAP based on local search methods. So we select two kinds of meta-heuristics based on (well known in the berth allocation problem, i.e., GA) and also lack of previous work in this field (i.e., VNS). This result present based on the quality of solution and computational time such that clarify which meta-heuristic is worked better on that.

Table 2. Chromosome representation.

| | | | | | | |
|---|---|---|---|---|---|---|
| No of vessels (X1) | 3 | 5 | 6 | 7 | 1 | 2 4 |
| No of berth (X2) | | 1 | | | 2 | |
| Order of service | 1 | 2 | 3 | 1 | 2 | 3 4 |

Table 3. Allocation representation.

| List A | | |
|---|---|---|
| No of vessel | Length of vessel | Allowable berth |
| 1 | 340 | 1 |
| 2 | 320 | 1 |
| 3 | 278 | 1,2 |
| 4 | 267 | 1,2 |
| 5 | 250 | 1,2 |
| 6 | 190 | 1,2,3,4 |
| 7 | 170 | 1,2,3,4 |
| 8 | 190 | 1,2,3,4 |

### 4.2 *Solution based on genetic algorithm*

The Genetic Algorithms (GA) are a particular class of evolutionary algorithms of global search heuristic techniques and are well established methods that mimic the process of natural selection to computationally find perfect or approximate solutions for optimization (see for instance Hansen et al., 1997). With regards to the genetic algorithm procedure, the main aspects are the representations of the solution and a fitness function to evaluate the solution. The algorithm begins with a set of randomly generated solutions and recombines pairs of them at random to produce offspring. Only sufficiently optimal offspring and parents are reserved to produce the next generation. The implementation steps of evolutionary algorithms are as follows.

#### 4.2.1 *Initialize population randomly*

It begins by creating a random population which consisted of a number of individual chromosomes that provides appropriate solutions for the problem. For example, if the number of population size is 50 and recalled the 30 vessels in quay, the population matrix (here mark as g-mat matrix) is formed of 50*60 chromosomes. In the present example the matrix rows illustrate the solution at hand (i.e., X1, X2) and the column corresponds to a random value for the vessel column (1:30), and no of berth column (31:60), respectively. After each loop iterations, the updated solution is used as a replacement in the population matrix.

4.2.1.1 Fitness evaluation
After generating the population randomly, the solution continues to find better values for the cost function. The cost function is calculated and compared with the initial cost function and if the new cost is better, the cost function is replaced by the better cost function.

4.2.1.2 Parents selection method
We used the elitism method for selecting of parents. It works by choosing the best fitness chromosomes in the current solution and replacing them by the next generation.

4.2.1.3 Crossover
The crossover is a recombination operator that is incorporated in the genetic algorithm that allows for the creation of the population of the next generation from the current generation. When crossover is applied, a section of the current generation's chromosomes (parents) are randomly selected, merging them, two new chromosomes (first and second children) are formed. Two types of operators have been used in our solution which is scattered crossover.

4.2.1.4 Mutation
Another modified operation that is used to create next generation in the solution is called a mutation. A combination approach was employed for the mutation process. At the first layer, the exchange method was used. In this method, a couple of genes, which were randomly selected, were exchanged. Moreover, for the next layer, the randomly value mutation method was used. In the randomly value mutation method, a single genome, which was randomly selected from the solution, was re-valued in the building blocks of the genome.

### 4.3 *Solution based on VNS*

The variable neighborhood search technique proposed by Hansen et al., (1997) is quickly becoming a fit established method in meta-heuristics for solving a set of combinatorial optimization and global optimization problems. As its name implies, the VNS concept is used to build set of neighborhoods within the local search space that tries to find promising solutions based on the moving strategy. In this paper, we adapt VNS to the berth allocation problem. To apply VNS in advance, it is required to design different types of neighborhoods that correspond to the problem. It is important to remark that the efficiency of local search meta-heuristics (i.e., the ability to generate feasible solutions, reduce objective functions and make a trade-off between solution quality and computational time to solve the problem) is proportional to the neighborhoods that are used. The VNS methodology is decomposed based on three steps: shaking, local search and an update of the solution. The basic concept of VNS is given in table 4.

Table 4. Basic variable neighborhood search.

```
Initialization
        Select a set of neighborhood structures N_k (K=1
,..., Kmax),
Find an initial solution x, choose a stopping condition
  K=1
   Repeat
     Perturb : x' ← x
     local search: x'' ← N_k(x')
      if x'' is better than x, then set x = x'' and k=1
      else
       k=k+1(Switch to Another Neighborhoods)
  Until k=k max
```

After generating an initial solution at random, as mentioned in section 4 (layer X1 and X2), in the first step, shaking is performed. The purpose of the shaking (perturbation) phase is to destroy a part of current solution (not completely) in such a way that helps to find good starting point for local search. Frequently, the perturbation phase is performed before the set of neighborhoods which are defined to improve the solution. In our study, the shaking step is executed before each neighborhood. In this step, from the list of existence berths, two berths are randomly selected. From the first berth, a certain percentage of the vessels is randomly removed (in our problem 20% of vessels in the selected first berth) and inserted them in the latter berth. It is important to notify that, the shaking step must be handled the feasible allocation of vessel to berth. The fact that due to applying different lengths of berths, the following rule must be checked. After selecting the two berths from the list randomly, the smaller berth must be considered berth one (meaning the berth that removing take places) and the latter must be considered the bigger one. Checking the berth length in this step implies that corresponding solution of the shaking step is feasible. The solution generated by the shaking step is submitted to the local search to find the local optimum solution. Two kinds of neighborhoods are used in the local search operation: 'remove—Insert and 'Swap.

1. *Swap:* two ships randomly selected and attempt to exchange together in the same or the different berths.
2. *Remove-Insert:* Removes a ship from a berth where it is served and inserts it in the same or a different berth.

Table 5. Parametric configuration of the proposed model.

| Parameter | Value |
|---|---|
| No of total vessel (*i*) | 25,50,75 |
| No of berth (*j*) | 3,5,7 |
| No of quay crane | A crane in each 50 m of quay |
| Rate and type of crane | Three types with 25, 35 and 45 moves per hour |
| Plan starting time $t_0$ | 360 |
| GA population size | 50 |
| Recovery time between two vessels | 20 minutes |
| Safety distance between two vessels at berth | 5% of length of vessel |

Before applying any kinds of moving, the feasibility of the allocation situation must be checked. If a feasible allocation is found, the moving is performed otherwise discarded. The difference between applying two kinds of moving in a same berth or between berths is important because of the vessel allocation problem. When a swap and remove movements are applied within a same berth no vessel allocation or length check must take place. Because the vessel is moved to another position in the same berth, the length of the berth is still the same. In step 3, the solution obtained has to be compared to the incumbent solution to be able to decide whether or not to accept it. The acceptance criterion in VNS is only based on improvement.

### 4.4 *Testing method*

In the present study, the optimal planning for all of the processes occurring in the berthing will be investigated in different values of vessel length (L_mat = 130 to 350), berth length (L_berth = 2850), number of vessels (N_v = 25,50,75) and number of berth (N_B = 3,5,7). Table 5 illustrates the values of the parameters assumed in our container terminal handling model (e.g., quay facilities, etc.).

## 5 VALIDATION

Table 5 provides the comparison of the objective function value for previous work (Cordeau et al., 2005) and our meta-heuristics (VNS and GA) with 500 iterations. The parameters, that are included in the comparison, are: Fraction time (0.084 to 0.5), ship size (25 and 35) and berth size (5, 7 and 10). The parameter F is defined as a fraction of arrival time of the first ships to the last incoming ship to the port. The last 2 rows demonstrate the percent difference between the results as reported by Cordeau et al. (2005) with GA and VNS, respectively. It can be observed that GA delivers better results than VNS for all instances. More precisely, GA demonstrates less deviation to best known results than VNS for all instances. Moreover, it is seen that both meta-heuristics outperform for instances that have large value of fraction time, i.e., instance f = 0.5 to 0.85.

Table 6. Validation results for generated instances: fraction time, ships and berths.

| Fraction | Ship* berth | GA | VNS | Validation (Cordeau et al. 2005) | Deviation-GA | Deviation-VNS |
|---|---|---|---|---|---|---|
| 0.5 | 25*5 | 1748 | 1749 | 1722 | 1.50987 | 1.5679 |
| 0.625 | 25*5 | 2214 | 2218 | 2187 | 1.23457 | 1.4175 |
| 0.75 | 25*5 | 2581 | 2583 | 2551 | 1.17601 | 1.2544 |
| 0.875 | 25*5 | 2911 | 2931 | 2877 | 1.18179 | 1.877 |
| 0.084 | 25*5 | 1072 | 1086 | 1058 | 1.32325 | 2.6465 |
| 0.084 | 25*7 | 1007 | 1011 | 990 | 1.71717 | 2.1212 |
| 0.084 | 25*10 | 1025 | 1031 | 1009 | 1.58573 | 2.1804 |
| 0.084 | 35*7 | 2087 | 2093 | 2040 | 2.30392 | 2.598 |
| 0.084 | 35*10 | 2099 | 2106 | 2036 | 3.0943 | 3.4381 |

## 6 RESULTS

All problem instances and experiments have been performed on a personal computer Intel (R) Core (TM) 3i CPU with 2.13 GB of RAM. All our methods have been coded by high-level technical computing language MATLAB version R2010a on SONY (VPCEB1E0E) computaer. As noted previously, the main contribution of our objective concentrates on the minimization of both total berth idle time and ship waiting time. Meeting this objective requires considering a number of parameters such as the optimal facilities and proper planning in port. In order to approach the above parameters, the effects of the number of available berth and the rate of crane have been investigated in this paper.

Table 7 reports both computation times and objective function for 500 iterations. All running times are given in seconds. In this table, the first row shows the problem size (number of vessels and berths). The next rows reveal that the value of computational time and objective function for both meta-heuristics, respectively. It is seen for all proposed instances for GA in the lower sample sizes, that the values of the objective function are less than VNS, whereas for the large sample sizes, this behavior improves for VNS. More than, from computational time point of view, VNS yielded considerably shorter times compared to those reported by GA from the small to the large instances.

Table 7. Computational results for generated instances.

| No of ships | GA-Time-second | GA-results | VNS-Time-second | VNS-results | Deviation—GA and VNS |
|---|---|---|---|---|---|
| 15*10 | 35 | 407 | 29 | 410 | -0.7371 |
| 20*10 | 37 | 1179 | 33 | 1185 | -0.5089 |
| 30*10 | 45 | 1376 | 34.3 | 1384 | -0.5814 |
| 40*10 | 47 | 4321 | 36 | 4375 | -1.2497 |
| 50*10 | 48.5 | 5047 | 36 | 5098 | -1.0105 |
| 60*10 | 49 | 5641 | 42 | 5635 | 0.10636 |
| 70*10 | 51 | 8371 | 43 | 8349 | 0.26281 |
| 80*10 | 54 | 13580 | 44 | 13541 | 0.28719 |
| 90*10 | 55 | 19707 | 44.2 | 19653 | 0.27401 |
| 100*10 | 56 | 22238 | 45.1 | 22198 | 0.17987 |

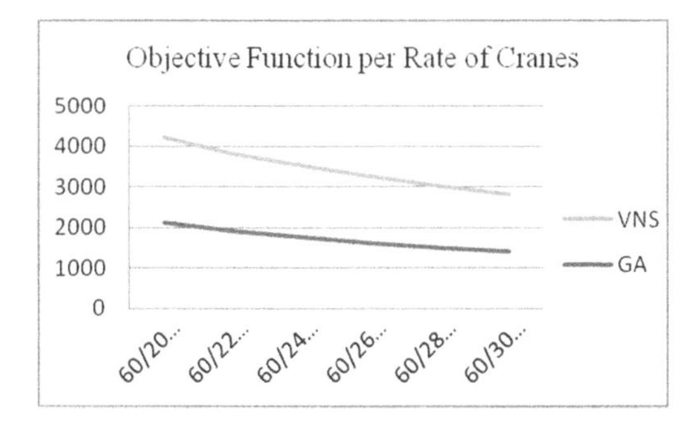


Figure 1. Relationship between objective function and rate of crane.

Figure 1 shows the relationship between objective function and rates of crane for both meta-heuristics. It is seen that the objective function value decreases as the rate of crane increases, which is expected. This is because of the increase of the rate of handling in the port process and as a consequence less service time for each ship. Form this figure it is also obvious that the trend of results for GA is much slightly than VNS.

## 7 CONCLUSIONS

In this paper we present two types of meta-heuristics for obtaining optimal ship berth allocation in the dynamic berthing plan. In this case, ships are arriving in the port while the port is in the process. We also presented the effect of facilities and planning parameters on the objective function by changing the rate of crane, fraction time and number of available berths. The results showed that the GA algorithm found better solution than VNS for lower sample sizes. Another important factor with respect to runtime shows that the VNS is quite well. More precisely, the results demonstrate that for the VNS meta-heuristic, the increase in running time is lower while the problem size rises in comparison with the GA meta-heuristic. We also made special efforts to compare both algorithms to problem instances (i.e. those with different fraction time). It is concluded that for the large value of fraction time, our results are compatible to previous work (f = 0.5 to 0.85). Future research may include considering another neighborhood operator in the local search steps within VNS. Apart from these statements about the algorithm, from the modelling point of view, we try to include the influence of cranes assignment and impact of the internal transportation on the operation procedure in the container terminals simultaneously.

## REFERENCES

A. Imaia, b, E. Nishimuraa, M. Hattoric, Berth allocation at indented berths for mega-containerships, European Journal of Operational Research Volume 179, Issue 2, Pages 579–593,2007.

A. Imai, E. Nishimura, S. Papadimitriou, The dynamic berth allocation problem for a container port. Transportation Research Part B volume: 35 pages: 401–417, 2001.

A. Imai, E. Nishimura, S. Papadimitriou. 2005. Berth allocation in a container port: using a continuous location space approach. Transportation Research Part B Volume: 39 pages: 199–221.

A. Miguel, R. Molins, F. Barber. 2011. Integrated intelligent techniques for remarshaling and berthing in maritime terminals, Advanced Engineering Informatics, Volume 25, pages: 435–451.

B. Raa, W. Dullaert and R.V. Schaeren. 2011. An enriched model for the integrated berth allocation and quay crane assignment problem, Expert Systems with Applications volume: 38 (11) pages: 14136–14147.

B. Skinner, S. Yuan, S. Han, D. Liu, B. Cai, G. Dissanayake, H. Lau. 2013. Optimisation for job scheduling at automated container terminals using genetic algorithm, Computers & Industrial Engineering, volume: 64 pages: 511–523.

C. Bierwirth, F. Meisel. 2010. A survey of berth allocation and quay crane scheduling problems in container terminals. Eur J Oper Res volume: 202 pages: 615–627.

C. Yang, X. Wang, Z. Li. 2012. An optimization approach for coupling problem of berth allocation and quay crane assignment in container terminal, Computers & Industrial Engineering volume: 63 pages: 243–253.

D.H. Lee, J.X. Cao, Q. Shi, and J.H. Chen. 2009. "A heuristic algorithm for yard truck scheduling and storage allocation problems", Transportation Research Part E: Logistics and Transportation Review, Vol. 45, pp. 810–820.

D. Lee, J. Hang Chen. 2010. The continuous Berth Allocation Problem: A Greedy Randomized Adaptive Search Solution, Transportation Research Part E, Volume: 46, pages: 1017–1029.

D. Xu, Chung-Lun Li, and Joseph Y.-T. Leuing. 2012. Berth allocation with time-dependent physical limitations on vessels. European Journal of Operation Research, 216(1):47–56.

E. Nishimura, A. Imai, S. Papadimitriou. 2005. Yard trailer routing at a maritime container terminal, Transportation Research Part E volume: 41 pages: 53–76.

G. Giallombardo, L. Moccia, M. Salani, I. Vacca. 2010. Modeling and solving the Tactical Berth Allocation Problem, Transportation Research Part B Volume: 44, pages: 232–245.

H. Javanshir, S.R. Seyedalizadeh Ganji. 2011. Yard crane scheduling in port container terminals using genetic algorithm, J. Ind. Eng. Int., 6 (11), 39–50, ISSN: 1735–5702.

J. Cordeau, G. Laporte, 2005. Models and Tabu Search heuristics for the Berth-Allocation Problem, Transportation Science, Vol. 39, No. 4, pp. 526–538.

M. Flavia Monaco, M. Sammarra. 2007. The Berth Allocation Problem: A Strong Formulation Solved by a Lagrangean Approach, Journal of Transportation Science Informs, Vol. 41, No. 2, pages: 265–280.

P. Hansen, C. Oğuz and N. Mladenovic. 1997. Variable neighborhood search for minimum cost berth allocation, European Journal of Operational Research, Volume 191, pages: 636–649.

S. Han, B. Skinner, S. Yuan, D. Liu, B. Cai, G. Dissanayake, H. Lau. 2013. Optimisation for job scheduling at automated container terminals using genetic algorithm, Computers & Industrial Engineering, volume: 64 pages: 511–523, 2013.

S.R. Seyedalizadeh Ganji. A. Babazadeh, N. Arabshahi, 2010. Analysis of the continuous berth allocation problem in container ports using a genetic algorithm, J Mar Sci Technol, Volume 15, pages: 408–416.

UNCTAD (United Nations Conference on Trade and Development) secretariat 2011 Review of Maritime Transport 2011 United Nations publication. http://wwwunctadorg/en/docs/rmt2011_enpdf. Accessed 26 Jan 2012.

Y. Dumas, Desrosiers, J., Soumis, F., 1991. The pickup and delivery problem with time windows. European Journal of Operational Research 54, 7–22.

Z. Pengfei, K. Hai-gui. 2008. Study on Berth and Quay-crane Allocation under Stochastic Environments in Container Terminal, Chinese language journal, Volume 28, Issue 1, pages: 161–169.

Z.X. Wang, Felix T.S. Chan, and S.H. Chung, 2013, 3rd International Conference on Intelligent Computational Systems (ICICS'2013) April 29–30, 2013 Singapore "Storage Allocation and Yard Trucks Scheduling in Container Terminals Using a Genetic Algorithm Approach".

*Maritime-Port Technology and Development – Ehlers et al. (Eds)*
*© 2015 Taylor & Francis Group, London, ISBN 978-1-138-02726-8*

# Ontology based management of maritime rules and compliance

M. Hagaseth & Ø.J. Rødseth
*MARINTEK, Trondheim, Norway*

P. Lohrmann & D. Griffiths
*BMT, Teddington, UK*

M. Seizou
*Temis, Paris, France*

ABSTRACT: This paper presents the current status of an ongoing project to develop a new knowledge management system for maritime rules and legislation. This includes a number of use cases, a preliminary system architecture as well as a first version of a new ontological framework for modelling the rules. The management system will use the ontology to guide semantic search in and lexical annotation of the legal texts as well as to define more accurate search criteria for the annotated rules texts. This will significantly simplify the rule creation as well as the compliance processes. The long term objective of the project is to use the ontology directly in rule creation, rule compliance and rule enforcement to ensure full unambiguity and consistency in management and use of rules through exact annotation and automatic processing of new rules. The project will also investigate ontological approaches to classification of Key Performance Indicators (KPIs), check-lists and other tools that can be used in rule compliance.

## 1 INTRODUCTION

### 1.1 *Background*

Commercial sea borne shipping is regulated by international, EU and national authorities and is subject to a number of commercial constraints. The long history and large number of organisations associated with regulation has led to a high level of complexity in managing the development of regulations, their implementation by transport operators, and their enforcement by authorities. The complexity of the situation is aggravated by the long lifetime of ships, the different phases of ship operations, and the number of parties in the operation and the interests of other stakeholders.

### 1.2 *The e-Compliance project*

In the e-Compliance project[1], we address the aforementioned problem by developing an IT system to support the creation, enforcement and compliance of regulations. This system consists of a number of parts as illustrated in Figure 1.

- A digital database of maritime regulations covering a broad set of regulations.
- A maritime ontology, encapsulating detailed knowledge about maritime regulations in a machine-understandable form. This is linked to a thesaurus used as basis for semantic searches and annotations.
- A set of core services used for semantic enrichment and exploitation of regulations (annotating, searching and reasoning) as well as other general services used in the front-end functions.
- A creation system to support the legislators during the creation of international and national regulations and of port bye-laws, and to help the readers of the regulations to correctly interpret the text.

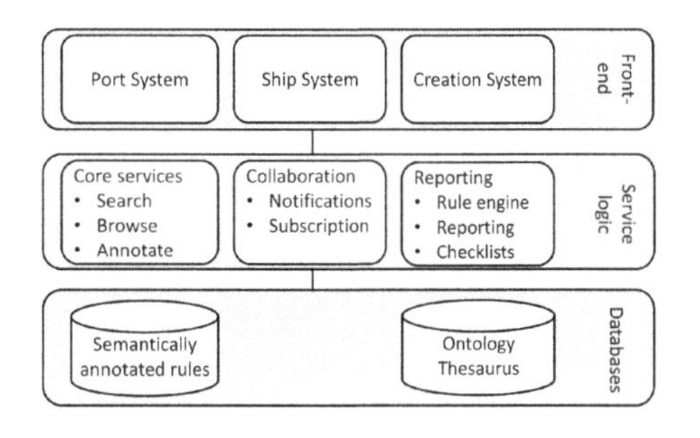


Figure 1. e-Compliance system.

[1]This work has been performed as part of the e-Compliance project which receives funding from the 7th Framework Programme of the European Commission under grant agreement MOVE/FP7/321606.

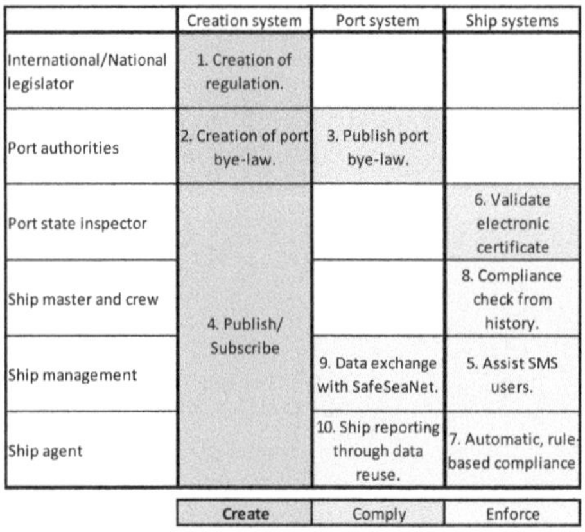

| | Creation system | Port system | Ship systems |
|---|---|---|---|
| International/National legislator | 1. Creation of regulation. | | |
| Port authorities | 2. Creation of port bye-law. | 3. Publish port bye-law. | |
| Port state inspector | | | 6. Validate electronic certificate |
| Ship master and crew | 4. Publish/ Subscribe | | 8. Compliance check from history. |
| Ship management | | 9. Data exchange with SafeSeaNet. | 5. Assist SMS users. |
| Ship agent | | 10. Ship reporting through data reuse. | 7. Automatic, rule-based compliance |
| | Create | Comply | Enforce |

Figure 2. e-Compliance use cases.

- A port system to support the ship agents and ship managers in complying with port specific regulations, and to support the port authorities in managing port specific regulations.
- A ship system to support the ship master and ship agent to comply with rules as well as implement their own compliance systems, such as the Ship Management System (SMS).

All the different requirements for the e-Compliance system from the maritime legislators, enforcers and reporting parties have been summarized in ten use cases that will form the basis for the further development of the e-Compliance system as shown in Figure 2.

The use cases focus on different aspects of regulation creation, enforcement and compliance to be able to improve the quality of the regulations and to reduce the burden for those that have to enforce the regulations, as well as for those that must fulfil them. In addition, the use cases also handle publish/subscribe functionalities to ensure that all parties involved are aware of the current active regulations and updates on them.

The next section will give the methodological background for the regulation management system. Section 3 will describe the e-Compliance ontology. Section 4 will give a few examples of application of the ontology and section 5 will summarize the conclusions so far in the project and an overview of the further work that will be undertaken.

## 2 METHODOLOGY FOR SETTING UP A MARITIME REGULATION ONTOLOGY

### 2.1 *Termino-ontological resource*

In e-Compliance, we develop two semantic tools: A thesaurus and an ontology.

A thesaurus can be viewed as a hierarchically structured controlled vocabulary with additional lexical information for each concept. This information consists of a preferred term, alternative terms (synonyms) associative relationships (related concepts) and definitions (definition and scope notes).

In contrast, an ontology is a formal, explicit conceptualisation of a domain, defining its concepts through object classes (objects) and their semantic properties (attributes and relationships). An ontology formally represents the "knowledge" about a domain in a structured and computer-readable way (Gruber, 1993). In the context of this project, we will express maritime concepts as classes in the ontology including their properties (both data values, the relationships to other classes, and annotated data). The ontology also contains a hierarchy of classes describing inheritance of properties from super-classes to sub-classes. We only use single inheritance, not multiple inheritances.

Both the thesaurus and the ontology play an important role in the e-Compliance use cases. In particular, they will support the following functionality:

- During the creation process, the thesaurus will be used to ensure that terms are used consistently throughout and across regulations; "things" should be referred to by their preferred term.
- The ontology will ensure that newly created regulations are structurally consistent, i.e. that they contain at least a "target" and a "requirement" (these terms will be explained in the next chapter).
- The ontology will provide the e-Compliance system with a certain "understanding" of the content of a regulation, thus allowing the system to search for related regulations and (potentially) even to recognise contradictions.
- Indexing and annotating regulations text will allow us to perform semantic search.

To enable this functionality, we will link the thesaurus and the ontology to create a "Termino-Ontological Resource" (TOR). This structure can be used for annotating and consistently indexing documents, searching and browsing as well as inferring and reasoning (Omrane et al., 2011). The linking of the Thesaurus to the ontology will be done via the "rdf:about" label. This label is a feature of the Resource Description Framework (W3C, 2014) and is used in both OWL (W3C, 2004) (for describing ontologies) and SKOS (W3C, 2012) (for describing thesauri). The links between the ontology, the thesaurus and the plain regulations text is depicted in Figure 3.

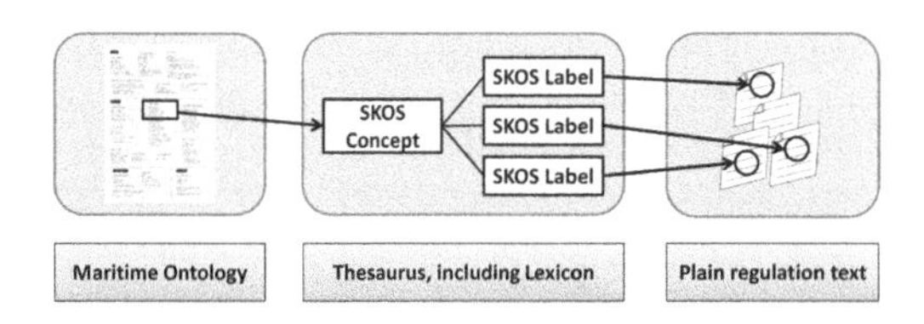


Figure 3. Lexicalised maritime ontology.

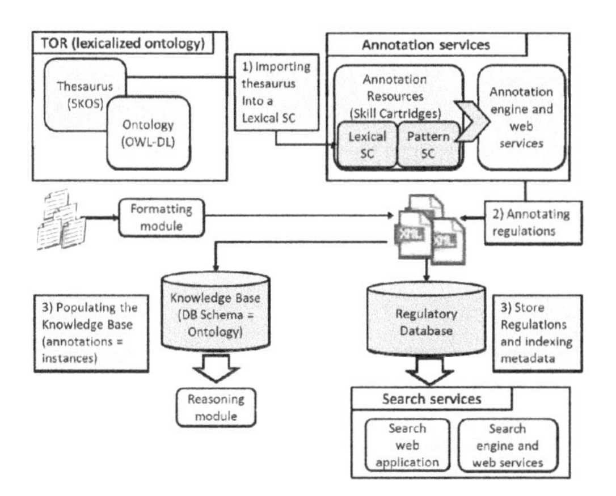


Figure 4. Core services and regulation enrichment process.

### 2.2 *Regulation enrichment process*

Figure 4 shows the overall enrichment process of the regulations and the different core services that is used. This shows how the TOR will be used as part of the e-Compliance system.

By interfacing with the Creation, Port and Ship systems, the core services will provide functionalities for the access and search of regulations, as well as support collaboration between stakeholders, in terms of creation, amendment and publication, and enabling efficient usage of the regulations (compliance).

Uniform creation, representation and management of regulations will improve the compliance of maritime regulations. Such uniformity increases transparency of the regulations and will ensure that the regulations are fully understood by both the compliers and enforcers. More importantly, the uniformity ensures that all users will interpret the regulations in the same way. The core services also act to ensure that all stakeholders are aware of the current rules and regulations and that the new regulations are available to all stakeholders in a timely manner.

The e-Compliance regulation database (library) will store both regulations and their indexing metadata (structural metadata and semantic annotations coming from the annotation service), for supporting the search and browse service. Some of the search and browse services are reused from the project Flagship (Flagship).

### 2.3 *Example queries*

In this section, we list some example queries that are relevant to the users of the e-Compliance system. Answering these queries will be possible through using the structured data contained in the Knowledge Base and Regulations Repository described in the previous section.

- Ship agent or ship manager searches regulations to find out which clauses apply to a ship carrying dry bulk.
- Ship agent or ship manager searches regulations to find out which clauses apply to their ship if the cargo is changed from one type to another.
- A ship manager searches for which clauses in the SOLAS instrument have to do with maintenance, as referred to in the ISM Code.
- The ship manager wants to search for regulations related to a certain activity (safety, maintenance, fire protection, training etc.).
- A legislator searches in existing regulations for what term is used for the vessel type of a ship of size 300BT, carrying dry bulk and powered by a single engine.
- The creator of a new regulation has entered the term "ships used for fishing". What term is the preferred term for this?
- Ship agent and ship manager want to know which port bye-laws apply to their RORO ship when entering Barcelona.
- The creator of Port bye laws wants to know which port bye-laws are valid for garbage handling in the neighbouring port.
- An enforcer wants to find out which certificates are mandatory for a given ship in the next port of call.

## 3 E-COMPLIANCE ONTOLOGY

### 3.1 *Class structure*

The e-Compliance ontology can roughly be split into four domains:

- **"Legal":** This domain covers the regulations and their parts, the documents they are published in (or refer to) and—most importantly—the "meaning" that can be extracted from a regulation.
- **"Maritime":** This domain focuses around ships and maritime activities and situations as the topic of regulations.
- **"Organisational":** This domain covers organisations and roles that are responsible for or addressed by regulations.

- **"Territorial":** This domain models the area (geographical, political or legal) in which regulations are applicable or in which an action takes place.

A class can have two types of properties:

- Attributes, which are basic data types (like strings, numbers, dates etc.);
- Relationships, which are instances of other classes.

The class structure of the ontology is given in Figure 5. The individual domains are colour coded as follows: "Legal" in red, "Maritime" in blue, "Organisational" in grey and "Territorial" in green. Mandatory properties are indicated in bold font; relationships are written with a leading "->".

### 3.2 *The legal domain*

The **Regulation** class is used to unambiguously identify the maritime regulation we wish to model. It has the attributes "title" (which is a unique identifier), "validityDate" (which denotes the date from which the regulation is effective), "objective" (which describes the 'topic' of the regulation) and "kpi" (which stands for a key performance indicator linked to the regulation). In addition, we use the attribute "type" to denote the category of regulation (guideline, procedure, check list etc.).

Regulations are published in documents. The **Document** class unambiguously describes the reference document (including its "title", "publisher" and "publicationDate"). This class is also used to describe legal documents that are mentioned in regulations. In particular, we have defined the two subclasses Certificate and Report. Certificates are linked to a ship and have a "validityPeriod".

Reports are submitted to a specific organisation (for example a local authority); in addition, they are due at a certain point in time. This point in time is described relative to an activity ("activityDefiningDeliveryTime"), for example the ship's arrival in port. The time for delivery is then described as the amount of hours before or after this event ("maxHoursDueAfterDefiningActivity" and "minHoursDueBeforeDefiningActivity").

Regulations consist of clauses. A clause is the smallest document unit that contains a complete requirement or definition; it is an "atomic" part of a regulation. Thus we always break a regulation we wish to model down into a set of clause objects. The **Clause** class is linked to one or more regulations ("isPartOf" and "hasReference") and contains an "objective" to capture its 'topic'. In addition, we note that some clauses do not contain rules, but rather *definitions* of terms. To capture this, we use the two attributes "definesTerm" and "definesThing". The former contains the term that is being defined (exactly as it appears in the clause). The latter specifies what type of object is being defined; a carefully controlled list of categories will be used to capture this information (like "ship", "role", "equipment" etc.).

The very heart of the e-Compliance ontology is the **Rule** class. This is where we capture the "meaning" of a clause and hence of a regulation. By definition, a rule is derived from a clause (which is identified by the "isDerivedFromClause" property). A clause can contain more than one rule.

A rule consists of three parts:

- A target ("hasTarget" attribute of the class): this is the subject of the rule, i.e. the "thing" of which a demand is made;
- A requirement ("hasRequirement" attribute): this is what the target has to "do" in order to be compliant with the rule;
- A context ("hasContext" attribute): these are conditions that further specify when the rule is applicable.

A well-formed rule needs to have a target and a requirement; context information is optional.

In the e-Compliance ontology, the target of a rule can be a ship, an organisation or a role; these are considered the "players" in the domain that can take action to ensure compliance with regulations. A requirement can be any of these three, or additionally a ship part, a piece of equipment, an activity, a document, a type of cargo or a pollutant. A context can be given by a maritime situation, a journey, an activity or a type of cargo. All of these classes will be described in more detail in the following sections.

In addition to these relationships, the rule class also contains a number of attributes. The Boolean "isMandatory" attribute indicates whether a rule is mandatory (i.e. legally binding) or optional. The "isDisjointRequirement" indicates if the rule's requirement contains an "OR" operator; this is difficult to model within an ontology and may require a user's special attention. We also provide an optional "type" attribute that can be used to further specify a rule if required.

Note that we will not create rule objects for clauses that contain definitions.

### 3.3 *The maritime domain*

Central to the Maritime Domain is the **Ship** class. An instance of this class does not describe an individual vessel but rather a set of ships with certain properties that are subject to a given maritime regulation. To specify this set we use attributes like ship type, tonnage, length, keel laid, draft, crew number, passenger number etc. As we are

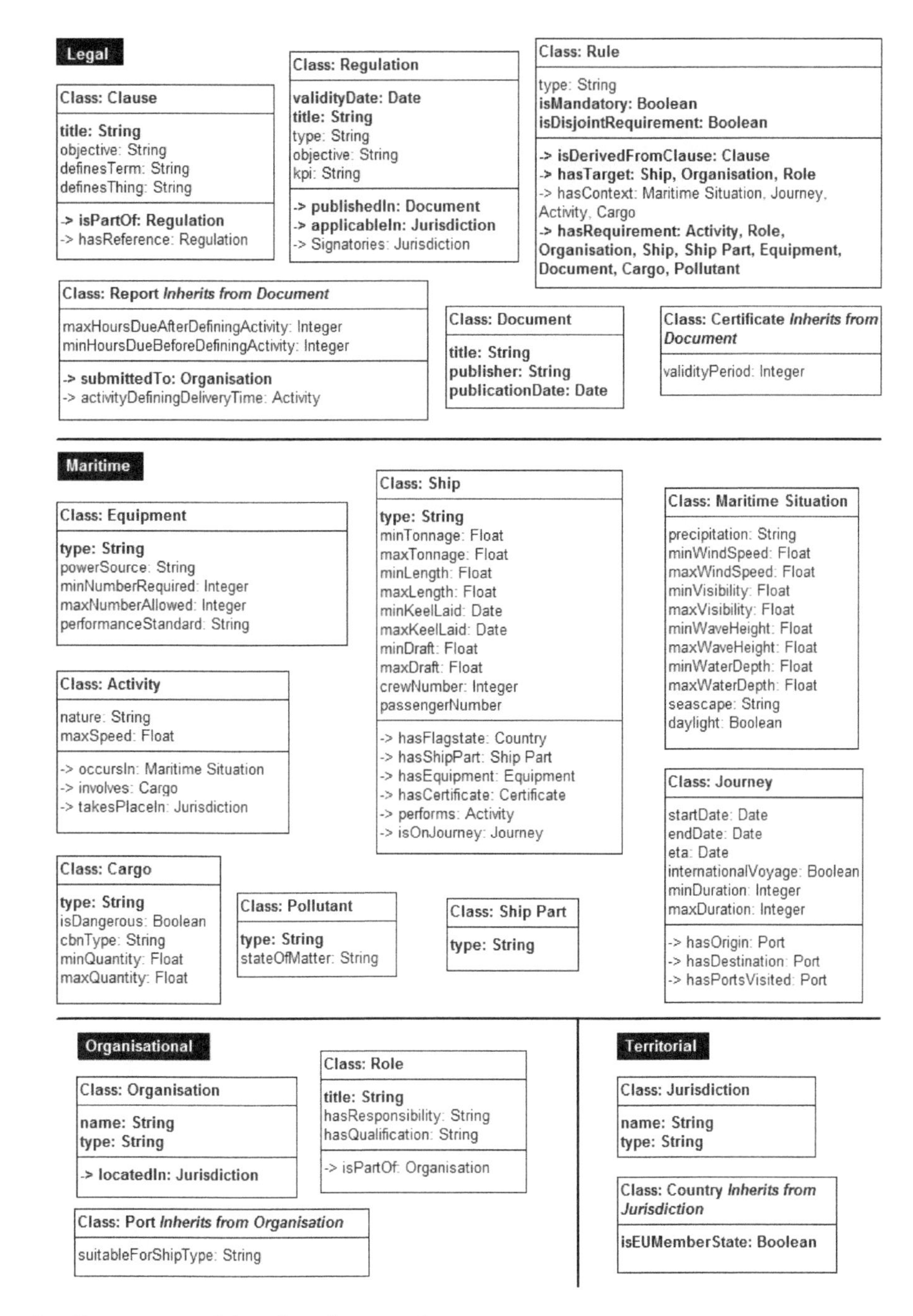


Figure 5. Class structure of the e-Compliance ontology.

describing sets of ships, we define ranges for the numerical attributes rather than exact figures; e.g. rather than "length" we define "minLength" and "maxLength". This allows us to define rules of the type "Cruise ships of more than 100 m length must…".

In addition, the ship class is linked to various other classes. In particular, a ship object can have a flag state, a ship part, a piece of equipment or a certificate. Moreover, it can perform an activity (like "berthing") or be on a journey. All these classes are explained in the following.

The **Ship Part** class describes structural parts of a vessel (like "hull", "engine" etc.). This class is currently somewhat generic and only contains the attribute "type". We expect to define more properties in future versions of this ontology in order to refine our modelling of regulations that target parts of a ship.

In contrast to ship parts, the **Equipment** class describes parts of a ship that are not structural but can be moved or replaced with relative ease. Examples of instances are lifesaving appliances, communication equipment, first aid kits or protective clothing. The "type" attribute specifies an equipment object; to define the number of objects that are required/allowed on board we use the attributes "minNumberRequired" and "maxNumberAllowed". To further describe equipment, we have defined the attributes "powerSource" and "performanceStandard". More attributes will be added as needed in future versions.

Similarly, the **Cargo** class describes the cargo that can be carried by a vessel. Objects of this class are specified by their "type"; in addition, we define the "minQuantity" and "maxQuantity" attributes to capture the restrictions regulations place on the amount of a certain cargo in a certain context. Dangerous cargo is flagged up by the Boolean attribute "isDangerous". Lastly, "cbnType" specifies if cargo falls under the categories chemical, biological or nuclear.

A key part of the e-Compliance ontology is the **Activity** class. Its instances model activities that a ship can perform (like "Berthing", "Sailing", "Entering port", "Submitting report", "Collecting meteorological data", "Avoiding collisions" etc.) and that can as such be the context or requirement of a maritime regulation. An activity instance has a "nature" and a "maxSpeed" at which it is performed (for example applicable to "Sailing"). In addition, an activity can occur in a specific Maritime Situation ("occursIn"), involve a certain type of cargo ("involves") and take place in a certain jurisdiction ("takesPlaceIn"). As this class covers a large subdomain, we expect to add more structure (in particular subclasses for certain types of activities) in the future development of this ontology.

The **Maritime Situation** class captures the hydrographical and meteorological context of a maritime regulation. It can describe minimum and/or maximum values for wind speed, visibility, wave height and depth. In addition, it captures precipitation, the type of seascape and daylight (yes/no).

The **Journey** class is used to describe the type of voyage that a regulation may refer to. The attributes are "startDate" and "endDate", expected time of arrival ("eta") as well as "minDuration" and "maxDuration". In addition, the ports visited during the journey are described by the properties "hasOrigin", "hasDestination" and "hasPortsVisited".

Finally, we define a **Pollutant** class specifically to model the regulations in MARPOL. The attribute "type" specifies the pollutant, whereas "stateOfMatter" describes whether it is solid, fluid or gaseous.

### 3.4 *The organizational domain*

This part of the ontology is used to describe the organisations and individual persons that are active "players" in the maritime world. Firstly, we define an **Organisation** class; objects of this class can be ship operators, governments, the IMO, the EU or local administrations, port authorities, etc. The "type" attribute specifies the organisation instance, whereas the "locatedIn" relationship describes the jurisdiction it is subject to.

The **Port** class is a subclass of Organisation; it takes account of the prominent role that ports play in the maritime regulation domain. Apart from the properties it inherits, the Port class also contains a "suitableForShipType" property.

The **Role** class describes jobs and responsibilities as they are addressed by regulations. Typical examples are "ship master", "ship agent", "pilot", "harbour master" etc. A role object has a "title", and it may be further specified by the attributes "hasResponsibility" and "hasQualification". In addition, a role may be linked to a particular organisation through the "isPartOf" property.

### 3.5 *The territorial domain*

This part of the ontology describes the "area" in which a regulation is applicable. We have defined a **Jurisdiction** class which captures this information in the most general form; its instances can be port areas, regions, parts of several countries (like the Baltic Sea area), continents, international territory or a hydrographical region (like the English Channel). We specify the jurisdiction instance through the attribute "type".

The **Country** class is a subclass of Jurisdiction. It contains the additional Boolean attribute "isEUMemberState".

## 4 EXAMPLES

To illustrate how the ontology is used to capture the content of "real-world" regulations, we will populate two examples. We will sketch the relevant properties of the rule objects and omit details on the underlying regulation and document objects.

### 4.1 *ILO*

The following is a slightly abridged extraction from an ILO regulation (C164 Health Protection and Medical Care (Seafarers) Convention, 1987—Article 8, Paragraph 1).

*"All ships carrying 100 or more seafarers and ordinarily engaged on international voyages of more than three days' duration shall carry a medical doctor as a member of the crew".*

This text consists of only a single clause; we only need a single rule to capture its content. The target of the rule is a Ship object (say, "TargetShip"); it is specified by the minimum number of seafarers (TargetShip.crewNumber > 99).

The requirement is described by a Role object (say, "MedicalDoctor"), which is something the ship must "have".

The rule's context in which this rule is applicable; it is specified by a Journey object (say, "OurJourney") which is further specified by its nature (OurJourney.internationalVoyage = true) and its duration (OurJourney.minDuration = 3 days).

### 4.2 *SOLAS*

The following text is taken from SOLAS Chapter III, Part B, Section I, Regulation 7, Paragraph 2.1:

*"A lifejacket complying with the requirements of paragraph 2.2.1 or 2.2.2 of the Code shall be provided for every person on board the ship".*

This clause contains a single rule. Its target is a ship ("OurShip") without further specification; hence the regulation applies to all vessels.

The rule's requirement is given by an Equipment object ("LifeJacket"); its type is given by the preferred term for "life jacket". In addition, the performance standard of this piece of equipment is specified in the referenced paragraphs of the regulation.

Finally, the minimum number of life jackets is specified by the number of people on board the target ship. Formally, this would be expressed as

LifeJacket.minNumberRequired = OurShip.crewNumber + OurShip.passengerNumber

This rule does not have a context. For further examples, refer to Lohrmann et al. (2014).

## 5 CONCLUSION AND FURTHER WORK

In this paper we have presented the first version of the e-Compliance ontology for the modelling of the maritime regulations domain. We have constructed a class structure that allows for the modelling of maritime regulations, in particular for describing the "who", "what" and "when" of a piece of legislation—who is it aimed at, what is required and under which circumstances. We have then proceeded to populate the ontology by creating instances for some example regulations taken from the ILO Convention and SOLAS. We have demonstrated that the e-Compliance ontology is capable of capturing the "meaning" of these regulations.

In addition, we have briefly discussed how the ontology will be linked to a thesaurus containing relevant maritime terms to form a termino-ontological resource. We have explained how this structure will be used as part of the e-Compliance system to assist users with the creation of regulations and to provide the basis for advanced search functionality.

The presented ontology is a first version; while modelling example regulations, several ways to make improvements were identified. In particular, we will refine the Activity class to allow for the description of various types of activities (like navigation, communication and cargo handling). In addition, we may introduce an "Exception" class to precisely model targets that are exempt from a given regulation. These refinements will be undertaken in future versions of the ontology, produced during the lifetime of the e-Compliance project.

## REFERENCES


Flagship. Flagship Project. Available at http://www.flagship.be/.

Gruber T.R. 1993. Towards principles for the design of ontologies used for knowledge sharing. *In Formal Ontology in conceptual Analysis and Lnowledge Representation.* Available as Stanford Knowledge Systems Laboratory Report, Kluwer Academic.

Lohrmann P., Hagaseth M., Seizou M., Griffiths D. 2014. "e-Compliance Deliverable 2.2: Ontology", e-Compliance Consortium.

Omrane N., Nazarenko A., Rosina P., Szulman S., Westphal C. 2011. Lexicalized Ontology for a Business Rules Management Platform: An automotive use case, In *5th International Symposium, RuleML,* Florida.

W3C 2004. OWL Web Ontology Language. Available at http://www.w3.org/TR/owl-features/.

W3C 2012. SKOS Simple Knowledge Organization System. Available at http://www.w3.org/2004/02/skos/.

W3C 2014. Resource Description Framework. Available at http://www.w3.org/RDF/.

*Maritime-Port Technology and Development – Ehlers et al. (Eds)*
*© 2015 Taylor & Francis Group, London, ISBN 978-1-138-02726-8*

# Maritime Single Windows: Lessons learned from the eMAR Project

I. Koliousis & P. Koliousis
*Decision Dynamics Limited, Lexington, SC, USA*

T. Katsoulakos
*Inlecom Systems Ltd., London, UK*

ABSTRACT: This paper addresses the maritime transport administrative complexities and reports the lessons learned from developing and implementing an innovative single window solution for maritime transport. The European Commission (EC) is supporting Maritime Single Windows (MSW) through different policies, as a solution to simplify and facilitate ship reporting formalities. The e-Maritime Strategic Framework and Simulation based Validation (eMAR) Project, a 3 Year EU funded project, develops and implements a Single Window that will be used seamlessly and efficiently by different stakeholders. In this context, the paper describes a conceptual model of the key features that a successfully user driven MSW should incorporate in addition to reporting feedback from the implementation process as well as identifying the key drivers that facilitate the implementation.

*Keywords*: eMAR; Maritime Single Window; policy recommendations

## 1 INTRODUCTION

### 1.1 *Introduction*

It is commonplace to suggest that the maritime industry is a globalized one in terms of functional, business and regulatory terms. Nevertheless, the former is also a barrier to the industry, since stakeholders must comply with a complex set of administrative procedures, including customs, taxation, immigration, safety and security, waste, health protection, to name but a few of these requirements. Competent authorities require even before a ship's arrival and/or after departure from a port, a number of documents, with mutually overlapping information. These formalities and the procedures to fulfil them are often considered duplicative and time consuming, resulting in costs and delays that could make maritime transport less attractive. It has to be noted though that there is no uniform way of practices competent authorities per EU country. In Norway, the National Competent Authority (NCA) covers all ports, airports etc. whereas in the Netherlands, the Port Authority of Rotterdam is the NCA and in Greece, the Ministry of Shipping is the NCA.

The Directive 2010/65/EU (European Union, 2010) guides the reporting formalities for ships arriving in and/or departing from ports of the EU Member States (the Reporting Formalities Directive) and simplifies and harmonises procedures by establishing a standard electronic transmission of information and by rationalising reporting formalities for ships. Essentially, EU MS (Member States) shall accept the fulfilment of reporting formalities in an electronic format and through submitting those via a national single window. The deadline that has been decided is no later than June 1st, 2015.

In support of this EU eMaritime concept, the eMAR project develops and tests systems that enable easy exchange of documents and information in support of improved networking for the maritime industry stakeholders. One of the key elements of these systems is the Maritime Single Window, which aims to undertake an important role in the trade and transport chains by evolving to the single point of data entry. The approach that the eMAR has developed will be further analysed in the following paragraphs.

## 2 MARITIME SINGLE WINDOWS

### 2.1 *Background*

Competent authorities including trade, transport, hygiene as well as port owners/operators have established an extensive range of case specific, authority specific and country specific regulatory requirements. Furthermore, the lack of common understanding and coordination amongst each other, even at the local level, leads to increased effort to comply with these requirements. Information and Communications Technologies (ICT) systems intend to solve this situation, nevertheless, most reengineering efforts have been geared towards the means of communication exchange rather than reengineering the processes and the informational needs in general.

Similarly, the compliance support systems have not considered the specific needs of Small and Medium Sized Enterprises (SME), producing complex, bureaucratic environments for them too. This only builds up on the recent security related requirements for advanced trade and transport notifications.

In response to these, a number of initiatives have been adopted. The concept of a single point interaction between businesses and authorities, commonly termed as the "Single Window" (SW) have been in development for the past decades, at different levels, including, local, national, European and International level. Although Single Windows were first developed by Customs for trade facilitation, recently, they have been heavily used for transport facilitation, especially in the maritime sector. This has been reinforced by the EU Directive 2010/65/EU (commonly known as the 'FAL Directive') which mandates Member States to accept the fulfilment of ship reporting formalities in electronic format and their transmission via a single window no later than 1 June 2015.

## 2.2 *The UN/CEFACT recommendation 33*

Single Windows were introduced by United Nations (UN) Centre for Trade Facilitation and Electronic Business (UN/CEFACT) to improve the exchange of information between trade and government. UNECE's definition for Single Windows (United Nations Economic Commission for Europe, 2005) states that they constitute a facility allowing parties involved in trade and transport to lodge standardized information and documents with a single entry point to fulfil all import, export, and transit-related regulatory requirements as shown in Figure 1.

The concept of the recommendation is that electronic information is submitted only once, simplifying and expediting the process. Additionally, in cross-border trade, specific security and trade rules should be incorporated into the system. The core UNECE recommendation is that the Single Window should be managed centrally by a lead agency, enabling the appropriate governmental authorities and agencies to receive or access the relevant information for their purpose and co-ordinate their controls. In addition to that, UNECE recommends that the most important

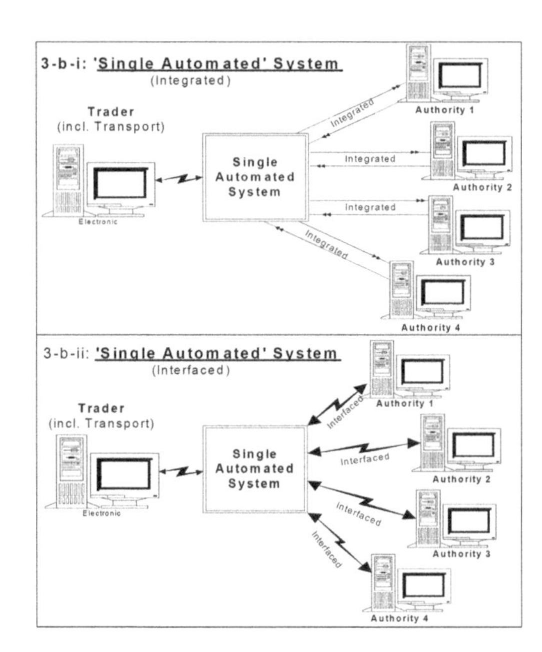


Figure 1. UNECE Recommendation No. 33 on Single Windows.

driver for the successful implementation of a Single Window facility is the will and preparedness of the government and the relevant authorities and the full support and participation of the business community.

Extending this argument, the national legal framework, including privacy laws and security rules in the exchange of information on "a needs basis" have to be carefully modified to enable the Single Windows.

## 2.3 *The National Data Set (NDS)*

National Single Windows (NSW) build on top a National Data Set (NDS) which sets the regulatory regime and should be aligned to trade and economic development policy. According to UNECE (United Nations Economic Commission for Europe, 2010), the simplification and standardisation reengineering activity should start with the Government's clear objective for the way in which the National Data Set will be used, particularly with respect to:

- interaction with other national, European and international systems,
- trade facilitation,
- safety, security and environmental risk management.

A generic Data Harmonization Process is shown in Figure 2.

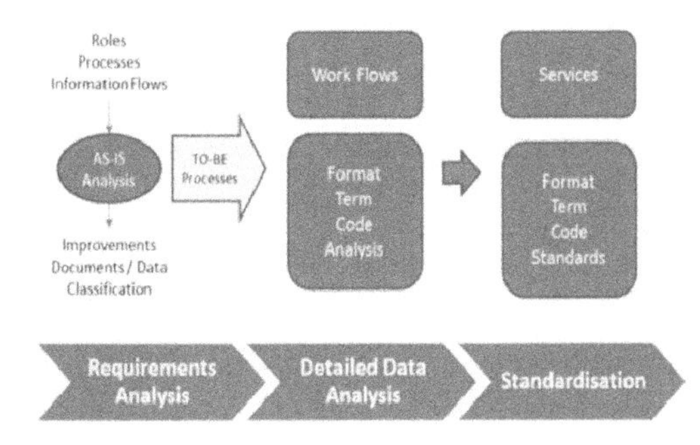


Figure 2. Data Harmonisation Process, adapted from (United Nations Economic Commission for Europe, 2010).

### 2.4 *Customs related initiatives*

The World Customs Organization's (WCO) perspective (World Customs Organisation, 2010) follows the UNECE definition of the Single Window. WCO recommends the following for a successful SW implementation:

- use of ICT and dataset standards commonly accepted by the relevant public and private stakeholders, particularly, a Harmonized System of Commodity Description and Coding, the WCO Data Model and the Unique Consignment Reference,
- agencies involved in Integrated Border Management should determine the essential data for their controls,
- in cases where inspection of goods is necessary, the Single Window should be used for the co-ordination of physical inspection amongst the relevant agencies.

WCO promoted the SW for harmonisation and the streamlining of cross-border customs procedures and in this context, the SAFE framework of standards for secure supply chains that was developed (Ireland, 2009) aimed at:

- promoting co-operation between the Customs and business communities,
- strengthening networking arrangements between Customs administrations to improve their capability to detect high-risk consignments,
- supporting the seamless movement of goods through secure international trade supply chains.

To this extent, Mutual Recognition (MR) is a broad concept embodied within the WCO SAFE Framework of Standards to Secure and Facilitate Global Trade (SAFE) whereby an action or decision taken or an authorization that has been properly granted by one Customs administration, is recognized and accepted by another Customs administration.

### 2.5 *Additional relevant initiatives*

The concept of Single Window has also been covered for some time by Port Single Windows to facilitate port state control. One of the main objectives of the Vessel Traffic Monitoring and Information System (VTMIS) Directive (Directive 2002/59/EC for vessel traffic monitoring, (European Union, 2007)) has been to guarantee that all Member States will be interconnected via the Community maritime information exchange system SafeSeaNet (SSN), in order to obtain a complete view of the movements of ships and dangerous or polluting cargoes in European waters. The VTMIS Directive mandated the development of National SSN applications which became operational by 2009.

National SSN applications differ from country to country, from basic low-end systems to high-end extended systems covering also Port Community Systems' needs and requirements.

Similarly, Port Community Systems (PCS) have become an essential component for the efficient operation of Ports. Typically PCSs have developed particular implementation guides for each Electronic Data Interchange Fact (EDIFACT) message, and support their Port Communities for interchange of Transport Orders, Bayplans, Gate Reports (Gate-in and Gate-Out), Bookings, Shipping Instructions, Custom Clearances, etc. Despite the value offered by PCSs to their communities the different implementation guides in each port create increased costs for the reporting parties.

Last but not least, the Electronic Port Clearance (EPC) ISO standard which was adopted by the International Maritime Organisation's Facilitation Committee (IMO FAL) in April 2013, lists ISO 28005 (ISO, 2013) as a reference for XML based electronic port clearance systems (EPC) in the FAL Compendium. Indicatively, ISO 28005 supports the following messages:

- All FAL standard declarations (FAL 1 to 7)
- International Ship and Port Facility Security (ISPS) reporting requirements
- All general ship reporting requirements
- Recommended reporting on ship generated waste
- Required reporting as defined in the bulk loading and unloading code
- Expected Time of Arrival (ETA) reporting to pilot station

In summary, it has to be noted that each system developed, has a specific focus, either trade, or safety or security and as such the content of the information and the dissemination level is different per system. For example, Port Community Systems are more trade oriented while the SSN is safety oriented. The SSN philosophy of a network of SSN systems around different geographical regions allows the (easy) exchanging of information among authorities (e.g. vessel movements). Private information on the other hand is oftentimes more difficult to be exchanged since the commercial character is more sensitive.

## 3 STATE OF PLAY IN THE MSW

### 3.1 *The WCO survey of single window implementation*

The WCO's survey of SW implementation (Choi, 2011) acknowledges five types of cargo clearance systems currently in operation:

- The Integrated Single Window: Individual data elements are submitted once to a single entry point (integrated automated system) to fulfil all import, export and transit-related regulatory requirements (i.e., enables multiple procedures to be performed from a single submission),
- The Interfaced Single Window: Individual data elements are submitted once to a single entry point but each regulatory agency maintains its own automated system and connects with other systems through custom-build interfaces,
- The Hybrid Single Window: A combination of the Integrated and the Interfaced approach,
- The One-Stop Service: Stakeholders are required to implement each procedure/declaration separately
- The Stand-alone system for Customs clearance.

The survey's most important insight is that 66% of the survey respondents (customs agencies) use a non-single window system, which duplicates efforts and increases compliance costs. Furthermore, the majority of the respondents (45%) indicated that they operate a stand-alone system, minimising the possibility of inter-agency communication and data fusion. Last but not least, the majority of Customs administrations indicated that they have harmonized single window data with internationally recognized standards and to this extent, the WCO Data Model and UN/EDIFACT were the ones widely adopted.

### 3.2 *The eMAR survey on the adoption of SW systems by EU ports*

The eMAR project has undertaken a comprehensive analysis (eMAR Consortium, 2013) of the different SW systems that are in use. The most important insights with regards to the Maritime Single Windows in European countries, include the following:

- Some kind of maritime IT systems exist practically in every EU country that has access to sea.
- All countries have established the EU mandatory SafeSeaNet system
- In some countries, SafeSeaNet system is the core element of Maritime Single Windows.

Several EU countries have established SSN systems that are tailored to specific country needs and interests.

Table 1 presents an overview of the different systems used by indicative ports across EU. Most of the ports surveyed, indicated that they have developed a Port Community system and they use this in their daily operations, acting as a Single Window.

Table 1. Overview of SW Systems used (eMAR Consortium, 2013).

| | Port single window | Port community system | National single window | Harbour authority system | Cargo community system | Polish harbours information & control system | Single point of contact |
|---|---|---|---|---|---|---|---|
| Antwerp | x | x | x | | | | x |
| Copenhagen Malmoe | x | x | | | | | |
| Cyprus Ports Authority | x | | | | | | |
| Dunkerque | | | | x | x | | |
| Bordeaux | x | | x | | | | |
| Esbjerg | | x | | | | | |
| Hamburg | | x | | | | | |
| Klaipeda | x | | | | | | |
| Livorno Port Authority | x | | | | | | |
| Luka Koper | x | | | | | | |
| Rauma Port | | | x | | | | |
| Southampton | | x | | | | | |
| Stockholm | x | | x | | | | |
| Szczecin and Świnoujście Seaports Authority SA | | | | | | x | |

Additionally, all of the respondents replied that these systems are bespoke and parameterized to the business models and the specific rules of each port. Although this is not a comprehensive nor representative list of ports, it serves as an indication of the focus and of the systems used by different ports. The authors selected ports that were offering cargo related services (e.g. containers, bulk, cars, etc) and Table 1 presents those that were finally interviewed. Based on the analysis, and on unreported feedback, no significant difference between public and private ports (within those interviewed) was identified. The difference among the systems used is primarily related to the services offered to the end users and to the port community in general. A comprehensive analysis of the services offered by the different systems identified in Table 1 is carried out by the eMAR Project (eMAR Consortium, 2013).

The main conclusion from this study is that all of the ports use a PCS or a variation of it. Such offerings are considered to be Value Added Services (VAS) for the ports and add to their competitive advantage. On the other hand, per the feedback collected shipping companies react positively to such systems and tend to select ports offering such VAS.

## 4 A PROPOSAL FOR A MARITIME SINGLE WINDOW: THE EMAR MSW

### 4.1 *A conceptual model*

Based on the eMAR recommendations (INLECOM Systems Ltd, 2014), EU Member States preparing for the implementation of the 2010/65/EU directive should consider the interactions between the different modes of transport (maritime, aviation, rail, road, inland waterways) as well as the cargo/passenger perspectives. Figure 3 indicates that National Single Windows need to address four fundamental aspects:

- Facilitation of business compliance to applicable regulations
- Exchange of information between national authorities, including coping with the different standards used by different authorities and/or different countries (i.e. IMDG, ADR, SSN, Customs, etc).
- Information exchange with external systems
- Value added Services for competiveness and growth.

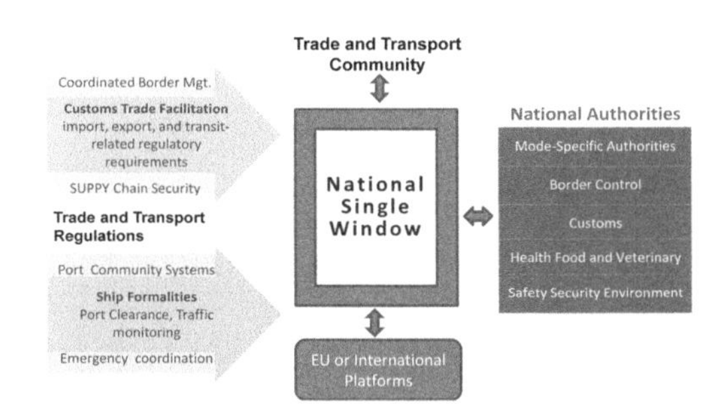


Figure 3. National Single Windows: the eMAR comprehensive approach.

### 4.2 *The eMAR MSW modules overview*

In the context of the eMAR Project, a set of tools, ontologies, methodologies and roadmaps were developed to facilitate rapid prototyping, development and deployment of Single Windows. The eMAR framework aims to upgrade the systems in producing intelligent MSWs that are interoperable and dependable. The Single Window Platform (SWP) is a solution for National Single Windows that allows member states to design and manage the MSW complying with 2010/65/EC directive, but allowing for sufficient autonomy for the different authority systems with respect to the information sharing and the access rights.

The eMAR MSW solution comprises of the following modules, which can be used either on their own or collectively:

1. Information Exchange between Reporting Parties and NSW
   - Receives an MSW message and converts it to Maritime Reporting Formalities (MRF)
   - stores MRF to NSW database or distributes to specified databases
   - sends a response message to Reporting Parties
2. Information Exchange between NSW and SafeSeaNet
3. Information Exchange between NSW and Authorities' systems (Systems Configuration)
   - Administrator can configure the Access Components for the authorities' systems.
   - Routing engine
4. Users Management
   - Central Administration management that can configure Access Rights for the Users—configuration specifies who will have access to the NSW and what permissions each user will have (e.g. View, Consult, Approve)
   - Migration/Synchronizer with existing authorisation systems.
5. Users Dashboard
   - Web interface for displaying information regarding Formalities, Notifications, etc. to different roles in different authorities.
   - Personalization assistants.
6. Intelligent Ship Reporting Gateways (ISRG), an innovative software application, enabling

both shipping Industry stakeholders to fulfil their reporting obligations to European Maritime & Custom Authorities, in accordance to the European Commission Directives as well as Authorities to effectively check and reference compliance.

### 4.3 *Key features*

The key features of the eMAR MSW Systems include the following:

1. MSWs provide a Single Interface between (trade and transport) establishments and authorities responsible for enforcing national, EU and international legislation.
2. MSWs are essentially (extended) eGov applications based on a National Data Set (and national data model) offering a Common Reporting Gateway to businesses for submitting Standard Messages (MSW-SM) which are stored by each MSW using the NDS.
3. MSWs contain basic services, including the Common Reporting Gateway, and are equipped with validation rules and User Management interfaces. User Management Systems implement common authentication rules allowing interlinking to user management of other NSWs and EU/multinational platforms.
4. Two main channels for the submission of reporting information/declarations are utilised:
   - Port Community Systems and similar systems that facilitate both commercial and regulatory information exchange. The interface between PCSs and NSW should be kept harmonised,
   - Business Reporting Gateways, that is commercial applications used by carriers, agents or aggregators to fulfil formalities
5. The MSW provides interfacing to external systems such as SSN. This implies that the NDS must contain all data elements to construct notifications as required by external platforms. Information exchange mechanisms should be developed for cooperation with other NSWs and additionally with platforms offering important information sources. The use of aggregated data available by EU and international databases should be considered early in the design of MSWs.
6. MSWs should include core services including:
   - Data Adapters primarily to link with existing authority systems but which could be extended to include adapters for linking commercial information sources to the MSW
   - Rules Engine to implement information sharing rules and event based information data flows
   - Data Quality Systems offering data integrity checks and monitoring quality indicators associated with submitting parties
   - Data Integration Systems offering common services to aggregate data according to specific requirements for authorities particularly for safety, security or environmental risk management
   - Change Management which can include observatory of legislation change, revisions to NDS and change propagation services helping user groups to implement changes in a timely and cost effective manner.
7. MSWs could also facilitate provision of Value Added Services. Provision can be made to allow service providers to offer VAS primarily for authorities but also to businesses when liability and confidentiality issues can be adequately resolved.
8. Last but not least, the systems under development should also offer intelligent services.

It has to be said though, that considering the state of the art for the maritime sector, intelligent MSW systems should primarily cover the fusion of relevant information to the relevant actors in its entirety, reducing the information submission duplication.

Figure 4 describes the relations between the proposed systems, including the normative and regulatory requirements.

Although utilizing Automatic Identification System/Long-Range Identification and Tracking (AIS/LRIT) information was not in the scope of this study, capturing and using both AIS and LRIT data in the maritime domain is a very promising area, especially for the NCAs. A forthcoming study from eMAR consortium reports on the value of using this source as data input for decision making (eMAR Consortium, 2014 (f)).

### 4.4 *eMAR Access Points*

The eMAR analysis has progressed beyond the state of the art by utilizing the Access Point technology

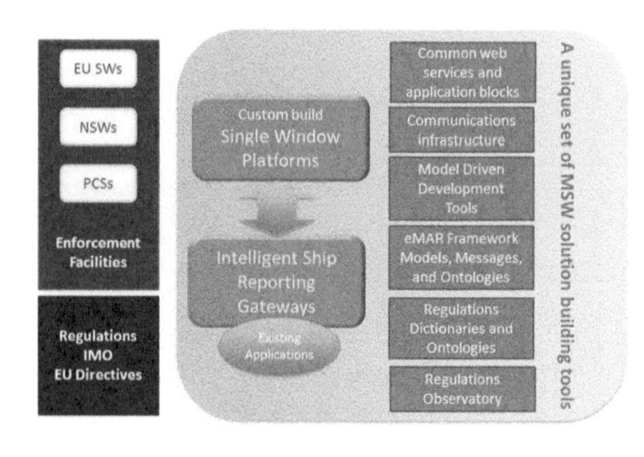


Figure 4. Relationships among the different SW systems: MSW driven approach.

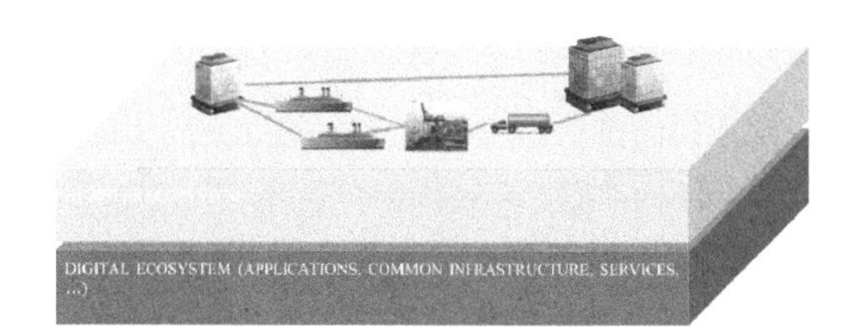


Figure 5. eMAR access points: an Ecosystem approach.

and combining this with an Ecosystem approach for the maritime industry. In an Ecosystem, a member is typically a company that participates in a commercial context and, in this process, connects with other members to exchange information. Ecosystem's members vary in characteristics, service provisions, objectives but they all have something in common: (a will for) sharing efficiently and effectively information, no matter if they are Shipping Service Providers, Logistics Services Providers, Governmental bodies, or anything else. Up to now, this communication is bi-directional, that is, the receiver gets the message from the sender(s) and this is replicated as many times as needed. The problems include multiplication of the same information and non-standardized information.

With an eMAR Access Point (AP), the exchange of documents and information between providers and consumers becomes easier. The "eMAR Ecosystem" supports a standard set of information types whereas the "eMAR Ecosystem Infrastructure" provides a robust gateway among Ecosystem members. A federated network of eMAR Access Points connected to each other using a common set of standards, within a community offer both push and pull services.

Maritime Logistics Service Providers offer their business as a value added service to their peer Maritime Logistics Service Consumers. The eMAR infrastructure supplements existing network services by providing the means for a participant to interconnect services instead of replacing existing services.

### 4.5 *Key drivers for success*

According to the above analysis, as well as to unreported feedback from the surveys carried out with competent authorities, the key drivers for the successful implementation of a Single Window include the following:

- A highly flexible and user-friendly tool for linking ship/shipping related information (e.g. voyage/ship cargo planning information) with port formalities reporting for use on-board and ashore
- Streamlining and reengineering of the reporting work-flows facilitating the exchange of ship and cargo information among all the actors involved in reporting, respecting their access rights on a "need-to-know" basis.
- Significant reduction of the reporting burden (in terms of duration and resource consumption) allowing the shipping/maritime company staff to focus on efficiency and safety of operations.
- Reduction of the overall cost of reporting by eliminating non-adding value intermediaries
- Reduction of IT complexity and enabling integration/sharing of information with other company "in-house" systems providing reference data for fulfilling reporting requirements and/or vessel tracking
- Compliance with international standards (e.g. ISO 28005, WCO, EDIFACT) and EU specific formats and requirements.

### 4.6 *Lessons learned*

Based on interviews from both government partners of the eMAR project (including The Royal Ministry of Fisheries and Coastal Affairs of Norway—Fiskeri—Og Kystdepartementet as well as the Valsts Akciju Sabiedriba Latvijas Juras Administracija—the Maritime Administration of Latvia) and from industry partners (including Danaos Shipping, DNV GL, European Community Shipowners' Association and the European Association for Forwarding, Transport, Logistics and Customs Service), the request to implement an MSW has very different origination, per the user group requiring it. However, all of the feedback collected so far agree on the following:

- The maritime and logistics procedures have to be simplified through the new system
- The shipping services have to be upgraded from a business perspective, and delay reduction is the most important element to this respect
- The new system has to decrease face-to-face interactions not only for cost reasons but also for improved transparency
- Upgraded security to the transactions has to be offered
- The data exchanging has to accommodate all the existing systems
- Paper based transactions have to be eliminated
- The new system will (have to) be universally a single point of data input, avoiding duplication
- The system will have to be as robust as possible to reduce risks from attacks, data leakage and virus
- The systems will have to be scalable with a simplified method of including new services for the users.

## 5 CONCLUSIONS

Maritime transport faces increased reporting formalities and procedures, which are often considered duplicative and time consuming, resulting in costs and delays that could render the industry less attractive. Maritime Single Windows (MSW) is promoted as a solution to simplify and facilitate ship reporting formalities. This paper addressed administrative complexities and reports on the lessons learned from developing and implementing an innovative single window solution for the maritime domain. Based on the experience and the analysis of the eMAR Project, a 3 Year EU funded project, the authors described a conceptual model of the MSW as well as the key features and the key drivers that a successfully user driven MSW should incorporate. The main objective of such systems is to significantly reduce compliance effort, enabling seamless trade within EU and abroad.

## ACKNOWLEDGEMENTS

The study is part of the eMAR Project, which receives funding from the European Commission under the No 265851 Grant Agreement in the context of the 7th Framework Program. The opinions of the authors do not necessarily reflect the opinions of the European Commission, the European Union nor the eMAR Consortium in its entirety.

The authors would also like to thank the anonymous reviewers for their comments s and suggestions to improve the quality of this paper.

## REFERENCES

Choi, J.Y. (2011). *A Survey of Single Window Implementation.*

eMAR Consortium. (2013). *D4.4 eMAR NSW Survey.*

eMAR Consortium. (2014 (f)). *D4.3 Interfacing e-Maritime with SSN and related developments.*

European Union. (2007). *Directive 2002/59/EC of the European Parliament and of the Council establishing a Community vessel traffic monitoring and information system and repealing Council Directive 93/75/EEC.*

European Union. (2010). *Directive 2010/65/EU of the European Parliament and of the Council on reporting formalities for ships arriving in and/or departing from ports of the Member States and repealing Directive 2002/6/EC.*

INLECOM Systems Ltd. (2014). *eMar delivering Intelligent Maritime Single Window Solutions.*

Ireland, R. (2009). *The WCO SAFE Framework of Standards: Avoiding Excess in Global Supply Chain Security Policy.*

ISO. (2013). ISO 28005, *"Security management systems for the supply chain—Electronic Port Clearance (EPC)".*

United Nations Economic Commission for Europe. (2005). *Recommendation and Guidelines on establishing a Single Windo.* New York & Geneva: UNITED NATIONS.

United Nations Economic Commission for Europe. (2010). *Data Simplification and Standardization for International Trade.* Geneva: UN.

World Customs Organisation. (2010). *The Single Window Concept: The World Customs Organization's Perspective.* Retrieved from www.wcoomd.org/en/topics/facilitation/activities-and-programmes/%7E/%7E/media/FA35ECDE953D4CDDA32A58D6F620B1FE.ashx.

*Maritime-Port Technology and Development – Ehlers et al. (Eds)*
*© 2015 Taylor & Francis Group, London, ISBN 978-1-138-02726-8*

# A study on estimation methodology of GHG emission from vessels by using energy efficiency index and time series monitoring data

T. Kano & S. Namie
*National Maritime Research Institute, Tokyo, Japan*

ABSTRACT: The National Maritime Research Institute of Japan (NMRI) has developed an Eco-Shipping Support System for domestic coastal shipping to provide energy saving navigation route and just-in-time speed plan. The operations by these services using weather routing are becoming a popular method combined with the rising fuel price. Accordingly, many simulation-based studies have been accomplished for comparative analysis. However, shipping company attempt to evaluate the effects of these operations based on actual data. The NMRI has developed methodologies to evaluate the amount of GHG emission reductions by using the speed plan services. One is based on Energy Efficiency Navigational Indicator (EENI), which is proposed to evaluate both loaded and unloaded conditions, in different size and speed of ships. The other is the estimate of operational performances of ship by using monitoring data. This paper presents applicability of these estimation methodologies to actual ships. Consequently, effectiveness and practical applicability was confirmed.

## 1 INTRODUCTION

The shipping industry has always taken a "hurry up and wait" approach, meaning merchant vessels would steam to meet a pre-agreed ship's speed regarding of fuel was burned with "full ahead" steaming leaving vessels often sitting idle at port awaiting berthing slots. To avoid risk of delay caused by weather and sea conditions these approaches are taken. If operated in this manner, any delay is due to weather, and the captain is absolved of responsibility. This operational style causes vast waiting loss time at port.

On the other hand, uses weather analysis and algorithm to calculate and agree a notional vessel arrival time, so that the ship will arrive "just-in-time". This radically reduces bunker fuel consumption and emissions, while easing congestion and enhancing safety. Shipping company seek to promote higher efficiency of ship's operation and also to evaluate effect of the operation.

The NMRI have developed an Eco-Shipping Support System for Japanese domestic shipping to provide energy-saving navigation route and speed plans, increase operational efficiency, improve fuel oil consumption efficiency, and reduce GHG emissions. The system uses the fruits of research and development of "Environmentally Friendly Shipping Support Systems for Coastal Vessels" (2006–2008) conducted by NMRI and supported by the New Energy and Industrial Technology Development Organization (NEDO). Trials conducted in Japan have shown that reductions in fuel consumption of 3–5% through route plans, and more than 20% through vessel optimum speed plans, can be expected, Kano et al. (2008) This trial results show that we could expect higher reduction in fuel consumption effects by optimum speed plan compared to route plan. And these operational means leads to GHG emission reduction.

Within the UN Framework Convention on Climate Change (UNFCCC), post-Kyoto Protocol discussions have included strong calls for $CO_2$ reduction measures in the area of the marine transport. The International Maritime Organization (IMO) promotes research into technical, operational, and economic means to reduce GHG emissions from ocean going ships. Under operational means, the organization seeks to promote higher efficiency of ship operations through improve voyage planning (IMO 2009, IMO 2011 & IMO 2012). The $CO_2$ emission regulation measures for international voyage vessels are examined by IMO. Regulations on energy efficiency (EEDI: Energy Efficiency Design Index (IMO 2011 & IMO 2012)) for ships entered into force on 1st of January 2013. Environment measurements for vessels are still under consideration by IMO. These indicators are expressed as the ratio of fuel consumption to transportation work (Akagi, S. 1995). The transportation work of EEDI is obtained from the product of deadweight and sailing distance. The product of cargo carried and distance corresponding cargo carried is transportation work of EEOI (Energy Efficiency Operational Index; IMO 2009). These indicators based on weight of cargo carried, is not able to evaluate at unloaded voyage. In the present study, the voyage work is defined by product of displacement and sailing distance in water since both loaded and unloaded conditions are targets for evaluation.

Speed planning of the system provides to reduce a vessel's speed on voyage to meet an arrival time at the destination port. The FOC per unit time is proportional to the cube of vessel speed. Therefore, speed reduction reduces an amount of fuel consumption and leads to GHG emission reduction. The methodologies to evaluate the effectiveness of reduction of GHG by using speed planning of this system was introduced, Kano et al. (2013, 2014).

In this paper, the estimation methodologies are introduced and applied to actual RoRo ships, corrected Energy Efficiency Navigational Indicator (EENI) of the ships at same speed was compared, applicability of the methodologies was confirmed and effectiveness confirmed.

## 2 ESTIMATION METHODOLOGY

### 2.1 *Evaluation methodology*

The effect of $CO_2$ emission reduction is based comparison between the case using the Eco-Shipping Support System and the case without using the system. The NMRI has developed methodologies to evaluate the effectiveness by using the speed plan services. One is based on Energy Efficiency Navigational Indicator (EENI) and the other is estimate of operational performances of ship by using monitoring data (Kano, T & T Seta, T. 2013, Kano, T & Namie, S 2014).

The following shows two evaluation methodologies.

### 2.2 *Methodology based on EENI*

To evaluate the vessel operational performance, $CO_2$ emission for vessel was considered and EEOI was proposed at IMO. However, EEOI was an indicator that was based on weight cargo carried and was not able to evaluate at empty voyage. EENI was proposed and applied as an indicator to evaluate both carried and empty conditions, in different size and speed of ships.

#### 2.2.1 *EENI*

The EENI for a voyage is the ratio between navigational work and $CO_2$ emission of a vessel with a displacement of $W$ tons and sailing distance $D$ shown as follows.

$$\text{EENI} = \frac{\Sigma_j FC_j \times C_{Fj}}{W \times D} \tag{1}$$

where $j$ is the fuel type; $FC_j$ is the fuel mass to $CO_2$ mass conversion factor for fuel $j$; $W$ is the displacement (tonnes); $D$ is the distance in nautical miles corresponding to the navigation.

By clarifying the physical meaning of EENI, it shows that it is proportional to the square of vessel speed and inversely with the third root of displacement (Kano, T & Namie, S 2014). Then the EENI is expressed as follows.

$$\text{EENI} = \frac{\Sigma_j (SFC \times C_{TV} \times v^2)_j \times C_{Fj}}{\rho \times g \times \mu \times \nabla^{\frac{1}{3}}} \tag{2}$$

where $C_{TV}$ is total resistance coefficient; $\rho$ is density of sea water; $R$ is total resistance in the sea conditions; $p$ is brake power; $\mu$ is propulsion efficiency.

It may be considered to be Froude number that uses displacement as a measure.

Assuming that the rate of loading is $Rp$, EENI has a relation with EEOI of the IMO guideline as follows.

$$\text{EEOI} = \frac{\text{EENI}}{Rp} \tag{3}$$

$Rp$ is shown as follows.

$$Rp = \frac{\text{Mcargo}}{W} \tag{4}$$

where Mcargo is cargo carried (tonnes).

$Rp$ is affected by the market and varies considerably. Therefore, EEOI for each voyage vary considerably too.

#### 2.2.2 *Evaluation*

First, EENI for each voyage is calculated for a certain period time X, and the following average is obtained at regular voyage without using Eco-Shipping Support service (BAU—Business As Usual—navigating the regular route with regular speed).

$$\text{Average EENI}_X = \frac{\Sigma_i \Sigma_j FC_j \times C_{Fj}}{\Sigma_i W_i \times D_i} \tag{5}$$

where $i$ is the voyage number.

EENI varies due to external forces such as wind and waves even when the same vessel navigates at same route. However it is considered to be convergent after taking average of navigations.

$CO_2$ emission that corresponds to total work becomes presumable by using Average EENIx obtained during period X (a period for the average to become constant). Defining it as reference $CO_2$, the value during period $Y$ could be obtained.

$$(\Sigma_i \Sigma_j FC_j \times C_{Fj})_{BAU:Y} = \text{Ave} \cdot \text{EENI}_X \times (\Sigma_i (W \times D)_i)_Y \quad (6)$$

Eventually, the $CO_2$ reduction $\Delta CO_2$ during period $Y$ with Eco-Shipping Support service could be shown as follows.

$$\Delta CO_2 = (\Sigma_i \Sigma_j FC_j \times C_{Fj})_{BAU:Y} - (\Sigma_i \Sigma_j FC_j \times C_{Fj})_{Project:Y} \quad (7)$$

The following shows the reduction ratio $Rd$.

$$Rd = 1 - \frac{\text{Ave} \cdot \text{EENI}_{Project:Y}}{\text{Ave} \cdot \text{EENI}_{BAU:X}} \quad (8)$$

EENI could be shown as follows and is proportional to the square of vessel speed. Therefore, $K_i$ is defined as follows.

$$K_i = \frac{\text{EENI}}{v_i^2} \quad (9)$$

The following shows the average $K_n$.

$$\text{Average } K_n = \frac{\Sigma_{i=1}^{n} \text{EENI}_i}{\Sigma_{i=1} v_i^2} \quad (10)$$

The effect of $CO_2$ emission reduction ratio using the Eco-Shipping Support service could be calculated by measuring average vessel speed $v_{BAU}$ (assumed as the support service did not use for the ship operation) and $v_{Project}$ (assumed as the support service used for the ship operation) during period $X$ and period $Y$.

$$Rd = 1 - \frac{\Sigma_i v_{project}^2}{\Sigma_i v_{BAU}^2} \quad (11)$$

Furthermore, $K_n$ makes the operational performance of different ships navigating at same speed comparable.

$$Rd = 1 - \frac{\text{Ave} \cdot K_{n:ShipA}}{\text{Ave} \cdot K_{n:ShipB}} \quad (12)$$

### 2.3 *Methodology based on time series monitoring data*

Supposing the case where actual navigation/engine data in support service uses for ship operation are obtained as follows; Power of main engine (kW); $P$, Fuel oil consumption in unit time (ton-fuel/hour); $f$, Specific Fuel oil Consumption; $SFC$, Logging speed (knot); $V$. The previous study (T. Kano & S. Namie (2014)) had proposed a methodology of the external force of power increase per unit time to be constant. Consequently, it was assumed that the additional power $\Delta P_{project}$ against external resistance $\Delta R$ is just same as that $\Delta P_{BAU}$ for different output operation passing through same waypoint on the route.

In this paper, the methodology was modified to improve the estimate accuracy that external resistance $\Delta R$ which is determined by sea and weather states, is equal to that for different output operation passing through same waypoint on the route, that means,

$$\Delta P_{project} \times \left( \frac{V_{BAU}}{V_{project}} \right) = \Delta P_{BAU} \quad (13)$$

$$\Delta P = \frac{\Delta R \times V}{\mu}. \quad (14)$$

After all rewriting the equation on referring to equation (13), it is estimated $CO_2$ emission when the ship operates in BAU on the same route in the same manner. For further detail refer to previous study.

## 3 APPLICATION TO THE ACTUAL SHIPS

### 3.1 *Target ships*

The route of the investigated ships is mainly round-trips from Tokyo to Tomakomai (Northern part of Japan). There are two schedule patterns, 33 and 27 hours from Tokyo to Tomakomai, and 34.5 and 27 hours from Tomakomai to Tokyo (Table 1).

Table 2 shows the particular principal and time schedule of the target Japanese coastal RoRo ships.

Following analysis was conducted using monitoring data of the Ship A and Ship B for the duration of 3 months from October to December 2013.

### 3.2 *Application of the methodology based on EENI*

#### 3.2.1 *EENI*

Figure 1 shows EENI obtained from each shipping route and each voyage displacement. Those

Table 1. Time schedule.

| | Tokyo to Tomakomai | | Tomakomai to Tokyo | |
|---|---|---|---|---|
| Sailing time | 27 hours | 33 hours | 27 hours | 34.5 hours |

Table 2. Principal particulars.

| | Ship A | Ship B |
|---|---|---|
| Length | 166.9 m | 161.15 m |
| Breadth | 27 m | 24 m |
| Depth | 12.5/6.7 m | 12.0/6.4 m |
| MCO | 13920 kW | 16920 kW |
| Gross tonnage | 10497 tons | 7323 tons |
| Nominal speed | 23.0 knots | 23.0 knots |

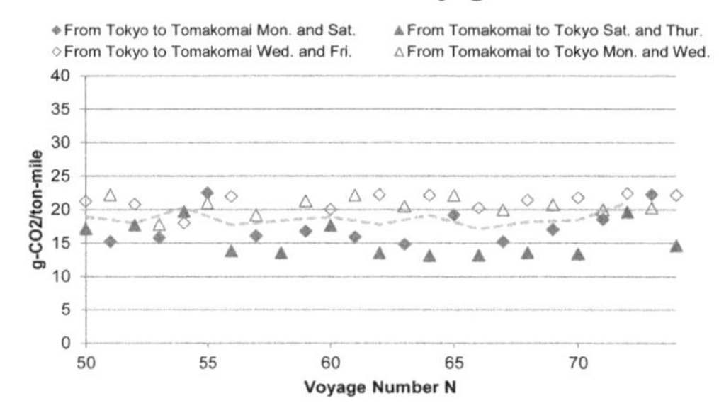


Figure 1. EENI for each voyage.

values vary widely, even though in the same shipping route. It seems that the values of longer sailing time are smaller.

Figure 2 shows average EENI through voyage number up to N. As the values of N increase, the differences decrease further. The average value over approximately 10 voyages approaches convergent value. This convergent value is smaller when the navigation time is longer. In other words, the value becomes smaller as the required speed increases.

1. $K_i$

Figure 3 shows $K_i$ obtained from each shipping route and each voyage displacement. Those values still vary widely according to each voyage but seem to be more coherent in comparison with that with no correction for ship's speed. It seems that the values of longer sailing time are smaller.

Figure 4 shows the average EENI $K_n$ through voyage number up to N. Irrespective of both shipping route and schedule, the values are convergent. It shows the vessel average performance in actual navigation under external forces might be convergent in sufficient voyage number N. The performance is proportional to the square of vessel average speed. The proportional constant $K_n$ becomes easier to evaluate the ship's performance and fuel consumption reduction or $CO_2$ reduction.

Table 3 shows the average EENI of each schedule of ship A and evaluation results. Assuming the Usual navigation condition (Business As Usual),

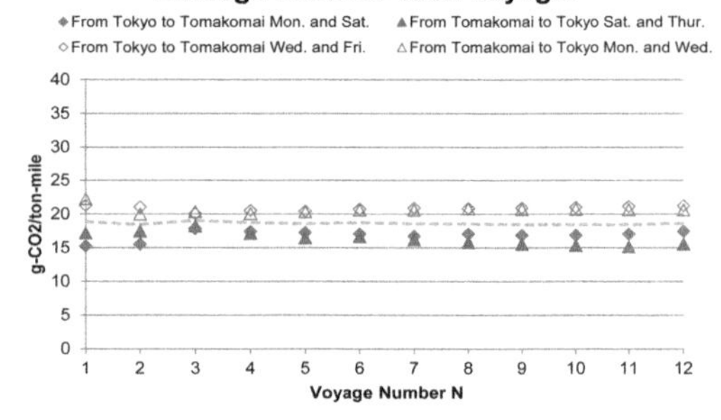


Figure 2. Average EENI for an obtained number of voyages.

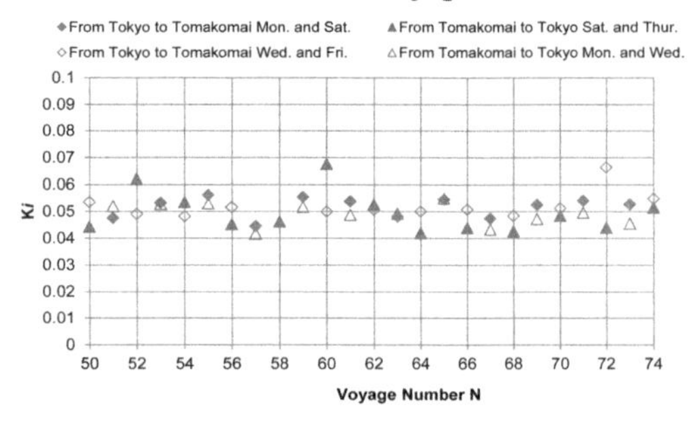


Figure 3. $K_i$ for each voyage.

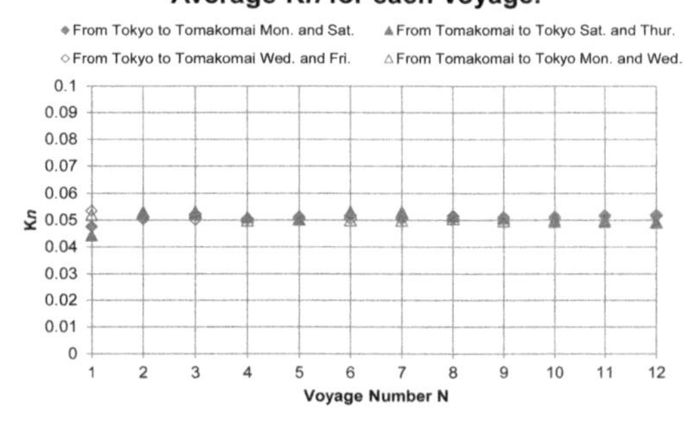


Figure 4. Average $K_i$ for an obtained number of voyages.

ship schedule 27 hours of Tokyo to Tomakomai and Tomakomai to Tokyo respectively.

The reduction ratio is 17.9% and 25.2% from equation (8). And, if average $Kn$ is constant, the reduction ratio is able to estimate using average of ship speed by equation (11). These values (17.9% and 25.5%) are coincides with the values from equation (8).

Table 3. Average EENI of each schedule and evaluation result (Ship A).

| | Tokyo to Tomakomai | | Tomakomai to Tokyo | |
|---|---|---|---|---|
| Sailing time (hours) | 27 | 33 | 27 | 34.5 |
| Num. of voyage | 12 | 12 | 12 | 13 |
| Ave. speed (knot) | 20.2 | 18.3 | 20.5 | 17.1 |
| Ave. EENI | 21.2 | 17.4 | 20.6 | 15.4 |
| Reduction ratio | – | 17.9% | – | 25.2% |
| Ave. K at 20 kts | 20.8 | 20.6 | 19.6 | 19.8 |
| $\frac{V_r^2 - V_p^2}{V_r^2}$ | – | 17.9% | – | 25.5% |

Table 4. Average EENI of each schedule and evaluation result (Ship B).

| | Tokyo to Tomakomai | | Tomakomai to Tokyo | |
|---|---|---|---|---|
| Sailing time (hours) | 27 | 33 | 27 | 34.5 |
| Num. of voyage | 13 | 10 | 14 | 8 |
| Ave. speed (knot) | 20.7 | 17.0 | 20.0 | 15.8 |
| Ave. EENI | 35.1 | 24.1 | 32.7 | 20.7 |
| Reduction ratio | – | 31.3% | – | 36.7% |
| Ave. K at 20 kts | 33.0 | 33.3 | 33.0 | 33.3 |
| $\frac{V_r^2 - V_p^2}{V_r^2}$ | – | 32.6% | – | 37.6% |

Table 4 shows the average EENI of each schedule of ship B and evaluation results with same assumption. The reduction ratio is 31.3% and 36.7% from equation (8). And, the reduction ratio is estimated by equation (11). These values (32.6% and 37.6%) are coincides with the values from equation (8).

The estimated values by two different methods almost agree. On the basis of the experimental result, we speculated that the equation (11) provides simple practical approach for estimation, because it requires monitoring ship speed only. Further experiments will be carried out in the future for verification.

The noteworthy point from the Table 3 is that average $K_n$ at 20 knots of ship A estimated by different time schedule of Tokyo to Tomakomai (20.8 and 20.6) or Tomakomai to Tokyo (19.6 and 19.8) convergent almost same value respectively. These values might be shown the character of the routes, although it is not clear from the Table 4.

2. Comparison EENI of these ships

Figure 5 shows corrected values of EENI corresponding to 20 knots for ship A and ship B by equation (10). We can found the value of EENI of ship A is superior to that of ship B at the same speed.

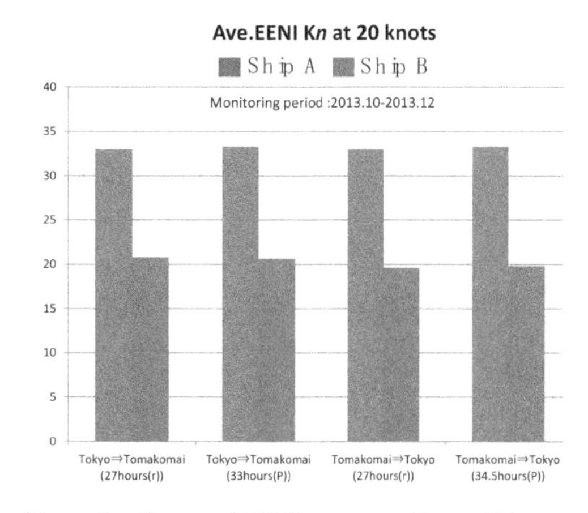


Figure 5. Corrected EENI corresponding to 20 knots.

### 3.3 *Application of the methodology based on time series monitoring data*

For example, investigated navigation distance and sailing time is respectively 324 miles and 15 hours. The logging speed of optimal navigation, $V_{log}$, is about 17.8 knots, which is compared with speed 21.5 knots of the reference navigation. We choose

Table 5. Evaluation result (Ship A & B).

| Ship A | | Ship B | |
|---|---|---|---|
| *$CO_2$ emission base (ton)* | | | |
| BAU | Reference | BAU | Reference |
| 78.8 | 56.2 | 175.2 | 130.5 |
| *$CO_2$ emission reduction* | | | |
| 22.6 (28.7%) | | 44.7 (25.5%) | |
| *$CO_2/T$ (ton—$CO_2$/hour)* | | | |
| BAU | Reference | BAU | Reference |
| 5.37 | 3.83 | 8.62 | 6.42 |
| *$CO_2/D$ (ton—$CO_2$/mile)* | | | |
| BAU | Reference | BAU | Reference |
| 0.243 | 0.173 | 0.367 | 0.273 |
| *EENI* | | | |
| BAU | Reference | BAU | Reference |
| 18.4 | 13.1 | 36.6 | 27.3 |
| *K* | | | |
| BAU | Reference | BAU | Reference |
| 0.0398 | 0.0284 | 0.0778 | 0.0579 |

ship A to mention the contents of Table 5. Table 5 indicates the result of evaluation.

Total $CO_2$ emission for the ordinary BAU navigation is 78.8 tons, while that for optimal nav. is 56.2 (ton), that means the $CO_2$ reduction is 22.6 (ton) and reduction rate is 28.7%. These results owes to the reduced logging speed Vlog in optimal navigation comparing to the BAU speed Vr. Another indicators for evaluation are $CO_2$/(sailing time) and $CO_2$/D, and EENI and K. These values are also indicated in Table 5 both for BAU navigation and for optimal navigation (Reference).

## 4 CONCLUSIONS

These operations using weather routing are becoming a popular method combined with the rising fuel price (Jukka, I. et al. 2014, Takashima, K. et al 2007, Yohan, G. & Claude, A. 2014). Accordingly, many simulation-based studies have been accomplished for comparative analysis. Shipping company seeks to promote higher efficiency of ship's operation and undertake the evaluation of the effect of these operations based on actual monitoring data.

The authors have developed two methodologies to evaluate the operation by using the speed plan services. First is the Energy Efficiency Navigational Indicator (EENI) is proposed to evaluate both loaded and unloaded conditions, in different size and speed of ships. The other is estimate of operational performances of ship by using monitoring data. Consequently, the applicability was confirmed in this paper.

The Summary of results is as follows.

First, average Energy Efficiency Navigational Indicator (EENI) was proposed as an indicator to evaluate the vessel performance at actual sea and was applied RORO ships. It had been confirmed that it becomes constant after taking the average of several voyage approximately 10 voyages.

Second, by clarifying the physical meaning of EENI, it shows that it is proportional to the square of vessel speed and inversely with the third root of displacement. $K_n$ was introduced to exclude ship speed effect, $K_n$ make the operational performance not only same ship and but also different ships comparable.

Third, regarding speed reduction, we have proposed a methodology to estimate $CO_2$ emission when the ship navigates the same route at different power by using monitoring data from actual navigations.

For the purpose of evaluating the effect of $CO_2$ emission reduction (FOC reduction effect) using the Eco-Shipping Support System, two methodologies were proposed and applied to the RoRo ship and practical applicability was confirmed.


## ACKNOWLEDGMENTS

We had started a project from 2013 to 2015 supported by Ministry of Environment (MOE) of Japan. We would like to express our sincere appreciation to the MOE, and Nippon shipping Co., Ltd for their kind and invaluable cooperation in collecting data. We are also indebted to Ms Kon and Ms Yamazaki (MTL) are gratefully acknowledged for their valuable suggestions and persevering job.


## REFERENCES


Akagi, S. (1995), Transportation Vehicles Engineering (Social Demands Technology), Corona Publishing Co., Ltd.

IMO 2009, Guidelines for Voluntary Use of the Ship Energy Efficiency Operational Indicator (EEOI), MEPC.1/Circ. 684, 17 August 2009.

IMO 2011, Amendments to the Annex of the Protocol of 1997 to Amend the International Convention for the Prevention of Pollution From Ships, 1973, As Modified by the Protocol of 1978 Relating Thereto, Annex 19, Resolution MEPC. 203(62), Adopted on 15 July 2011.

IMO 2012, Guidelines for the Development of A Ship Energy Efficiency Management Plan (SEEMP), Annex 9, Resolution Mepc. 213(63), Adopted on 2 March 2012.

Jukka, I. et al. (2014), A Comprehensive Performance Management Solution, 13th Conference on Computer and IT Applications in the Maritime Industry (COMPIT), 559–568.
Kano, T., Kobayashi, M., Toriumi, S. (2008), Energy saving operation for coastal vessel, Papers of National Maritime Research Institute of Japan 7(4), 425–472.
Kano, T., Seta, T. (2013), A study on the Eco-Shipping support system for keeping regularity of ship's schedule, 15th Int. Congr. Int. Maritime Assoc. Mediterranean (IMAM), 667–674.
Kano, T. & Seta, T. (2013), Application of set partition based fleet scheduling system to Japanese coastal oil tanker fleet, 15th Int. Congr. Int. Maritime Assoc. Mediterranean (IMAM), 691–696.
Kano, T. & Namie, S. (2014), A study on Estimation of GHG Emission for Speed Planning Operation Using Energy Efficiency Index and Time-series Monitoring Data, 13th Conference on Computer and IT Applications in the Maritime Industry (COMPIT), 167–180.
Takashima, K. et al. (2007), Energy Saving Operation for Coastal Ships based on Precise Environmental Forecast, the Journal of Japan Institute of Navigation 118, 99–106.
Yohan, G. & Claude, A. (2014), E2-An Advanced Ship Performance Monitoring Tool, 13th Conference on Computer and IT Applications in the Maritime Industry (COMPIT), 238–244.

*Maritime-Port Technology and Development – Ehlers et al. (Eds)*
 *ISBN 978-1-138-02726-8*

# A pathway towards more sustainable shipping in 2050: A possible future for automation and remote operations

Gabriele Manno
*DNV GL, Strategic Research and Innovation, Høvik, Norway*

ABSTRACT: Developments in ICT will have a profound effect on the shipping industry, providing new ways to analyze ship functions to significantly improve efficiency and safety performance. Similarly, advances in automation will shape the ways ships are designed, built, and operated. Sensor technologies and monitoring systems, combined with seamless ship-shore connectivity and software-enabled decision support tools, will create a more data-centric, responsive and flexible industry that is fully integrated with global transportation networks. This paper describes briefly the state of the art on ship automation and remote operations and presents a possible future scenario for shipping where onshore control centers will be responsible for the management of the ship. Main drivers, barriers and business opportunities are investigated alongside with the possible changes these developments may impose on the current business models of the main stakeholders involved in ship design and operation.

## 1 INTRODUCTION

Today, we live in a world that is continuously becoming more data-driven and automated, where physical systems and people are increasingly connected and mirrored into a virtual space. Key developments in ICT include sensor technologies, improved ship-shore connectivity, advanced software tools and algorithms, increased computing power and faster processing times. ICT has also enabled more far-reaching concepts (e.g. Big Data, "Internet of Things", Cloud Computing, etc.) which will provide the shipping industry with new ways to collect, store and process valuable data (Miorandi et al., 2012).

Advances in ICT have occurred so rapidly that they have outpaced existing systems used by the shipping industry to manage a broad range of challenges. Indeed, as more and more land-based industries adopt ICT systems to improve performance, the shipping industry will be compelled to do the same. In this way, the technology—not the demand—will drive change in the shipping industry.

ICT will have a dramatic effect on how the industry manages information and operates its assets. Most systems and components will be linked to the Internet, making them accessible from almost any location. This connection enables a virtual reality made up of data, models, and algorithms, embedded in software, databases and information management systems. At the same time, by combining data streams from multiple sources, the sheer volume of information available will enable the industry to make more informed decisions, faster, leading to more efficient and responsive organizations. In time, these databases will be accessible through vast information management systems combined with fast computing and advanced software via distributed networks.

To realize these potentials, different stakeholders must manage a broad range of challenges, including the growing complexity of systems, data networks, sensor technologies, systems integration, tools to manage increasingly large volumes of data, and processes to ensure software integrity and data security (Atzori et al., 2010). Furthermore, the adoption of any new technology requires users to change existing behaviors and develop the right competencies. In our view, the impact of ICT will be far-reaching and develop quickly. However, it will most likely take some time before legacy systems now used by the shipping industry are replaced. ICT may also challenge traditional competitive business models, which often act as a barrier to the sharing of information. Certainly, owners will have to invest in systems to protect and secure sensitive data and the integrity of software systems, but the full benefits of this technology cannot be fully realized unless the industry learns to be more transparent and cooperative (Bughin et al., 2010).

For the shipping industry, ICT will change how ships are designed and built, what materials are used, how ships are operated and how shipping fits into the global supply-chain logistics network. In this work, we will focus on one area where we believe ICT is likely to have a big impact: Automation

& Remote Operations. As sensor technologies and connectivity become more robust, remotely operated vessels, or even unmanned vessels, could become a reality. We are also likely to see many of the traditional activities performed on board shifted to shore-based centers, responsible for vessel health management, operations management and logistics.

The remainder of this paper is organized as follows: Section 2 describes the main developments in ship automation of the last 50 years; Section 3 describes a possible future scenario where onshore control centers are responsible for the management of the shipping transportation value-chain; Section 4 outlines the main drivers and barriers to the development of automation and technologies for remote operations, as well as the possible impacts on the existing business models; Section 5 presents a future storytelling; and, finally, Section 6 concludes the paper.

## 2 AUTOMATION AND REMOTE OPERATIONS IN SHIPPING

Over the last decade, expanding computing power and faster processing times have outpaced the ability of humans to manage complex systems effectively. Indeed, computers are much more effective at managing low-level intensity situations than humans, and have proven effective in supporting personnel during high-intensity events. In other industries, more and more systems are automated or controlled from remote locations. Over the next decade, it seems likely that the shipping industry will increasingly look to the offshore industry, which has developed a number of automated systems with marine applications, to improve performance.

### 2.1 *A brief history of automation in shipping*

Following WWII many other industries than shipping were introducing innovative automation solutions at a rapid rate and achieving major improvements in productivity and manpower utilization (Saveriano, 2010).

The first milestone in ship automation can be tracked back to 1961 when Japan announced to the maritime world the first successful automated seagoing ship, the KINKASAN MARU. The ship was fitted with bridge control of the main propulsion plant and a centralized control system for all engine room machinery. Progresses towards increasing automation in ships continued through the 60s and resulted in the production of a series of highly automated merchant ships during the period 1969–1973. The main force driving these developments was the quest for a reduction in ship manpower requirements, since the Japanese industry was facing a severe shortage of qualified seagoing professionals.

Also in the Scandinavian countries efforts directed towards the development of ship automated systems were initiated in the late 60s. Norway has been in the forefront of development and has successfully introduced innovative automation solutions to most of the operating functions of modern seagoing ships. Scandinavian countries were amongst the firsts that recognized that: (i) automation could have a major impact on almost all aspects of the maritime industry; (ii) that the cost and risks of introducing electronic hardware to ships would be high; and (iii) that, nevertheless, the payback to the whole industry would be very attractive. Moreover, differently from Japan, the Scandinavian countries recognized immediately that ship automation could lead to improved operating efficiencies, a reduction in maintenance costs, unscheduled repairs and downtimes, improved job satisfaction and increased safety.

Indeed, although some reduction in crew size has resulted from automation, many automated ships are still not operated at minimum crew levels authorized by statutory rules and regulations. This is mainly due to the fact that planned maintenance programs are generally carried out during sea voyages in order to reduce waiting times at ports. Automation was instead seen as a means to improve the sociological aspects of the seagoing profession rather than just focusing on technical improvements.

### 2.2 *The evolution of the enabling technologies*

The prime catalyst supporting the introduction of automation solutions in maritime and non-maritime industries has been the availability of intelligent electronics. Computers, processors, sensing devices and associated control circuitry permit the monitoring and control of complex functions. Since the 60s these developments have been further enhanced through the introduction of more reliable solid state components that have resulted in automated shipboard systems that can perform at sea for extended periods in a relatively hostile environment, isolated from specialized personnel and facilities.

In this way, a journey started with analogue electronics systems and relay-based logic, continued through the use of dedicated micro-controllers and Programmable Control Logics (PLC), distributed network control systems, and, finally, through the use of fast, low-cost and mass produced Windows-based PCs as workstations. Software development in the early PCs was far from complete, resulting in unreliable control systems, but thanks to the involvement of IACS and the leading class societies continual rule updating based on compute science and automation expertise has helped ensure the required reliability of today's versions.

Automation is today an integral and essential part of the ship and its operation. Heavier loadings, more complex operations and smaller tolerances require the adoption of advanced control and monitoring systems. For instance, many of the traditional mechanical parts of a ship propulsion system are being replaced by electronic and electrical equipment. Modern machinery cannot function without automation.

Finally, further developments of ICT within ship navigation focus on the integration of bridge system, ranging from simple position-fixing equipment, through automatic navigation and course-keeping systems to complete network-based workstation arrangements serving all the primary bridge functions. Nowadays, the total bridge system achieved through these advancements in automation and information and communication technology enables the officer of the watch, under normal conditions, to perform all the bridge functions without extra assistance.

### 2.3 *The increasing role of software*

At the start of the IT era, hardware was the significant cost factor and few talked about commodities. Today, the hardware portion of the total cost of a system is marginal, and most programs are either freeware or commodity-priced. Perhaps, software characteristics are a major importance factor in modern business. Used correctly, software solutions enforce common standards and operational consistency; disseminate past and present knowledge; and expand a company's capability. The complexity of today's software systems requires the use of abstract modelling techniques to be able to handle all the necessary data.

Software holds great promise. With its help, high-speed ships can come to a stop more quickly, it can automatically position itself, and it can release fewer emissions to air and water. The performance of computers and software could well continue to double every year. But it has to be recognized that only information, not knowledge, can be retrieved or re-used by ITC. Hence it is vital to establish the right workflow to enable people and software solutions to best exploit and complement each other' capabilities.

### 2.4 *Looking ahead: The ship gets a nervous system that is connected to land-based control centers*

The use of sophisticated robotics and automation are now commonplace for many land-based industries, particularly in manufacturing. In the past decade, we have seen the deployment of a number of unmanned autonomous and remotely operated vehicles, including Unmanned Aerial Vehicles (UAVs), Remote Operated Vehicles (ROVs), and the development of driverless trucks and autonomous cars (Naeem et al., 2012; Munin, 2013).

For shipping, remote operations will require automation of the engine and other integrated systems, alongside advanced navigation systems and sophisticated software that can manage smart sensor and actuator networks, maintain a vessel's course in changing sea and weather conditions, avoid collisions, and operate the ship efficiently, within specified safety parameters (Bhaskar et al., 2007; Guan et al., 2012; Vigna, 2013). This will increase the amount of data generated from shipping, thus, in the future new technologies for handling Big Data will become even more necessary. The "Big Data" provided from various levels of collaborative sharing and properly structured, will provide new insight. We can see this already in fleet operational efficiencies as wind/current data is mapped to AIS positioning data for ships—resulting in efficiencies in route planning and thus reduced fuel use. The data is not from one source/provider but collected through shared collaborate data models in structured form. In Remote operations, systems will need to be able to integrate data from different sources as from onshore control centers, other ships, weather and oceanography stations, ports, etc. Therefore, advanced systems for analyzing the enormous amount of data that will be generated by Remote operations will be needed. These systems need to be deployed both on board the ship and onshore. This will require advances in data storage and computing, mathematical modeling and software with optimized response that will be embedded in computer systems (Halevy et al., 2005). Finally, the software should offer a high level representation of the main risks and parameters as well as the possibility of full control over automated systems from the onshore control centers.

This system will also rely on robust and secure communications via satellite and land-based systems (Dark, 2012). Therefore, IT connectivity and reliable communication links as well as robust communication architecture between ship and onshore centers, that ensure that the onshore and offshore components are appropriately connected, will be a fundamental enabling technology for the diffusion of remote operated vessels. In this context issues related to bandwidth and security will be the most important to handle. The on board ship control and decision management system can be adjusted to allow different levels of autonomy, but with further advances in these enabling technologies, we can imagine a completely autonomous ship that reports to shore-based operators only when human input is needed, or if emergency situations arise.

Finally, while the idea of remotely operated vessels remains controversial, the development of such systems will not be limited by technology. Rather, the industry will have to weigh the benefits of remote operation, which include reduced manning costs, increased safety and improved vessel condition, against their perceived risks (see Section 4).

## 3 TOWARDS 2050: THE ADVENT OF ONSHORE CONTROL CENTERS

Shipping will benefit from developments in the offshore, aviation, aerospace, and automotive industries, which have been the primary drivers for advances in automation and remote operations. Shipping will likely apply these technologies to instrumented machinery first and then gradually to vessel navigation, which might be operated remotely from shore-based centers. These solutions will increasingly rely on sensor technologies and computers to manage onboard systems from remote locations. As more onboard systems become automated, the number of onboard personnel will be reduced, and more decisions will be made from shore-based control centers.

Onshore control centers will be responsible for the condition management of the ship and risk related to the failure of onboard equipment or broken communication links. These control centers will be responsible for operating vessels in congested sea-lanes, or in proximity to ports and terminals, and in emergency situations. To manage these tasks, control centers will be equipped with system simulators designed to select optimal routing procedures and interfaces with land-based supply chain networks. As with many emerging technologies, the ability of the system to manage the interaction between man and machine will be critical. Such systems should provide accurate representations of risk and allow humans to take full control of vessels from a remote location, when necessary.

### 3.1 *Ship health monitoring centers*

As advanced real-time condition monitoring becomes a reality, asset maintenance will broaden to allow owners to assess vessels in a life-cycle perspective. Today, some engine manufacturers have systems in place to collect maintenance data, which they can analyze and use to recommend actions. In time, manufacturers, system integrators and related service providers will be able to support owners with real-time critical diagnostic and prognostic information about the conditions of various on board systems, providing specific guidance to maintenance crews via virtual-space software and hardware.

Therefore, a Health Monitoring Center will have the task to monitoring the main parameters of a ship in order to take optimal actions on the reliability, safety and risk related to the vessel. Such analyses can be run e.g., at the system integrator/maintenance provider center, but results need to be sent to the control center that is responsible for the operation of the vessel.

### 3.2 *Ship operation control centers*

The operation control centers will be responsible of the pilotage, traffic services and overall risk management of the ship. There could be different centers depending on e.g., sea areas, type of traffic, ship segments, clusters of companies. System simulators will be used in many practical cases to choose optimal solutions.

Control centers cover not only the generic concept of ensuring that the mission is properly planned and considered, but also the specific matter of uploading mission instructions and related information into the remotely operated onboard computers. Control centers are fundamental to the successful and safe operation of an autonomous ship. To ensure this, personnel engaged in control centers should be appropriately trained, competent and fully conversant with mission planning software and all other aspects of unmanned operations.

Finally, the human-machine interface, with ergonomic control levers and joysticks, visual displays and alarms, plays a central role, whereas an important aspect for assuring safety and control of such system is not to leave the human out of the loop. However, towards 2050 many optimization decision would be taken by computers/the vessel itself, and the onshore operators will be responsible only of managing emergency situations and critical planning and routing activities.

### 3.3 *Vessel logistic control center*

Information Management offers to businesses the possibility to learn the ability to view physical operations more effectively. This means that the foundation for the virtual value chain is used to coordinate the activities of the physical value chain. Furthermore, with the assistance of IT, it is then fully possible to plan, implement, and assess events with greater precision and speed. In this way, the physical value chain is mirrored into the virtual one, giving the possibility to move also some of the value-adding activities into the market virtual space. This last step allows businesses to present value to the customer by new means and in new fashions, whereas the new relationship between business and customer is based on IT. However, this has to be structured in such a manner that all the involved actors are providing their own specialization skills, capabilities and data. Moreover, a common agreed integration platforms must be defined, so that the information is accessible to all the parties involved, while it should allow to maintaining the unique capabilities (competitive advantages) of the different actors.

# 4 DRIVERS, BARRIERS AND IMPACTS ON BUSINESS MODELS

## 4.1 *Drivers*

### 4.1.1 *Sustainable performances*

Performance management will become increasingly important to meet demands for transparency, particularly on sustainability performance and emissions. Companies will need tools for decision support and optimization, and for documenting fuel consumption and reduction for charterers so as to justify investments in energy efficiency. Moreover, an increasing legislation attention towards transparency and control will drive developments towards more data-shared based decisions (Baumgarten et al., 2009).

- **Monitoring, Reporting and Verification:** By the means of more sophisticated data analysis and automated transmission of data and reporting from ship to shore and the widespread implementation of e-formats, data and information exchange will become less challenging. In addition, there is a potential to increase transparency of operations and ensure compliance with port state control requirements using semi-automated procedures (Boccaletti et al., 2008).
- **Quality management:** Shore support can engage the crew in a continuous improvement process. Therefore, while (perhaps) fewer skills will be needed for the crew, at same time, the crew has the possibility of learning from the experience gained on the job with the support of the more skilled operators onshore. Also the availability of Big-Data will offer a tool for continuous improvement where benchmarked data can be used as a feedback loop to improve operations and design.

### 4.1.2 *Safe and reliable operations*

Reliable and safe operations contribute to the overall system performances. The request for solutions providing control over the status of degradable systems, increase in situational awareness and human reliability, support in the definition of corrective actions and reduction of operational risk will drive the adoption of new digital and automation technologies in the shipping industry.

- **Increase Human reliability:** In terms of advancing maritime safety, Remote operations will provide much improved surveillance technology that can be used in the vicinity of the ship. This will allow for better object identification and relieve conventional ship' crews from tedious and repetitive tasks. Watch-keeping at sea and monitoring machine performance are two tasks where automation is possible. Furthermore, highly functional detection and situation assessment capabilities will aid human operators in dealing with complex situations. Finally, there is a potential for a continuous learning of crew from onshore support.
- **Reduce Operational Risk:** Fatigue and attention deficit is also caused by monotonous sea passages, short and busy port stays and lengthy periods away from the social environment at home. Thus, Remote operations shift the human reliability issue from the ship to shore, where enhanced control can be achieved by the support of opportune technological and organizational settings. Enhanced awareness of ship condition is another of the main factors increasing system safety. In fact, as a system ages, through use or even inactivity, the actual system capability is most likely less than the nominal capability assumed at the design stage. When the current capability does not fully contain the operating space required by the mission requirements, the mission must be replanted or additional or alternate resources assigned to avoid potential system failures.

### 4.1.3 *Smart and efficient operations*

The ability to make smarter decisions based on accurate and timely information provides a competitive edge. Those who are able collect information leading to improved maintenance procedures, increased safety and availability in addition to lowered fuel consumption and better logistic planning and economies of scale achieved through more integrated and remote operations will have lower operating costs and are hence better positioned for both complying to changing regulations and competing in the market.

- **Manning costs:** With Remote operation the bridge is moved onshore, with a consequent reduction for the need of control functions on board. Moreover, the technology of the future (e.g., robots, smart materials), coupled with increased reliability and more use of redundancy, could allow for better condition management and life prognostic of systems, and reduction of manpower on board. In addition, organizational solutions could eliminate the problem, e.g., high speed crafts reaching the ship for embarking crew for maintenance.

- **Efficiency:** While supporting the crew in time-consuming and often undemanding tasks, Remote operation may also improve general performance: Computers may use sensors systems that can identify small objects which would otherwise go unnoticed, the sea and loading profiles, and computer analysis can often detect machinery degradation long before a human can. This information is readily available at onshore control centers where teams of experts and computer based system experts can optimize the operation of the overall enterprise on the base of real-time, highly informative information. Moreover, support from shore can engage the crew in a continuous improvement process, leading to increased performances and reduced errors.
- **Scale of economics:** The onshore control center will allow cutting down most of the manning and decisional costs and centralizing activities to achieve more standardization leading to even a greater potential in cost savings. Spare management can be optimized based on timely and accurate information. Moreover, changes in the business models like more outsourcing and vertical integration of value chains, will allow cutting down many operational costs.

### 4.2 *Barriers*

#### 4.2.1 *Change management*

The adoption of new digital technologies poses a challenge to the way things are usually done, and therefore to the mindset of people. New competencies will be required and new ways for addressing old problems, based on new types of information, will have to be defined.

#### 4.2.2 *Data and system complexity management*

Among the major technical issues there is the management of data and of the complexity of systems. The new systems are data centric in the fact that the focus is on liberating latent value from data in the context of sustaining ships in an operational environment. Data needs to be stored, maintained, processed, analyzed, integrated, and made secure. Systems will need to be integrated with many software and hardware components. Industry fragmentation may pose a challenge to the development of effective solutions as the focus in system design should be on overall system levels rather than on individual components.

#### 4.2.3 *Regulation, security and society*

New solutions need new regulations and standards to be defined, as well as the willingness to change old practices and the need for communicating the new risks to society.

- **Hacking:** One of the principal threats for Remote operation is the risk of hacking. A secure link needs to be established in order to ensure safe operations.
- **Main crew and competencies:** Currently several regulations exist regarding the composition of the main crew and the required competency. As many functions regarding the bridge of a remotely operated ship are performed onshore, regulations need to be changed. Governments have long played a key role in stimulating the development and deployment of technology with potential benefits to the economy, environment and society. It is likely that also for the deployment of Remote operations, potential benefits and shortcoming will need to be addressed at regulatory levels.
- **Public perception:** The dynamic relationship of technology and society is increasingly critical for businesses and policymakers to understand. Perceived risks associated with technologies can stall or halt their deployment, and companies perceived as untrustworthy by the public can struggle to secure a social license to operate. These barriers are very present also for Remote operations, where risks need to be correctly managed and communicated.

### 4.3 *Impact on business models*

The steady advance of ICT and access to vast amounts of data will continue to drive unprecedented human connectivity. For the shipping industry, the digital age will open up a new landscape of opportunities to "get smarter". In the short term, several relatively minor subsystems in ships are expected to grow more automated. One such important system is instrumented machinery, which can be monitored from a centralized, shore-based data centers. At first, maintenance and logistics planning may be performed by human analysts, but over time, these tasks will increasingly be handled by computers, which will make decisions on maintenance, ordering parts and scheduling work.

Today, some manufacturers offer systems to monitor on board conditions. This process is likely to become more mainstream in the next decade and, by 2020, data collection from machinery will be performed on advanced ships, such as offshore vessels. Data collected on board will be used for diagnostic testing to determine the condition of various components and if they need to be inspected, overhauled, or replaced.

The first prototype of a fully autonomous ship may appear as early as 2015, with fully automated ships entering the market by 2025. In 2035, many types of ships may routinely be delivered with autonomous operation capabilities. At the same time, ports will have more automated systems for the loading and unloading of cargo. If so, it is conceivable that some segments, like container transportation, may be fully automated by 2050.

Other expected developments include collaborative software tools to enable seamless coordination between various stakeholders, on board robots, modular designs, Autonomous Decision Support Systems, and tools for Virtual Operations, such as virtual surveys, virtual guidance from land-based operators, etc.

While the deployment of ICT in shipping is likely to reshape established business models through more data-centric and more collaborative, extended value chains (Malone, 2004), we believe that these technologies will enable safer, smarter and greener operations and maintenance procedures. In the following we revise some of the main possible changes that might influence the existing business models.

#### 4.3.1 *Towards risk based maintenance*

The overall objective of the maintenance process is to increase the profitability of the operation and optimize the total life cycle cost without compromising safety or environmental issues. Risk assessment integrates reliability with safety and environmental issues and therefore can be used as a decision tool for real time maintenance planning when combined with real-time data from condition monitoring. Maintenance planning based on risk analysis minimizes the probability of system failure and its consequences (related to safety, economic, and environment). It helps management in making correct decisions concerning investment in maintenance and related field. This will, in turn, result in better asset and capital utilization (Arunraj et al., 2005; Campbell et al., 2011).

The concept of real-time risk-based maintenance integrates the condition of individual component at the system level and uses reliability and risk indicators evaluated real-time to prioritize maintenance actions. Such an integration of condition monitoring with risk based models requires considering the maintainability of an asset already in the design phase. In fact, at this stage, risk and reliability models of ship systems must be developed for future usability in operations and become part of the decision support and information management system of an enterprise.

Therefore, Risk based maintenance will have an impact on how the ship design is carried out. This includes new documentation contents and the development of opportune computer models for single equipment that must be integrated in a computer model at the system level. Computer models will include information about the physical behavior of components, failure modes, where sensors are located on equipment, which characteristics they measure, the functional and physical relations between components, etc.

During operation, risk based maintenance reflects in a significant way an important conceptual departure from the past. In a traditionally regulated industry environment, the emphasis is on cost minimization subject to achieving a certain required level of reliability. In contrast, the problem today is driven more by the business decision to allocate a certain level of resources to maintaining the equipment with the objective to achieve the best level of reliability subject to a constraint on cost (or resources). Thus, the maintenance manager is provided with a certain budget and personnel to do the work. Yet, it is inevitably the case that in the coming years, the maintenance tasks that the manager would like to perform require resources that exceed what is available. Thus, the task is to find the one way among the very large number of possible ways to utilize those resources so that the "good" that comes from those resources is maximized. Risk based maintenance allows quantifying that "good" as cumulative risk reduction that can be evaluated in real-time (Manno et al. 2014).

#### 4.3.2 *System integrators*

Systems Integration engineering is growing in importance as ship systems and operations become more complex and interconnected. It is only in recent years that systems have been deployed that can interconnect with each other; most systems were designed as 'stovepipe' designs with no thought to future connectivity.

Systems integrators function as a designer/engineer, bringing together a wide array of components from various manufacturers to accomplish the goal of creating a unified functioning system that meets the needs of the client. Systems Integrators are usually involved in the selection of instruments and control components from among various OEMs to determine the specific mix of output, function, interconnection, program storage, controls, and user interfaces required for specific projects.

System integrators will facilitate the shipbuilding and operation phases by simplifying the hierarchical levels of suppliers involved in the procurement of parts and spares. A system integration provider could also offer different consultancy services or even integrate vertically into the value chain of its customers and provide a service based on the usage of a proprietary asset.

- **System integrators in shipbuilding:** Building a ship is an enormous project management effort, one of the reasons laying in the large amount of hardware and software components. These components are organized hierarchically in order to form equipment, subsystems, systems, systems of systems and ultimately the ship. The figure of the system integrator simplifies this tree structure by comprising a large branch of it and providing an integrated solution fitting the requirements of higher-level clients. They generate value by researching and developing hardware and software components and by designing or building a customized architecture or application, or integrating it with new or existing hardware and software. Examples can already be found in in the Offshore sector where ship designers are ordering complete packages from sub suppliers including all components of a system. Another solution is for ship yards to extend their value chains also in component manufacturing. Some Korean shipyards are considering this solution to compete with the low cost Chinese yards. Finally, ship yards could also think to move into operations in order to control the whole supply chain from the manufacture of components and ships, and the management of spares and maintenance (e.g., Edison Chouest).
- **System integrators in ship operation:** It represents that kind of arrangement where the system integrator offers onshore support for diagnostics & prognostics, and guidance in emergency situations, while the responsibility of the decisions is left to the shipping operator. Nowadays, there are some OEMs providing this kind of service already, where remote diagnostics centers, managed by the OEMs themselves, have a direct link with the ship and can suggest preventive or recovery actions to the crew on board (e.g., ABB Marine services). In general these kinds of agreements allows increasing the competence of crew as they are directly supported by experts onshore and can, thus, be engaged in a learning process. Moreover, by enhancing the guidance of onshore based experts to the crew (e.g., by virtual spaces, vision-goggle), there will be less need for global presence, travelling to remote locations, and thus achieve cost savings. Finally, diagnostic and prognostic support offers a valid means to cut down procurement costs of spares due to larger lead times (i.e., prognostics provides a forecast of the condition of systems) and improving the overall resources allocation.
- **System integrators as service providers:** It is the increased viability for performance-based arrangements, where comprehensive aftercare services are offered to end users who actually pay a flat rate for set level of ship performance. System integrator providers may shift their business models to embrace aftermarket services associated with their product lines. Service revenues may be based on guarantying availability of systems. To mitigate risk, however, the System integrator must have access to product data as it operates, so that he can understand the system' behavior, predicts events, and avoids failure consequences. Some of the actual challenges are: the definition of an opportune business model and legal agreements around the service that is being sold, which comprises changes in the strategy and organization of the enterprises of the shipping sector (e.g., EPD).
- **New analytics and IT platforms providers:** While system integrators are responsible for delivering an engineering systems, other players can have a major role in just combining data from different sources and provide solutions for the most effective maintainability of an asset. These players will leverage on predictive analytics technology to analyze data on multiple ship parts, components and systems, and make recommendation to optimize maintenance and operations. These providers look also at the information system of an organization, therefore can enable organizations to openly share the information to other business areas to break down data silos (e.g., ESRG, in the maritime industry; and Taleris, in the aviation industry).

#### 4.3.3 *Onshore control centers*

Onshore control centers will perform several functions to ensure resilient and safe, cost-effective and environmental friendly shipping operations. The functions of onshore centers will be related to ensure efficient and safe logistic chains, comprising tasks such as traffic control, controlling the condition of equipment and ensure availability of maintenance crew and spare, weather routing and performance optimization, etc.

One of the main issues with the establishment of onshore control centers is related to who should be in control of the operation of such centers, which functions the center should perform, and how responsibilities should be re-assigned. It seems unfeasible that each shipping company will establish a control center for themselves. This may have an impact on safety, as the whole system may results poorly integrated with the risk of poor and erroneous communication as well as the missed opportunity of exploiting industry-wide economies of scale.

#### 4.3.4 *Onshore based maintenance*

Traditionally most of maintenance activities are carried out onboard in order to save waiting times at ports or shipyards and there is still much resistance to change this setting. However, as more automation is adopted onboard ships, one has to think how this major barrier can be overcome. Several possibilities exist and more will be available. New materials, the use of robotics and modular design may make it possible to move these activities onshore. New materials will allow reducing the need for maintenance (e.g., self-recovering materials); robots may be used onboard for repairing or replacing faulty components (e.g., automotive industry assembly and paint, aerospace robotic arms for operating outside the space station); and modular design will allow to replace easily and quickly entire subsystems and units to functioning ones. Successively the faulty units can be repaired onshore and reinstalled on another ship when able to operate again.

#### 4.3.5 *Virtual classification, certification and verification*

Data, computer models, and risk based approaches may challenge the classification paradigm with certification and verification approaches based on computer-automatic algorithms for model checking (Manno, 2012).

## 5 FUTURE STORYTELLING: TOWARDS 2050

This section gives a summary of the above chapters in a future storytelling fashion. The definition of the timeline is based on the projection into the future of the deployment of the technologies and business models presented above, starting from the current status of the shipping sector, and has no pretense of being exact or complete; the scope being more to illustrate the progressive developments needed to achieve the complete deployment of the technologies object of this work (Fig. 1).

### 5.1 *Developments in 2020*

Several minor subsystems in ships are expected to grow more automated. An important system is the instrumented machinery with a centralized data analysis center on board the ship that can be accessed upon request from onshore via satellite or internet access points in ports. Data will be stored in proprietary databases with no possibility of access from users external to shipping companies, unless authorized.

Data will be used for condition and performance monitoring as well as for optimizing power distribution and consumption.

In particular, condition monitoring is performed by several machinery manufacturers today and towards 2020 data collection from machinery will be performed on a majority of newly built advanced ships like offshore vessels. Condition monitoring data will be analyzed by onboard computers to perform diagnostics and determine the health state of the component, and whether it should be inspected, overhauled or discarded. Algorithms will estimate the health status of machinery based on the collected data, and will be able to give warnings of imminent failures. The output of the algorithms will be checked by human personnel to validate the results. Moreover, the crew onboard will be supported by remote diagnostic systems at the OEMs facilities, which will be contacted for troubleshooting.

Performance monitoring will be poorly integrated with onshore systems, and, thus, performed by the personnel onboard with the aid of opportune software, or used in retrospective analyses for optimizing future operations. Onshore data analyses will be performed by the shipping companies and in some cases by IT-analytics provides that will support in optimizing performances and energy consumption, and will support in maintenance, ship voyage, and fleet logistic planning.

Maintenance will still be mainly scheduled, but condition based maintenance based on the current status of equipment will start to be implemented for some machinery components. However, at this stage, algorithms will not be capable to forecast future degradations, therefore, maintenance will be still reactive based on the identification of developing faults. In order to develop a correct strategy for condition based maintenance, however, we will see first developments in the integration of maintenance in the ship design phase (Nowlan et al., 1978). At this stage of implementation focus is put on determining what the most critical failure modes of a ship system are and to equip the related components with sensors that can detect related physical parameters.

Finally, some of the OEMs will provide complete integrated solutions for many of the ship functions. This will be positive for shipyards, which will outsource much of the complexity due to the several layers of interconnected systems, and the ship operations where procurement of spare parts will be made easier. However, this could lead to rise of costs due to reduced competition. Shipyards could start to integrate vertically in the value chain, both upward and downward. The upward integration is consequence of the cost-based competition of Chinese shipyards. Moreover, many Korean shipyards are large enterprises with manufacturing facilities that are generally used to produce equipment in other sectors; thus leading to a horizontal integration of the value chains of different industries. Downward integration will be searched due to the possibility of making use of the

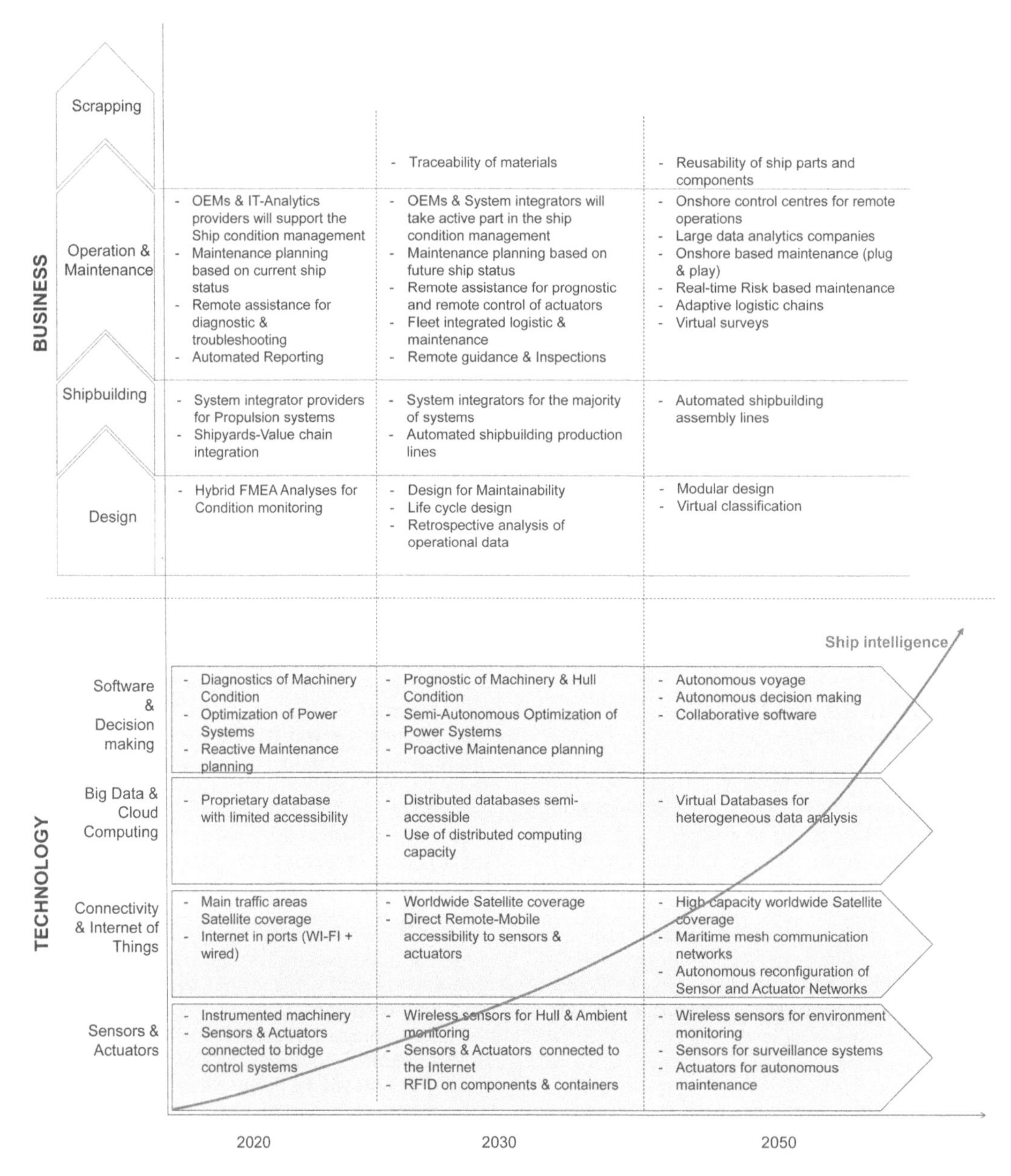


Figure 1. Future development in the technological and business arenas.

yards facilities that can be used for maintenance. Moreover, this will lead to economies of scale for shipyards with respect to inventory management of spare components, and allocation of workforce.

The benefits of this first stage are the capability to prevent failures and reduce costs and safety concerns associated with the prevented failures. The development of failure diagnosis software will result in two-fold benefits. First, it will improve the consistency and accuracy of automated failure diagnostics. Second, it will reduce the labor required to assess equipment condition and perform the appropriate maintenance tasks.

The companies who were early adopters of these new technologies start gaining momentum and as the benefits of a smarter maintenance become clearer, the costs related to maintenance will start falling and it will become increasingly difficult for competitors without such experience to compete. This will lead to an increased rate of retrofitting legacy ships, hence setting the stage for large-scale implementation of smarter maintenance practices throughout shipping.

Finally, we will see first deployments of automated reporting in some of the more advanced countries where internet access point in ports have been deployed.

### 5.2 *Developments in 2030*

From 2030, integrated ship health management systems will, to a large degree, become important in the shipping industry. These kinds of systems consider the entire ship (or even fleet) with associated operations and maintenance costs, health state and reliability of vessel, logistics, lifecycle design, emissions and environmental impact (Benedettini et al., 2008).

Different types of smart and wireless sensor networks will be installed on many systems. Examples are: easy to install hull monitoring systems, smart wireless sensors with mature energy harvesting methods, multiple-redundant gas and fire-alarm systems in engine rooms, high resolution gas sensor networks on LNG carriers, easily deployable sensor networks in ports, such that an upgrade to autonomous docking infrastructure will be made cheaper, RFID sensors that will be installed on components and materials will allow the traceability of materials during to improve the environmental-friendliness of scrapping operations. Moreover, RFID and other kinds of sensors will be used for the traceability and monitoring of containerized cargo (Knutsen, 2013).

This stage includes the use of prognostics to determine the Remaining Useful Life (RUL) of components and gain an understanding of when maintenance should be performed and plan for that. Prognostics uses statistical information about failures together with continuously measured parameters that determines health state to determine RUL. The benefit of having access to this information is that the reliability of the ship is estimated real-time, and actions can be taken if needed. The ability to detect, diagnose and isolate faults as well as the capability of predicting the residual useful life of critical components reduces the probability of technical risks, and thereby enhances system safety. This ability allows replacement of faulty components and adaptive adjustment of mission profiles, e.g., engine loads, activation of stand-by components, before the development of an accident.

While in the preceding decade planning activities were dominated by human analysts, from 2030 tasks will be gradually handed over to computers that make the decision to perform maintenance and order the production of parts within a timing interval. The end result of this is increased availability and lowered maintenance costs. As experience with such systems grow and the related mathematical knowledge develops, the accuracy of both diagnostic and prognostic algorithms will improve and give a more accurate health state and RUL estimate, thus providing better decision support information which will lead to proactive maintenance planning, based on the future prevision of system functionalities degradation.

System integrators and OEMs could change their business model and embrace the aftermarket stage with an active role. Performance-based agreements where the equipment supplier provides a service based on usage hours and guarantees a defined availability level will become more common. This will lead to new modalities of conducting operations where risks, competencies, costs and revenues are shared by the different actors.

This will also demand new class requirements, as maintenance depart from conventional intervals to maintenance on demand as determined from sensor data. Moreover, Design for maintainability will be integrated into the Lifecycle design of ships, thus, leading to new rules and methods for the design of ships (Manno, 2012). Also systems for inspection can be expected to take on a more autonomous character, for example aerial or submersible robots that are released into empty/full water ballast tanks to inspect wall condition. Finally, we will see the implementation of more automation in shipbuilding.

Performance management will also be much improved, where systems combining weather, sea status, traffic, and availability of ports will be integrated with the monitoring and actuator systems of the ship, which will allow to optimize energy consumption and other parameters like trim, etc.

Systems will be semi-autonomous, in the sense that notification will be given to the human operator that will confirm or not the undertaking of the suggested action. These systems will be connected to onshore facilities via global satellite coverage and ship operators can control the ship and its components through mobile devices.

As more experience is gained, databases of historical process data of failures and near misses as well as related condition, process, performance and environmental monitoring data will be built and accurate and systematic ways of handling data will be needed. Database will start to have a distributed architecture and some of the data will be accessible for users external to the shipping companies. This will gradually lead to a new innovation paradigm, where the design of ships and the optimization of shipping operations will be improved by collaborative networks of individuals that can communicate through data sharing platforms and advanced software which can run on distributed computing resources. With more operational data, empirical methods will strike back (e.g., statistical analysis). Moreover, model validation will become more viable from available information.

New applications with Multi-touch and sensors which provide high user interface responsiveness. Integrations of platforms between different industry' actors will allow to screen concepts in the early design stage w.r.t. intended functional performances and regulations. Finally, tools for virtual training and virtual guidance to support the crew onboard the vessel will start widespread.

### 5.3 *Developments in 2050*

In 2050 fully autonomous ships may have become reality. The first demonstrator may appear in 2015, while test ships in commercial operation may become reality in 2025. In 2035, many types of ships may routinely be delivered with autonomous operation capabilities at sea. However, on-shore there is still a lot of activity when un-loading and loading. In 2040 automated dock facilities for containers may be present and, in 2050, a fully automated sea going container transportation system may be in place.

A fundamental enabler will be a high capacity worldwide coverage satellite communication network that will be integrated by maritime mesh communication networks, which will allow enhanced, resilient and robust ship-ship and ship-shore communication.

The rapid implementation of highly sophisticated sensor and actuators networks, combined with advances in robotics, computer vision and artificial intelligence, has rendered ship crews redundant in most sea operations, with the most notable exception in advanced off-shore and deep-sea operations like construction and maintenance. These systems will allow the deployment of surveillance technology in the ship proximity that can identify nearby objects, environmental monitoring for several applications (e.g., oceanography research), as well as virtual spaces for the inspection and survey of ships.

Autonomous reconfiguration of smart sensors and actuators, routing and voyage planning will be possible by advanced software based on artificial intelligence algorithms that will be able to read present conditions, forecast future ones, integrate ship goals with traffic, weather and other logistic information, and act in order to maintain optimal safety and performance levels. These systems will allow logistic supply chain to become highly adaptive to varying conditions.

Robots will be used for many maintenance tasks, like painting and faulty components replacement. This will be made possible by advances in modular design, where components can be easily plugged in and out by robotic arms onboard the ship as well as onshore. Onshore based maintenance will be possible due to these advances, which have made traditional time consuming activities, simple and rapid. Substituted components can be, then, repaired onshore and installed on other ships trough opportune standardized interfaces. This will be enhanced also by the deployment of more smart material which may have self-recovering properties.

Maintenance will be based on accurate real-time data that will be integrated into risk models (Real-time risk based maintenance). This will allow to perform maintenance cost-effectively only on the actual most critical components.

In 2050, large data analysis and logistics centers will employ the largest number of people in maritime operations. Onshore control centers will be responsible for the condition management of the ship and the intrinsic risk due to the failure of onboard equipment and communication links; for the operation of the vessel especially in busy waters, in proximity of ports and in case of emergency situations; and finally for the overall logistic supply chain that will be mirrored into a virtual web-based space that will allow to establish new way of interaction between shipping operators, logistic providers and end-customers.

Collaborative software tools will increase due to the actual distribution of the work among individuals possibly scattered around the world and working with different design software and analysis tools (peer-to-peer communication infrastructure); and due to the diffusion of virtual databases that can merge heterogeneous data from different sources, and the large availability of distributed and fast computing (Bughin et al., 2008).

Class societies may become highly data proficient companies. They may have larger divisions for software, analysis and data collection than for traditional surveys and engineering disciplines. A new classification paradigm based on virtual tools for the certification of components will challenge the current paradigm.

Finally, the efficient use of Big Data has significantly improved the information available to regulators and engineers. Both of these focus heavily on using the available information to make new regulations and technologies to achieve highest impact on safety and sustainability performance.

## 6 CONCLUSIONS

In this paper we have revised the main developments and challenges for the deployment of ICT and automation solutions in shipping with major focus on the future of maintenance and remote and autonomous operations. The paper outlines the main technological, sociological and economic barriers, based on the current review of the main literature and nowadays' expertise level on the emerging technologies in the context of shipping. The main findings are that:

- The application of ICT on ships will have a positive impact on safety at sea. In fact, ICT solutions can provide control over the status of degradable systems, increase situational awareness and human reliability, support in the definition of corrective actions, and the reduction of operational risk.
- More automated operation will help reduce human errors, while remote operations may lead to a reduction of the number of people serving at sea.
- Finally, while enhancing safety and efficiency, ICT will also answer the need for more transparent operations and help build trust and collaboration between various industry stakeholders, based on the collection of objective facts.

This paper summarizes the work conducted during the "Future of Shipping" project, which was run within DNV GL for the 150 years jubilee of DNV. In particular it is the extract of a part of the project which had the goal to investigate the effect of digitalization in shipping. The interested reader can find more information at www.thefutureofshipping.com.

## REFERENCES


Arunraj, N.S., and Maiti, J. 2005. Risk Based Maintenance-Techniques and applications. Journal of Hazardous Material: 142(3), 653–61.

Atzori, L., Iera, A., and Morabito G. 2010. The Internet of things: A survey. Computer Networks: 54(15), 2787–2805.

Baumgarten, J., and Chui, M. 2009. E-governement 2.0. McKinsey on Government: 4.

Benedettini, O., Baines, T.S., Lightfoot, H.W., and Greenhough, R.M. 2008. State of the art in integrated vehicle health management. Procedings of IMechE: 223(G), 157–70.

Bhaskar, A.K., and Menaka, S. 2007. The state of the art of MEMS in Automation. International Society of Automation.

Boccaletti, G., Loffler, M., and Oppenheim, J.M. 2008. How IT can help cut carbon emissions.

Bughin, J., Chui, M., and Johnson, B. 2008. The next step in open innovation. McKinsey Quarterly: 3.

Bughin, J., Chui, M., and Manyika J. 2010. Clouds, Big data, and Smart Assets: Ten tech-enabled business trends to watch. McKinsey Quartely.

Campbell, J.D., Jardine, A.K.S., and McGlynn, J. 2011. Asset Management Excellence. CRC Press, Taylor & Francis.

Dark J., 2012. Marine Satellite Communication. Globar Marine Networks Blog.

Guan, S., Hansen, K., and Ayello, F. 2012. Applications of MEMS Sensors and Opportunities for DNV. DNV Research & Innovation, Internal report: 2012–9653.

Halevy, A.Y., Ashish, N., Bitton, D., Carey, M., Draper, D., Pollok, J., Rosenthal, A., and Sikka, V. 2005. Enterprise information Integration: Successes, Challenges and Controversies. SIGMOD: 778–87.

Knutsen, K.E. 2013. Condition monitoring in various industries and transport systems. DNV Internal Report 2013–1048.

Malone, T.W. 2004. The future of work: How the new order of business will shape your organization, your management style, and your life. Cambridge, MA: Harvard Business Press.

Manno, G. 2012. Next Generation Reliability Framework for Ship Machinery Systems. DNV Internal Report, 2012–1694.

Manno G., Knutsen K.E., and vartdal, B.J. 2014. An importance measure approach to system level condition monitoring of ship machinery systems. CM 2014 and MFTP 2014, Manchester.

Miorandi, D., Sicari, S., De Pellegerini, F., and Chlamthac, I. 2012. Internet of things: Vision, applications and research challenges. Ad Hok Networks: 10(7), 1497–1516.

MUNIN Project. 2013. http://www.unmanned-ship.org/munin/.

Naeem, W., Sutton, R., and Xy, T. 2012. An automatic control and fault-tolerant multi-sensor navigation system design for an unmanned maritime vehicle. Further advances in unmanned marine vehicles. IET, London.

Nowlan, F.S., and Heap, H.F. 1978. Relibility-centered maintenance. United Airlines, San Fransisco.

Saveriano, J.W. 2010. Remotely operated vehicles: history. Waldos & Howard Hughes.

*Maritime-Port Technology and Development – Ehlers et al. (Eds)*
*© 2015 Taylor & Francis Group, London, ISBN 978-1-138-02726-8*

# Ship retrofit solutions: Economic, energy and environmental impacts

R. Aronietis, C. Sys & T. Vanelslander
*Department of Transport and Regional Economics, University of Antwerp, Antwerp, Belgium*

ABSTRACT: Retrofitting existing ships with "greener" technologies currently has become common practice due to different economic and environmental considerations. This research was sparked by the need for the policy makers and ship owners to evaluate the economic, emission and environmental performance of the different available retrofit solutions. For the policy makers, the choice of appropriate retrofit solutions to support is welfare-based, while for the ship owners the industrial-economic aspects are the most important. In practice, viable policies need to balance the interests of both sides. The research uses a three-stage approach to examine the performance of the retrofit solutions. In the first stage, an aggregate simulation model is developed. It allows testing various scenarios that simulate the introduction of retrofit solutions on the European scale. In the second stage, the simulations are run to test the impacts of such currently often discussed technologies and practices as dual-fuel engines, wind propulsion, LNG-powered generators, speed reduction, variable speed operation of propeller, voyage optimization, SCR systems and scrubber systems. The third stage of the research focuses on the assessment of the impacts. It allows comparing the different retrofit solutions based on economic, emission and energy performance. The comparison of the performance of different retrofit solutions enables the choice of the best-performing technological option. The results of this research are most relevant for the ship owners and for policy makers, but other stakeholders might be interested as well. For ship owners, this research shows which retrofit solutions are economically viable and worth considering. For policy makers, this research shows the welfare benefits and costs for ship owners of the tested retrofit solutions, which is applicable for development of efficient and viable policies.

## 1 INTRODUCTION

International maritime organizations, the European Commission and national governments are currently in the process of developing and enforcing different policies that would reduce the emissions generated by shipping.[1] Taking into account the age of the shipping fleet and the average lifetime of ships, retrofitting ships with "green" technologies is one of the realistic options considered for achieving the goals set by the policies. There is no shortage of retrofit technology options. Some studies have compiled lists of up to 50 or more technological options, although just part of those are in the stage of commercial development. Technologies suitable for retrofitting include propeller optimization techniques, machinery enhancements, alternative energy adoption (e.g. LNG) and a long list of others (Yella et al. (2012), Stevens et al. (2014)). The vast range of available technologies is one of the reasons why retrofitting ships presents a challenge.

This research comes in to support the process of ship retrofitting by proposing an approach for evaluating the impacts that the introduction of the retrofit solution would have. It aims at helping the policymakers selecting the best technological solutions that are available for reaching their policy goals, and also at helping the industry to reach their environmental obligations in balance with their economic goals.

To assess the impacts of the retrofit solutions, this paper uses simulation. In the modeling, the cost and benefit implications of the retrofit solutions are important. A review of literature on retrofit cost components shows three major private cost categories and one benefit centre in the cost algorithm (Yella et al., 2012).

The *Capital cost*, sometimes referred to as the initial cost or investment cost, is an upfront cost that constitutes the fixed cost component of an investment. Once there is a decision on a measure, these costs must be incurred before the ship can start running on the new technology. Typically, it will include design, equipment, installation, commissioning and transaction costs. Most of the expenditures will become sunk costs and cannot be recovered in case the technology is removed from the ship.

[1]A good review of the upcoming environmental regulations can be found in DNV report shipping 2020, (Det Norske Veritas, 2012).

The *Lost service cost*, sometimes referred to as opportunity cost, is primarily the lead time cost and cost of space lost due to retrofitting. Some technologies like solar panels can take up space that will otherwise be used for carrying cargo. That results in lost deadweight, which is tonnage carrying ability, or deck space for handling volume. An estimate of the forgone revenue due to giving up this space must be made for the lifetime of the measure or the ship, whichever is shorter. Lead time refers to the amount of time the ship spends when docked for installation and maintenance of a retrofit.

The *Service or running cost* typically includes repair and maintenance costs, and in some cases the cost of alternative fuel, extra personnel, and training. This cost category is not always stable, because it depends on variables that include for instance fuel prices.

There are also benefits that arise from introducing retrofit solutions. According to Yella et al. (2012), *internal or economic benefit* during the operational lifetime of the measure is basically savings for the shipping line that come from retrofitting.

Externalities from retrofitting are those costs and benefits that are incurred to society because of a certain technology adopted in the market. Examples of externalities include reduction of pollution (benefit), or extra $CO_2$ emissions (cost).

In order to test the performance of ship retrofit solutions, in the *first stage* this research develops an aggregate simulation model, which allows testing scenarios that simulate introduction of retrofit solutions on the European scale. The described internal and external cost and benefit components are used when creating the model in section 2. In the *second stage* of this research, in section 3, the model runs are done. The *third stage* of the research, in section 4, assesses the obtained results.

## 2 AGGREGATE SIMULATION MODEL

The aggregate simulation model focusses on Ro-ro and Ro-pax ships.[2] This ship type is interesting, because their sailing pattern has specific characteristics that are particularly favourable to the implementation of retrofit solutions: relatively high number of port calls and sailing close to the shores where higher external costs are incurred.

### 2.1 *The model*

The model allows calculating various parameters that describe the impacts, including direct and external, the introduction of a tested retrofit solution would have. In general, for the model, the impacts of a retrofit solution can be defined with several variables. For a shipping company introducing a retrofit solution the following can be defined as a result of innovation:

- $\Delta R_p$—change in private revenues;
- $\Delta C_p$—change in private costs.

For society, as a result of introducing a retrofit solution, the following can be defined:

- $\Delta B_s$—change in social benefit;
- $\Delta C_s$—change in social cost.

The impacts of introducing a ship retrofit solution for the shipping company would therefore be: $\mathbf{\Delta R_p} - \Delta C_p$, and the impacts for society: $\mathbf{\Delta B_s} - \Delta C_s$.

Figure 1 summarizes the structure of the aggregate simulation model. The formula that describes the impacts of a retrofit solution in general is given on top. The impacts of the reference scenario, which describes "business as usual" situation, are shown: $\Delta R_p = 0$, $\Delta C_p = 0$, $\Delta B_s = 0$ and $\Delta C_s = 0$. The modeled retrofit scenarios are shown below, showing the impacts they have and the model outputs that are generated. This allows assessing the economic, energy and emission performance of the modeled retrofit solution.

### 2.2 *Scenarios*

There are two types of scenarios in the model: the reference scenario and the retrofit scenario (see Fig. 1). The reference scenario simulates the "business as usual" situation with the aim of providing a comparison reference for the modeled retrofit scenario. In the reference scenario, it is assumed that no ship retrofit technology is introduced in the market. Furthermore, the economy follows normal historic development patterns.

The retrofit scenario assumes that a retrofit solution is introduced in the market. The retrofit solution tested within the scenario is one of those solutions that are selected in the project. Each of the selected solutions has certain economic parameters, which serve as inputs for the model at the second data level. The model allows for a number of such retrofit scenarios to be tested, depending on the inputs and requirements from other tasks in work packages of the project. The scenarios are detailed in section 3.1.

[2]The developed model is flexible and can be extended for use on other shipping segments (bulk, tanker, etc.). The input data regarding energy use, emissions, transport activity and other characteristics of those segments should then be added in the model to allow these calculations.

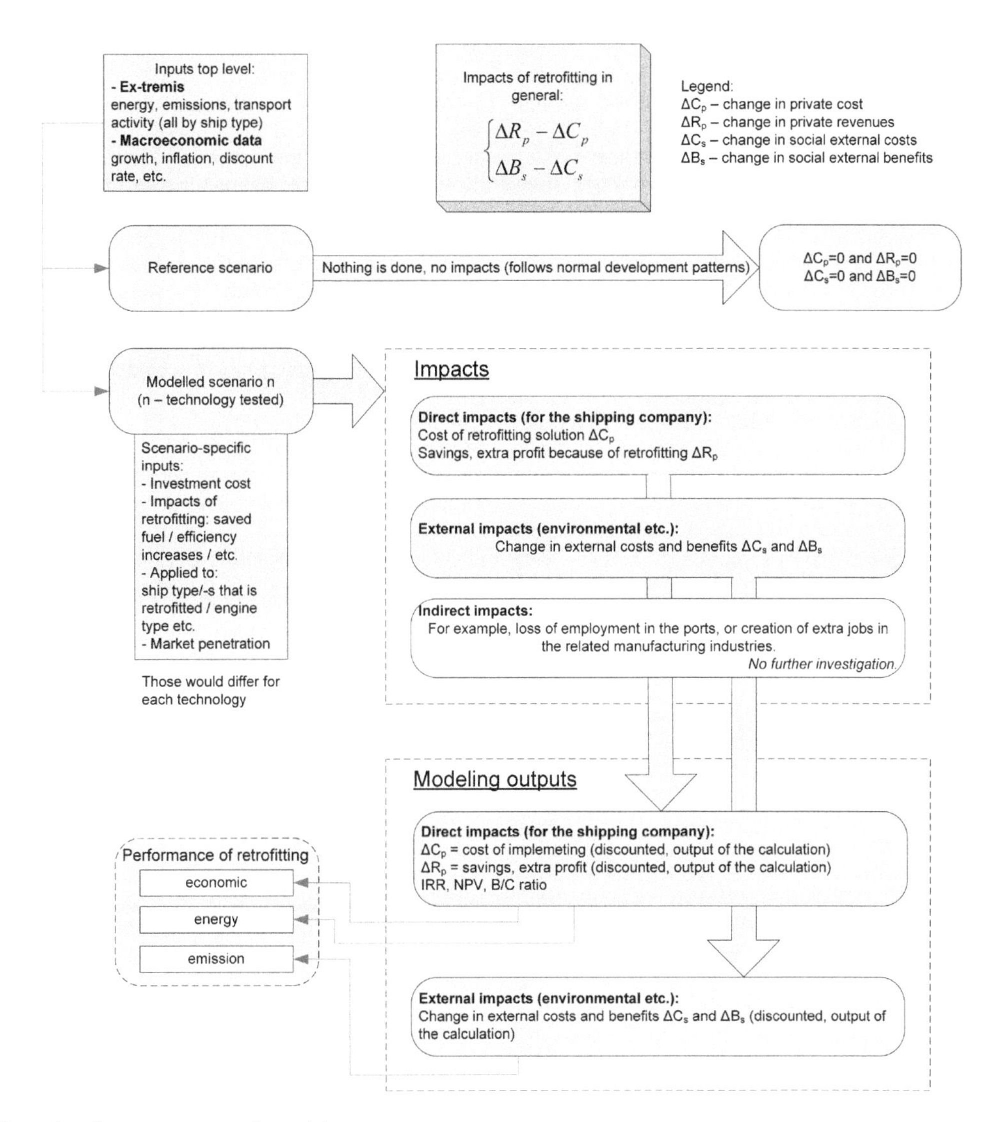


Figure 1. Aggregate economic model.

### 2.3 *Input data*

Two data levels are used in the model. Top level data are used in all calculations and do not change, but the scenario-specific data are changed to reflect the characteristics of the modeled scenarios.

#### 2.3.1 *Top level data*

The top data level includes data that are permanently used in the model and that do not change with the scenarios. They include a range of macroeconomic data, but also specific energy, emission, and transport activity data.

In literature, the following sources can be consulted for macroeconomic data and short-term forecasts. European Commission (2012), through its DG ECFIN provides short-term economic forecasts for the EU member states and some non-EU countries. The OECD (2012) Economic Outlook Database is a comprehensive and consistent set of macroeconomic data. It contains the bi-annual macroeconomic forecasts for each OECD country and the OECD area as a whole. The World Economic Outlook 2012 database by the International Monetary Fund (2012) also provides short-term forecasts of main economic indicators. Longer-term forecasts can be consulted in the European Energy and Transport Trends to 2030—update 2009 (European Commission, 2009).

With respect to the social discount rate, some controversy exists over the appropriate social discount rate to be used for calculation of the net present value for a project (Anthony et al., 2006). Cruz Rambaud & Munoz Torrecillas (2006) describe different possible options. For the purpose of the modeling, the exponential discount rate of 5% is chosen.

The data on the energy used in shipping is available from the EX~TREMIS database developed in an EU-financed project (TRT Trasporti e Territorio Srl et al., 2007). Extremis Maritime is an activity-based emission model for sea-going ships engaged in EU seaborne trade. The model is built upon three modules: a fleet module, an activity module, and an emission module. Emission estimates given in Extremis Maritime are for ship movements in European waters—EU 27 countries.

The number of Ro-ro ships that call European ports is not known. Based on data from ISL (2010) and Lloyds List Intelligence (2013), the passenger/Ro-ro cargo fleet for use in the model is estimated.

#### 2.3.2 *Scenario-specific data*

At the second level come the scenario-specific data. They include such items as investment cost, impacts of retrofitting and market data.

Investment costs are the costs that are needed to put the retrofitting solution into operation. These costs would usually be covered by the shipping companies that would benefit from retrofitting.

Here, the direct impacts of retrofitting are considered. For each of the scenarios, it means that quantified data on saved fuel, reduced or increased costs, and change in emissions of different pollutants is used.

For any product, including a retrofitting solution, market penetration is important. It shows the degree to which the product has been adopted by the market. In this research, market penetration is defined as a percentage of the ships fit for retrofitting that are actually retrofitted. Estimation based on expert evaluation of the probable market penetration is used here.

### 2.4 *Model outputs*

The outputs of the model include calculations of various parameters. Those parameters describe the impacts that the introduction of the tested retrofit solution would have. Three types of impacts are distinguished and described per type of impact as they appear in the model.

#### 2.4.1 *Direct impacts*

The direct impacts of the retrofit solution are the ones that the shipping companies introducing the retrofit solution would generate. The introduction of a retrofit solution for a shipping company is associated with a certain cost $\Delta C_p$, which in turn is compensated by the financial benefits $\Delta R_p$ that the retrofit solution brings. For each individual retrofit solution those costs and benefits differ due to the technological characteristics of the retrofit solution.

The outputs of the model at the direct impacts level provide the discounted values of costs $\Delta C_p$ and revenues $\Delta R_p$ caused by retrofitting. Also, other substantial financial parameters like IRR (internal rate of return), NPV (net present value) and B/C (benefit/cost) ratio are calculated.

#### 2.4.2 *External impacts*

When retrofitting a ship, external impacts are present. Those are costs or benefits imposed on others that are not taken into account by the person taking the action. In retrofitting a ship, an increase of social benefit, for example, could be related to the reduction of harmful emissions.

The outputs of the model at the external impacts level provide the discounted values of the change in social benefit $\Delta B_s$ and cost $\Delta C_s$ caused by retrofitting.

#### 2.4.3 *Indirect impacts*

The multitude of various indirect impacts that an introduction of a retrofit solution brings is acknowledged by the researchers working on this research. The indirect impacts of a retrofit solution could be, for example, price changes in related markets, increased sales, profits or (un-)employment in related industries (Starrett, 2011). The quantification of the indirect impacts at aggregate level is beyond the scope of this research.

## 3 SIMULATION AND RESULTS

In the *second stage* of this research, the simulations are run. In this section, the tested scenarios are described in detail and examples of modeling results are shown.

### 3.1 *Scenarios tested*

Several scenarios are constructed to model the impacts of each of those retrofit solutions. Every scenario is a combination of input data that characterize the technology investigated. These include associated investment costs and generated savings, achieved energy savings or waste, and assumptions on the environmental performance that the retrofit solutions bring. This is done for the following nine technologies, which were selected from a long list of technologies based on their commercial availability:

- Dual-fuel engine (LNG/Diesel);
- Wind propulsion;
- LNG-powered generator (for port use);
- Variable speed operation of propeller;
- Enable PTO/PTI (power take-off/power take-in) to improve loading of the engine;
- Speed reduction;
- Voyage optimization (weather, waves, current, speed);
- SCR (selective catalytic reduction) system;
- Scrubbing (SOx).

Tables 1 and 2 give a summary of the economic and environmental characteristics of the nine modeled technologies. The number of scenarios calculated differs for each technology according to the shown assumptions. If a cell contains several values related to the characteristics of the technology, a separate scenario is calculated for each possibility. With such approach, 117 scenarios are generated and calculated. As shown in Table 2, for some scenarios the specific values of emissions are not available for each emission type. In those cases the total change in emissions is used.

### 3.2 *Exhaust gas scrubbing*

The choice is made to present calculation results for scenario 9, exhaust gas scrubbing, which is an often considered technology. An exhaust gas scrubbing system can reduce the level of sulphur dioxide in the exhausts of ship engines, but at the same time the energy consumption increases. Three main principles of exhaust gas scrubbing exist: open-loop, closed-loop and dry. Their characteristics are detailed below.

*Open-loop seawater scrubbers.* Spray jets similar to the design of shower heads drench the exhaust gas with sea water just before the flue. Water and sulphur react to form sulphuric acid, which is neutralised with alkaline components in the sea water.

Table 1. Economic, lifetime and market characteristics of the modeled technologies.

| Scenario | Average investment cost, m € | Savings, % | Lifetime, years | Market penetration, % | Market size, number of ships |
|---|---|---|---|---|---|
| 1. Dual-fuel engine (LNG/Diesel) | 10; 15; 20 | 2; 3; 4 | 27 | 5; 10; 15 | 1560 |
| 2. Wind propulsion | 0.34 | 4; 8; 12 | 10 | 0.5; 1; 2; 3 | 1560 |
| 3. LNG-powered generator (for port use) | 1; 2; 3 | 0 | 20 | 1; 2; 3 | 1560 |
| 4. Variable speed operation of propeller | 0.5; 0.7; 1 | 1; 3; 5 | 27 | 5; 10; 15 | 1560 |
| 5. Enable PTO/PTI to improve loading of the engine | 0.1 | 1; 2; 3 | 27 | 10; 20; 30 | 1560 |
| 6. Speed reduction | 0.1 | 15 | 30 | 20; 30; 40 | 1560 |
| 7. Voyage optimization (weather, waves, current, speed) | 0.1 | 4 | 10 | 20; 30; 40 | 1560 |
| 8. SCR system (catalyst) | 2; 3; 4 | 0 | 7 | 5; 10; 15 | 1560 |
| 9. Scrubbing (SOx) | 2; 3; 4 | 4 | 10 | 5; 10; 15 | 1560 |

Table 2. Environmental characteristics of the modeled technologies.

| Scenario | Total Δ of emissions, % | $\Delta CO_2$, % | ΔNOx, % | ΔPM, % | $\Delta SO_2$, % | Δ energy consumption, % |
|---|---|---|---|---|---|---|
| 1. Dual-fuel engine (LNG/Diesel) | | –25 | –85 | –45 | –100 | –2; –3; –4 |
| 2. Wind propulsion | –4; –8; –12 | | | | | –4; –8; –12 |
| 3. LNG-powered generator (for port use) | | –0.125 | –0.425 | | –0.5 | 0,0 |
| 4. Variable speed operation of propeller | –5 | | | | | –5 |
| 5. Enable PTO/PTI to improve loading of the engine | –3 | | | | | –3 |
| 6. Speed reduction | –15 | | | | | –15 |
| 7. Voyage optimization (weather, waves, current, speed) | –4 | | | | | –4 |
| 8. SCR system (catalyst) | | | –80 | | | 0 |
| 9. Scrubbing (SOx) | | | | –45 | –97; –80 | 2 |

Filters separate particles and oil from the mixture before the cleaned water is sent back into the sea. The disadvantage of this scrubber technology is its relatively large space requirements on board. Its operation requires a capacity of 40 to 50 cubic metres of sea water per Megawatt hour of engine power.

*Closed-loop scrubbers.* In closed-loop scrubbers fresh water is used in combination with caustic soda as the neutralising additive. The scrubber requires less space than open-loop ones and its water requirements drop to 0.1 cubic metre per Megawatt hour output, and virtually no wash-water is produced that would have to be pumped into the sea.

*Dry scrubbers.* In dry scrubbers, the exhaust gas flows through granulated limestone. It combines with the sulphur to form gypsum, which can then be disposed of on land. The advantage is that the sulphur is locked in, meaning it cannot burden the biosphere at sea any more. The disadvantage is that a storage room has to be created on board for granulate, which reduces cargo capacity.

The differences between the three types of scrubbers are mostly technical, therefore in the aggregate simulation model scrubbers are tackled as one technology. As shown in Table 1, the investment costs for implementing a scrubber on a ship range from €2 to 4 million and the predicted

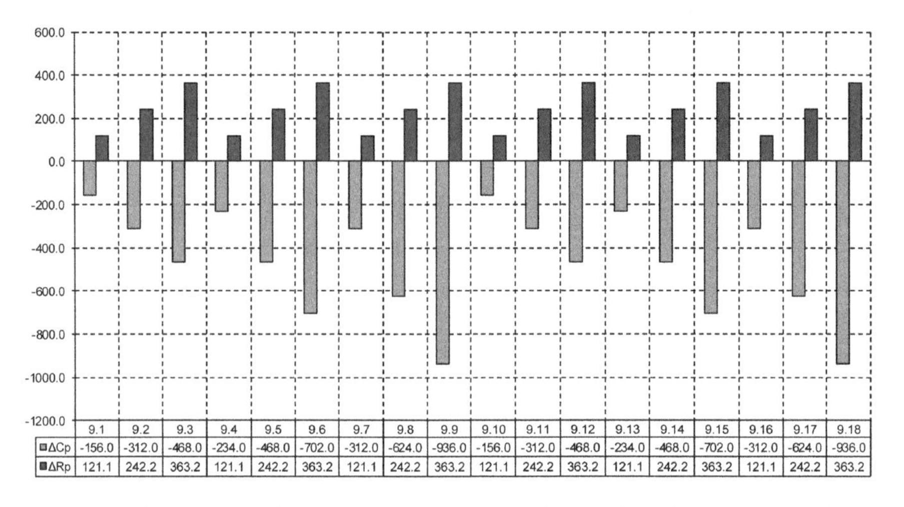


Figure 2. Impacts of the scrubbing solution on private costs $\Delta C_p$ and private benefits $\Delta R_p$ per scenario, € mn.

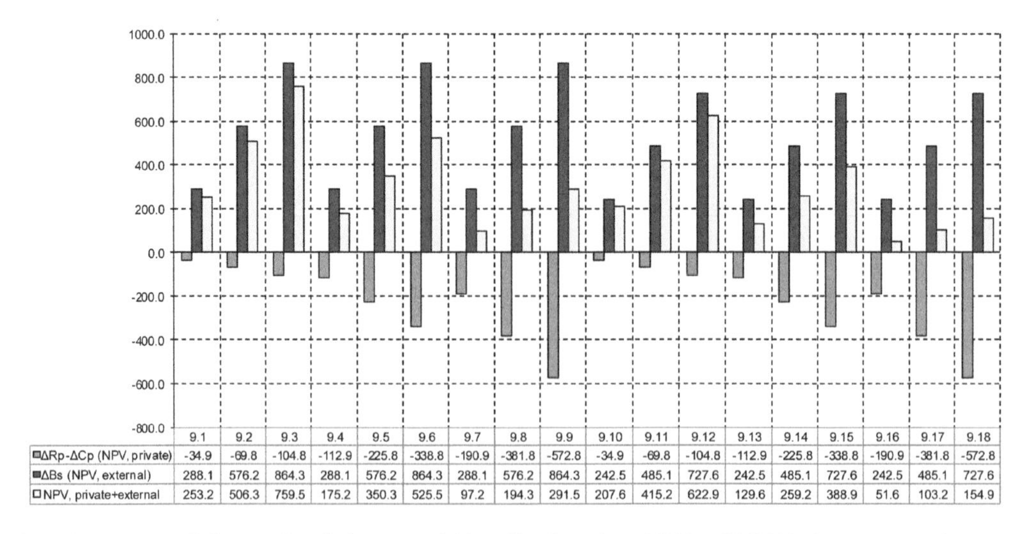


Figure 3. Impacts of the retrofit solution on social benefit ΔBs, private NPV and NPV (private + external) per scenario, € mn.

average lifetime of the scrubbers is 10 years. The savings come from using cheaper high-sulphur heavy fuel oil, and there are extra operating costs and increased fuel consumption. The market penetration of scrubbers is estimated to be between 5 and 15%. The reduction of PM emissions amounts to 45%, and $SO_2$ emissions are reduced by 80 to 97%, as shown in Table 2.

### 3.3 *Modeling results*

The results apply to the whole Ro-ro and Ro-pax market in Europe, under the assumption that a certain part of the fleet adopts the innovation (column "market penetration" in Table 1). The results are indicative and demonstrate the likely outcomes under the assumptions that are used in the model.

Figure 2 shows the impacts of the scrubbers on the private costs $\Delta C_p$ and private benefits $\Delta R_p$ for every scenario from 9.1 to 9.18. It can be observed that in all the investigated scenarios the discounted benefits or savings for the ship owners $\Delta R_p$ do not outweigh the costs $\Delta C_p$ that are associated with the introduction of scrubbers in the market.

Figure 3 shows the impacts of scrubbers for the ship owners (the difference between private benefits $\Delta R_p$ and private costs $\Delta C_p$), the impacts for the society $\Delta B_s$ and the sum of those—NPV (private + external). It can be observed that for all the scenarios, the NPV (private + external) is positive.

The benefit/cost ratio indicates the viability of a project, and it has values above 1 for viable projects. In none of the scenarios for scrubbers, this threshold is reached. Also, the return on investment is negative for all the scenarios that test the impacts of scrubbers (Table 3).

The calculations show that, as a result of the introduction of scrubbers, reduction of emissions will occur. At the same time, an increase of fuel consumption is also incurred (Fig. 4).

The results shown in this section are a good example of the calculation results that were obtained for the other scenarios. Based on these results, in the next step of the research, the assessment of the retrofit solutions is conducted to compare the different retrofit solutions based on economic, emission and energy performance.

Table 3. Benefit/Cost (B/C) ratio and Return On Investment (ROI) per scenario.

| | B/C ratio | ROI |
|---|---|---|
| 9.1 | 0.776 | –0.224 |
| 9.2 | 0.776 | –0.224 |
| 9.3 | 0.776 | –0.224 |
| 9.4 | 0.517 | –0.483 |
| 9.5 | 0.517 | –0.483 |
| 9.6 | 0.517 | –0.483 |
| 9.7 | 0.388 | –0.612 |
| 9.8 | 0.388 | –0.612 |
| 9.9 | 0.388 | –0.612 |
| 9.10 | 0.776 | –0.224 |
| 9.11 | 0.776 | –0.224 |
| 9.12 | 0.776 | –0.224 |
| 9.13 | 0.517 | –0.483 |
| 9.14 | 0.517 | –0.483 |
| 9.15 | 0.517 | –0.483 |
| 9.16 | 0.388 | –0.612 |
| 9.17 | 0.388 | –0.612 |
| 9.18 | 0.388 | –0.612 |

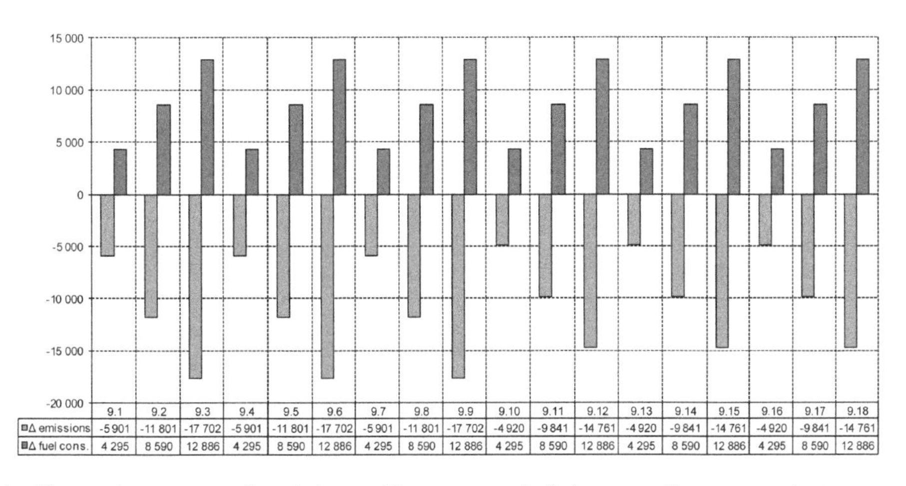


Figure 4. Changes in average yearly emissions and in average yearly fuel consumption per scenario, t.

## 4 ASSESSMENT

To compare the different retrofit solutions based on modeling results and assist in choosing the best-performing technological option, an assessment needs to be done.

### 4.1 *Assessment approach*

An assessment of the impacts is conducted to compare the different retrofit solutions based on economic, emission and energy performance. The assessment criteria are shown in Figure 5.

To perform the assessment in practice and make the assessment criteria comparable, a technology assessment index is used. This index is calculated for each of the assessment criteria in comparison with the other scenarios calculated in the model. The value '1' is assigned to the best or most desired value of the criterion amongst those achieved in all the model runs. And the value '0' is assigned to the least desired scenario. The other index values are calculated depending on the values obtained for the criteria in the model runs. This allows assigning three indexes to each of the calculated scenarios. An example of technology assessment index calculation for economic performance of exhaust gas scrubbing is shown in Table 4.

### 4.2 *Assessment of the retrofit technologies*

For each of the scenarios in the model, the index values for economic, energy and emission performance are calculated. Figure 6 shows graphically the obtained technology assessment index with average values for each technology. The value for each of the performance characteristics is marked on the corresponding axis, creating triangles that characterize the performance of each technological option.

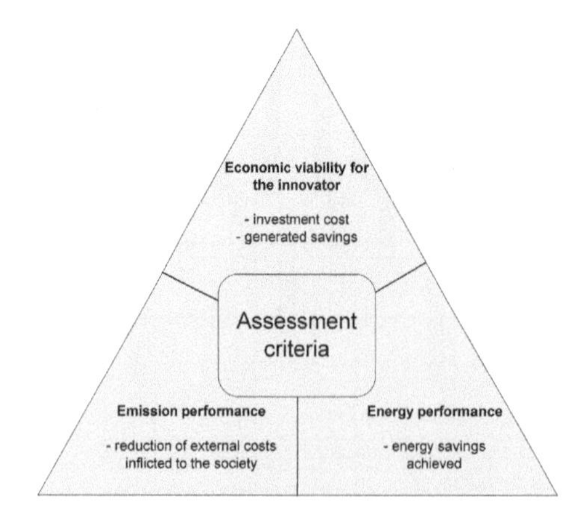


Figure 5. Assessment criteria.

Table 4. Calculation of technology assessment index value for economic performance of exhaust gas scrubbing.

| | Scenario | Economic performance, m € | Index value |
|---|---|---|---|
| Worst performing scenario | 1.21 | –4274.4 | 0 |
| ... | ... | ... | ... |
| Average performance of exhaust gas scrubbing | 9 | –225.8 | **0.3128** |
| ... | ... | ... | ... |
| Best performing scenario | 6.3 | 8667.5 | 1 |

Figure 6 shows which of the tested technology options are best for achieving specific policy targets, and which have the best overall performance. The technologies or approaches that have the best performance are those with the largest triangle areas, like speed reduction in this case. The shape of the triangle describes in which of the performance areas the technology gives best results. For example, scrubbers perform well economically, but the emission and energy performance is worse than that of the other technologies.

For comparison of the technologies, the values of the technology assessment index can also be summarized in a radar graph (see Fig. 7). It can be seen that speed reduction (technology 6) brings most benefits. The economic benefits (in dashed line here), that can be observed for this technology are probably one of the main reasons why speed reduction is a commonly used practice. The emission performance of the dual-fuel engines (technology 1), which in normal operation are assumed to run on LNG, is also very good. For dual-fuel engines, the high investment costs are the reason for the bad economic performance, although the emission performance is very good. The economic performance of voyage optimization (technology 7) and enabling PTO/PTI to improve loading of the engine (technology 5) are good due to low investment costs.

It is not surprising that for maritime transport, which is by definition a slow transport mode, the approach of speed reduction shows high performance. The shipping lines are aware of this and this approach, known as slow steaming, is commonly used. However, it must be taken into account that there are practical limitations to the application of the speed reduction. For example, for Ro-ro traffic, speed reduction often cannot be used, because of ships weekly schedules.

To put the results into perspective without the bias that the speed reduction as an outlier creates,

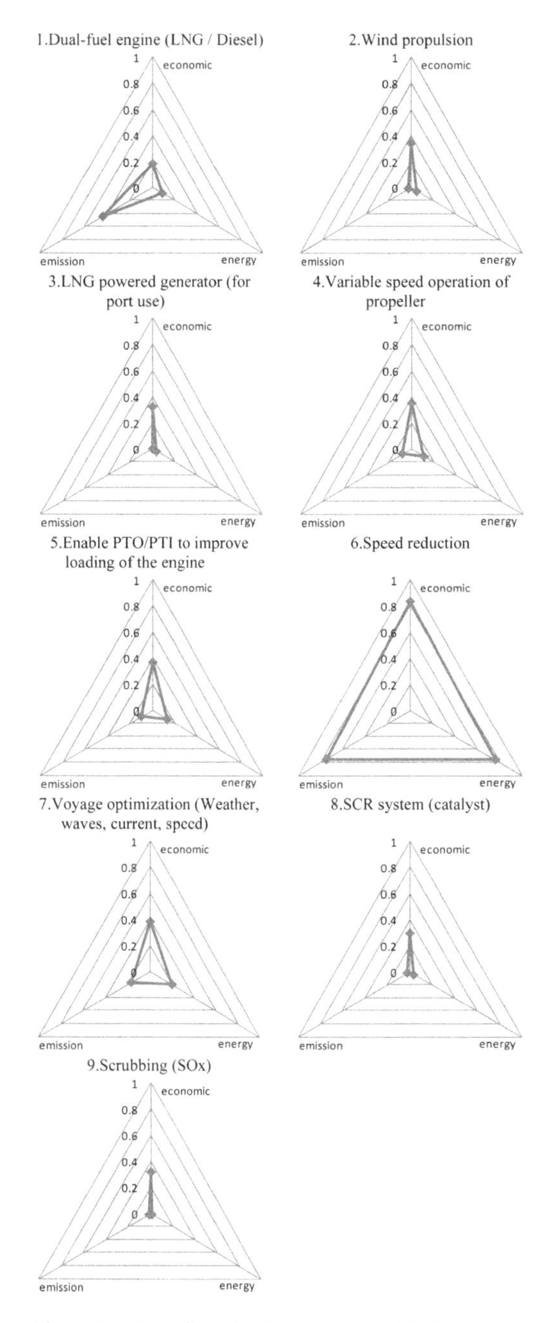


Figure 6. Retrofit technology assessment index.

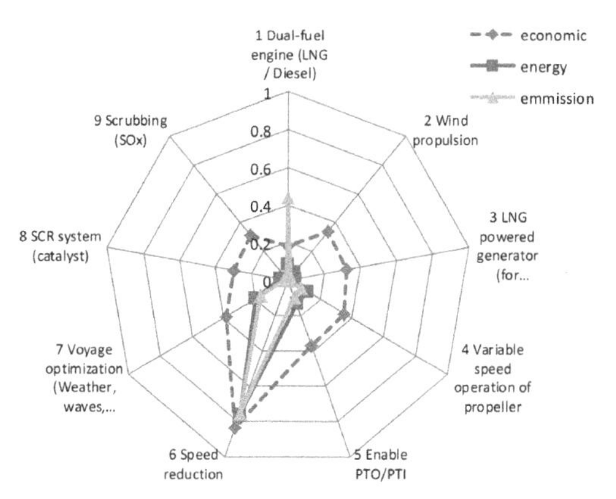


Figure 7. Performance comparison of the tested technologies.

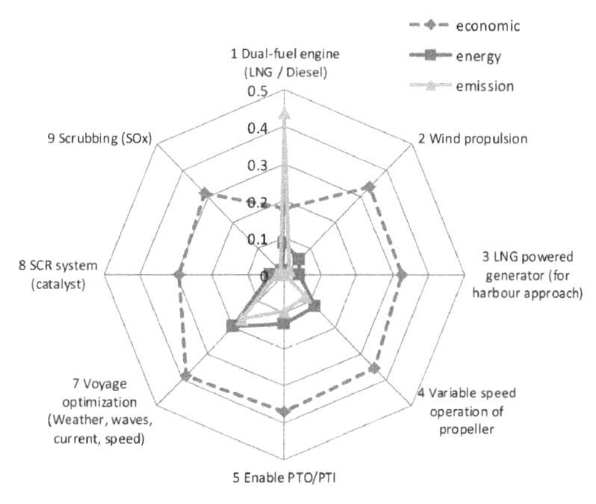


Figure 8. Performance comparison of the tested technologies with speed reduction excluded.

Figure 8 shows the performance comparison that excludes speed reduction. The performance of the other technologies can be clearly distinguished.

It must be noted that combinations of technologies are not investigated here because of the data unavailability. If sufficient data were available on the impacts of combinations of technologies, the model would permit the calculation of such results as well.

## 5 CONCLUSIONS

This research set out to investigate the selection of the best retrofit solutions for the Ro-ro and Ro-pax ships. It was done by creating an aggregate simulation model that simulates the impacts retrofit solutions would have if they were implemented in the market. The model was run for a set of nine commercially available technologies.

The developed assessment approach allows evaluating the impacts of each technology to compare the alternatives based on their economic, emission and energy performance. The evaluation of relative performance of the technologies allows choosing the best-performing technological option depending on the goals that a specific actor has. If the actor is a shipping company, the economic performance, and also existing and future regulation would be taken into account. For a government, the most important would be emission and overall performance of the technology.

The results of the assessment confirm what can be observed in the market. Those technologies that are performing better at the economic and/or energy level are the ones that are most likely to be invested in by shipping companies.

Speed reduction seems to be substantially outperforming all the other tested technologies. While the calculation results seem valid, certain aspects should be taken into account when interpreting these results. The relatively good performance is mainly due to the small investment that is required for application of this approach. At the same time this measure is often not applicable, especially for Ro-ro ships, but also in general. This is due to the technological limitations, sailing schedules and other issues that shipping lines face.

For policy makers, next to private cost/benefit analysis, this research also shows the welfare benefits and costs of the tested retrofit solutions. It also shows that a good economic performance of a technology is not always in line with good welfare characteristics. This should be taken into account for the development of efficient and viable policies.

It seems that out of the three performance criteria (economic, energy and emission) the economic performance of the technology chosen could be left for the market, as an investor will always choose the technology with the best economic performance. At the same time, developed policies could look into setting benchmarks for the energy and emission performance of the retrofitted technologies.

The developed model is flexible and can be extended for use on other shipping segments (bulk, tanker, etc.). The input data regarding energy use, emissions, transport activity and other characteristics of those segments should then be added in the model to allow these calculations.

The results of this research are most relevant for the ship owners and for policy makers, but other stakeholders might be interested as well. For ship owners this research shows which retrofit solutions are economically viable and worth considering. For policy makers, this research shows the welfare benefits and costs for ship owners of the tested retrofit solutions, which is applicable for development of efficient and viable policies.

## REFERENCES


Anthony E., David H., & Aidan R. (2006) *Cost-benefit analysis: concepts and practice*. Vol. 3 ed. (Upper Saddle River, N.J.: Pearson Education). Available at http://anet.ua.ac.be/ record/opacua/ c:lvd:6875424/N.

Cruz Rambaud S., & Munoz Torrecillas M.J. (2006) Social discount rate: a revision. *Anales de estudios economicos y empresariales*, pp. 75–98.

Det Norske Veritas (2012) *Shipping 2020, Technology Outlook 2020*. (Oslo).

European Commission (2009) European Energy and Transport Trends to 2030—update 2009. Available at http://www.energy.eu/publications/Energy-trends_to_2030.php, accessed 29 October 2012.

European Commission (2012) Economic forecasts. Available at http://ec.europa.eu/economy_finance/eu/forecasts/index_ en.htm, accessed 26 October 2012.

International Monetary Fund (2012) World Economic Outlook Database October 2012. Available at http://www.imf.org/external/pubs/ft/weo/2012/02/weodata/index.aspx, accessed 29 October 2012.

ISL (2010) *Shipping Statistics and Market Review*. Statistical Publications. (Bremen, Germany).

Lloyds List Intelligence (2013) Ship Sailings Ro-Ro 01-Mar-13. Available at http://www.lloydslist.com/ll/marketdata/containers/shipRoroPage.htm, accessed 13 June 2013.

OECD (2012) Economic Outlook. *Economic outlook, analysis and forecasts.* Available at http://www.oecd.org/economy/economicoutlookanalysisandforcasts/economicoutlook.htm, accessed 26 October 2012.

Starrett D.A. (2011) Economic Externalities, in, *Fundamental Economics*. Vol. 1. (EOLSS). Available at http://users.ictp.it/~eee/workshops/smr1597/Starrett%20-%20externalities. palfrey.doc, accessed 21 January 2014.

Stevens L., Sys C., Vanelslander T., & van Hassel E. (2014) Is New Emission Legislation Stimulating the Implementation of Sustainable (Retrofitting) Maritime Technologies?, in, IFSPA 2014 Conference Proceedings. (Hong Kong).

TRT Trasporti e Territorio Srl, Flemish Institute of Technological Research (VITO), & Institute for Prospective Technological Studies (IPTS) (2007) EX~TREMIS. *Exploring non-road transport emissions in Europe*. Available at http://www.ex-tremis.eu, accessed 26 October 2012.

Yella G., Sys C., Vanelslander T., & Frouws J. (2012) An Economic Analysis of the Costs Effectiveness Function for Measuring Ships Technology Abatement Potential, in, *Proceedings of NAV 2012 17th International Conference on Ships and Shipping Research.* (Naples: Centro Congressi Università di Napoli Federico II).

*Maritime-Port Technology and Development – Ehlers et al. (Eds)*
 *ISBN 978-1-138-02726-8*

# Model and simulation of operational energy efficiency for inland river ships

X. Sun, X.P. Yan & Q.Z. Yin
*Reliability Engineering Institute, School of Energy and Power Engineering, Wuhan University of Technology, Wuhan, Hubei, China*
*Key Laboratory of Marine Power Engineering and Technology, Ministry of Transportation, Wuhan, Hubei, China*

ABSTRACT: A greater understanding of energy efficiency of ships is vital to reduce $CO_2$ emissions. Previous studies have put more emphasis on ocean ships than inland river ones. In order to analyze the operational energy efficiency of inland river ships, and based on the resistance characteristics of different navigation environment factors, a main engine operational energy efficiency model was developed using the Energy Efficiency Operational Indicator (EEOI) as monitoring tool. The modeling and simulation, taking advantage of MATLAB/Simulink, were verified by the data onboard. The EEOI model was simulated under different navigation environment conditions. The results showed that the EEOI of the case ship is lower than its sea-going counterpart, and it varies with the main engine speed significantly, the simulation also showed how the EEOI is influenced by navigation environment such as wind, wave and water current.

## 1 INTRODUCTION

Greenhouse Gas (GHG) emission and the subsequent global warming attracted more and more attentions. As one of the biggest emission sources, shipping industry is under greater emission reduction pressure from the public (Lindstad et al. 2013). How to meet the energy consumption and pollution reduction targets, without affecting voyage safety, has been a realistic urgent problem for the shipping industry.

For international ships, mandatory measures from International Marine Organization (IMO) to reduce GHG had been entered into force since 1 January 2013. The amendments to MARPOL Annex VI Regulations added a new chapter 4 to Annex VI on energy efficiency for ships to make mandatory the Energy Efficiency Design Index (EEDI), for new ships, and the Ship Energy Efficiency Management Plan (SEEMP) for all ships. Generally, most of the current researches focused on the technology improvement for new ships rather than the energy efficiency analysis for existing ships. Tzannatos & Papadimitriou (2013) have expressed the energy and carbon efficiency of domestic passenger shipping in Greece during the decade 2001–2010, and discussed its seasonal characteristics. Based on the study on energy efficiency operational indicator of container ships, Ni & Zhao (2010) stated that a larger size could improve the energy efficiency for seagoing ships.

The inland rivers are playing an important role in the modern comprehensive transport system in some countries like China (Yan et al. 2010). However, China has not settled clearly regulation on exhaust emissions restriction for inland river ships until now, and the relevant researches are at the beginning stages. Based on a case study of container shipping on the Yangtze River, Sun and colleagues (2013) calculated the EEOI for inland river ships. However, this previous study did not build a model to describe the EEOI. In this work, a main engine operational energy efficiency model was developed, with the integrated consideration of environment-hull-propeller-shaft-engine. The simulation was performed on Matlab/Simulink to analyze the influence from navigation environment on energy efficiency.

## 2 ENERGY EFFICIENCY EVALUATION TOOL AND CASE SHIP

### 2.1 *EEOI*

According to the IMO regulations, the EEOI could be considered as the primary monitoring tool for operational energy efficiency, which is defined as the ratio of mass of $CO_2$ emitted per unit of transport work:

$$EEOI = \frac{\sum_i F_i \times C_{carbon}}{\sum_i m_{cargo,i} \times D_i} \tag{1}$$

Table 1. Parameters of the case ship.

| | | | |
|---|---|---|---|
| Length m | 107.00 | Block coefficient | 0.86 |
| Breadth m | 17.20 | Mid-ship section coefficient | 0.91 |
| Depth m | 5.2 | Water plane coefficient | 0.72 |
| Propeller diameter m | 2.38 | Pitch ratio | 0.73 |
| Number of propeller blade | 5 | Blade area ratio | 0.65 |
| Load capacity TEU | 220 | Engine power kW | 518 |
| Number of propeller | 2 | Number of engine | 2 |

where $F_i$ = the mass of consumed fuel at voyage $i$; $C_{carbon}$ = the fuel mass to $CO_2$ mass conversion factor; $m_{cargo}$ = cargo carried or work done; $D_i$ = the distance.

### 2.2 *Case ship*

There are some differences between inland river ships and seagoing ships as for the operation energy efficiency, they are to be seen in the following:

1. The inland river ships' displacements are relatively small, they are mainly flat-bottomed with a large breadth depth ratio. These characteristics bring difference on the resistance prediction methods;
2. Due to the winding navigation channel, the turning of an inland river ship would be very sharp, so the double engines & double propellers design style is usually adopted in the Yangtze River;
3. Low-power diesel engines are often equipped for propulsion, and most of them burn light diesel oil only.

A container ship in Yangtze River was selected as the case study in this work, the main parameters of which is shown in Table 1.

## 3 MODEL OF SHIP OPERATIONAL ENERGY EFFICIENCY

The main engine is the largest energy consumption facility onboard, and it is the only equipment taken into consideration for energy efficiency analysis in this work. As mentioned before, two 4-stroke middle-speed diesel engines would generally be adopted by an inland river ship, and the engines work on propeller characteristics, so the engine service power could be calculated based on rotation speed and torque. On the other hand, as the linkage between the main engine and ship hull, the propeller becomes an energy converter. So, the working condition of hull, propeller and engine is interrelated (Feng et al. 2012).

### 3.1 *Main engine*

The statistical model of the diesel engine is usually employed in ship emission research to describe the relationship between rotation speed, torque and specific oil consumption, such as (Shi. 2008):

$$M_E^* = 1 + a(1 - N_E^*) + b(1 - N_E^*)^2 + c(1 - m_{fuel}^*) + d(1 - m_{fuel}^*)^2 + e(1 - N_E^*)(1 - m_{fuel}^*) \quad (2)$$

where $M_E$ = engine output torque; $N_E$ = engine speed; $m_{fuel}$ = mass of fuel consumption; ()* means the ration of working condition value to rated condition value, and *a, b, c, d, e* are undetermined coefficients.

Five sets of data (shown in Table 2) were collected onboard to determine the coefficients.

According to the above parameters, the oil specific consumption would easily be fitted as:

$$\mathrm{m}_{fuel}^* = 1.1970 \times N_E^* + 0.0173 \times (-215.2934 \times N_E^{*2} + 143.2159 \times N_E^* + 115.7828 \times M_E^* - 44.7114)^{1/2} - 0.2002 \quad (3)$$

### 3.2 *Transmission system*

Along the shaft system from main engines to propellers the transmission losses should be taken into account. To simplify the model, the torque losses could be considered as a liner relationship:

$$Q_{loss} = 0.04 * M_E \quad (4)$$

where $Q_{loss}$ = the torque losses on transmission system.

As for speed relationship, the speed reducing ratio of the gear box in the shaft system is 3.7391, thus:

$$n = N_E/3.7391 \quad (5)$$

Table 2. Data for engine torque prediction.

| Item | $M_E^*$ | $N_E^*$ | $\mathrm{m}_{fuel}^*$ |
|---|---|---|---|
| 1 | 0.35 | 0.56 | 0.42 |
| 2 | 0.44 | 0.64 | 0.51 |
| 3 | 0.60 | 0.76 | 0.66 |
| 4 | 0.70 | 0.83 | 0.75 |
| 5 | 0.95 | 0.97 | 0.99 |

where $n$ = the shaft speed after gear box, that is the propeller rotate speed.

### 3.3 *Propeller model*

The thrust force and torque generated by propeller can be represented by coefficients $K_T$ and $K_Q$ (Ying 2007):

$$T = K_T \rho n^2 D_P^4 \quad Q = K_Q \rho n^2 D_P^5 \tag{6}$$

where $T$ = thrust force; $Q$ = torque; $\rho$ = water density; $D_P$ = propeller diameter.

Based on the propeller open water map (Ying, 2007), $K_T$ and $K_Q$ could be fitted as the function of advanced coefficient $J$:

$$K_T = -0.1429 * J^2 - 0.3286J + 0.3486 \tag{7}$$

$$K_Q = -0.0123 * J^2 - 0.0311J + 0.0402 \tag{8}$$

### 3.4 *Ship motion*

The ship motion should be described as a 3-dimentional function, which could be simplified as Equation 8 when considering the 1-dimentional motion along ship hull only (Ying, 2007).

$$K_w m \frac{dV_s}{dt} = 2(1-t)T - R \tag{9}$$

where $K_w$ = water coefficient; $m$ = weight; $V_S$ = voyage speed to water; $t$ = thrust deduction coefficient; $R$: ship resistance.

### 3.5 *Calm water resistance*

The calm water resistance can be divided into frictional resistance $R_f$ and residual resistance $R_r$.

According to "International Towing Tank Conference (ITTC) 1978", the frictional-resistance is (Georgakaki & Sorenson, 2004):

$$R_f = \left[\frac{0.0776}{(\log_{10}^{Rn} - 1.88)^2} + \frac{60}{Rn}\right]\frac{1}{2}\rho S V_S^2 \tag{10}$$

where $R_n$ = Reynolds number; $S$ = wet surface.

The residual resistance is given by:

$$R_r = \frac{1}{2} C_r \rho S V_S^2 \tag{11}$$

where $C_r$ = residual resistant coefficient, which is related with Froude number, length-breadth ratio, longitudinal coefficient, and it was estimated according the method of Georgakaki (2004).

### 3.6 *Additional resistance under service conditions*

#### 3.6.1 *Wind*

Wind resistance is important for ships with large structure areas above the water level, including containerships. When only the wind projected to the direction of the ship course is taken into consideration, the following wind resistance calculation method is adopted (Ying, 2007):

$$R_{wind} = 47.94 k_{wind} A_w V_w^2 \tag{12}$$

where $k_{wind}$ = wind resistance coefficient; $A_w$ = front face area; $V_w$ = relative wind speed.

#### 3.6.2 *Wave*

The prediction of wave added resistance is a quite difficult problem. Here the following model is employed for inland river ships (Ying, 2007):

$$R_{\mathrm{wave}} = \frac{0.065}{(F_r)^2}\left(\frac{h}{L_{\mathrm{wl}}}\right)^2 \frac{\rho}{2} S V_{\mathrm{s}}^2 \tag{13}$$

where $h$ = wave height; $L_{wl}$ = length of water line; $F_r$ = Froude number.

### 3.7 *Water current*

Unlike sea going ships, where the current direction is varied with time, the inland river ships will always be influenced by longitudinal upstream or downstream water current. The water current would change the voyage speed directly:

$$V = V_S + V_{current} \tag{14}$$

where $V_{current}$ = water current speed; $V$ = ship speed to ground.

### 3.8 *EEOI*

The case ship was voyaged with the capacity of 220 TEUs during the data collection process, and the same assumption in simulation will be adopted to keep an equal draft, so the EEOI could be calculated by:

$$EEOI = \frac{FC_i \times C_{carbon}}{m_{cargo} \times D_i} = \frac{Fph \times 2 \times 3.206 \times 1.852}{220 \times V \times 3.6 \times 1000} \tag{15}$$

## 4 MODEL SIMULATION

Based on the above analysis, the model was simulated on MATLAB/Simulink to function the energy efficiency system and sub-systems, as shown in Figure 1.

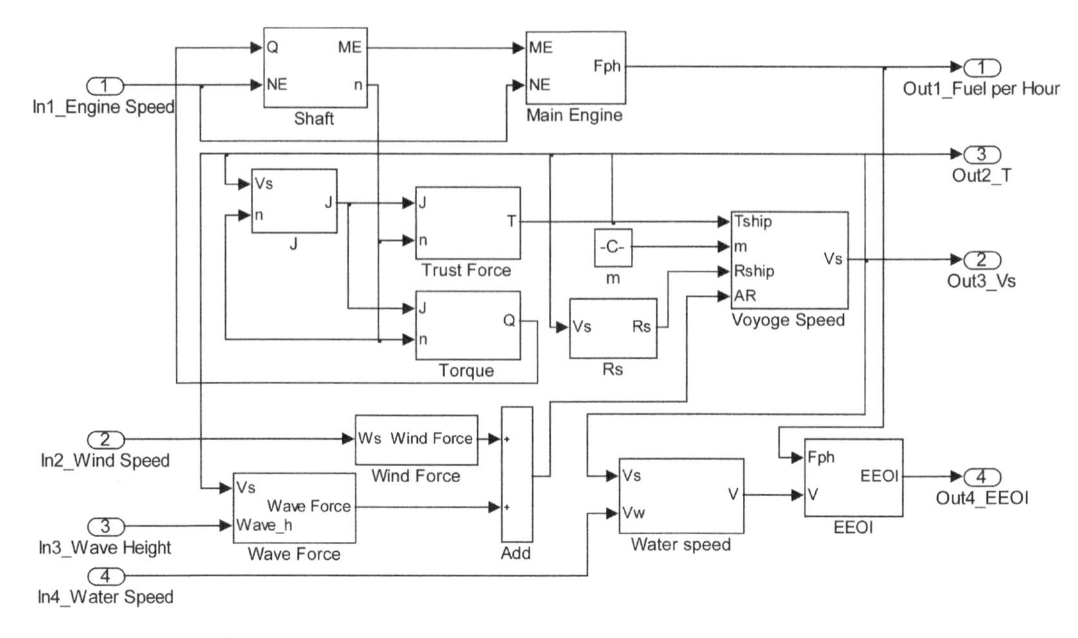


Figure 1. The operation energy efficiency model on Simulink.

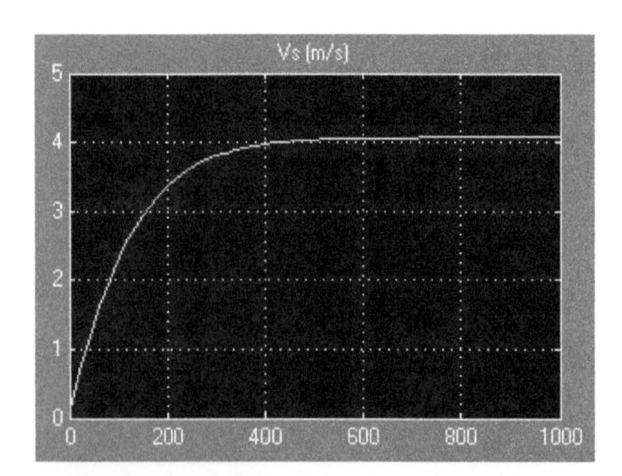


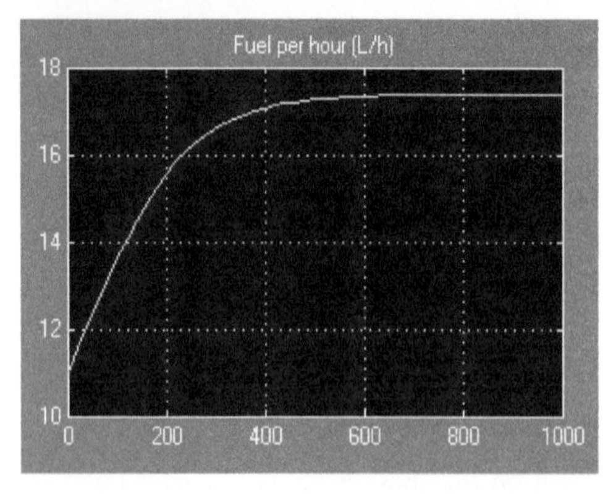


Figure 2. The simulation results of voyage speed and fuel consumption.

Table 3. Validation of voyage speed and oil consumption.

| Engine speed (r/min) | Voyage speed (m/s) | | | Oil consumption (L/h) | | |
|---|---|---|---|---|---|---|
| | M | S | D | M | S | D |
| 400 | 3.7 | 4.1 | 0.11 | 20.3 | 17.4 | 0.14 |
| 460 | 4.2 | 4.7 | 0.12 | 28.5 | 24.0 | 0.16 |
| 550 | 4.9 | 5.6 | 0.14 | 43.7 | 38.0 | 0.13 |
| 600 | 5.6 | 6.1 | 0.09 | 54.2 | 46.7 | 0.14 |
| 700 | 6.3 | 7.1 | 0.13 | 83.9 | 67.0 | 0.20 |

In order to validate the model simulation, the voyage speed and oil consumption data under the engine speed of 400, 460, 550, 600, 700 r/min were collected onboard under calm conditions.

Simulation processes and results under 400 r/min are shown in Figure 2. It can be seen from the graph that after a short time of increase, the voyage speed and oil consumption remained stable in 4.1 m/s and 17.4 L/h separately, where X-axis is the operation time in "oscilloscope". The simulation processes under other rotation speed (460, 550, 600 and 700) are similar with Figure 2. All of the results are shown in Table 3.

The comparison and deviation analysis between real data and simulation results are shown in Table 3. In this table, "M" is short for measure-

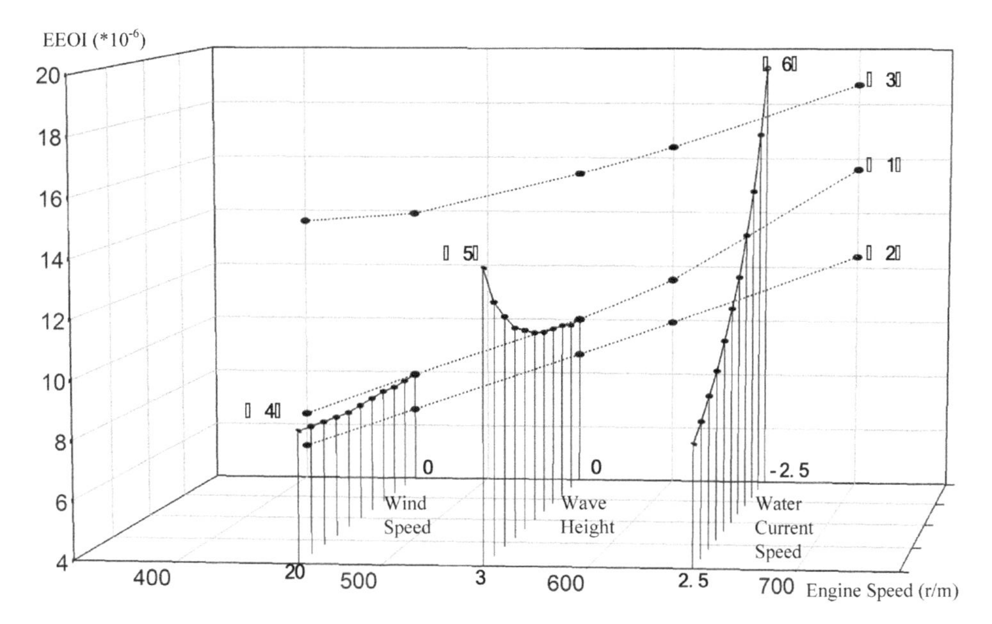


Figure 3. The simulation results of EEOI under different conditions.

ment, "S" for simulation and "D" for deviation, which is defined as:

$$D = \frac{|S - M|}{M} \tag{16}$$

As the two most important input parameters for energy efficiency model, voyage speed and oil consumption showed a deviation around 10%. Considering the complexity of ship resistance prediction, a lot of simplified methods are adopted to qualitatively analyze the EEOI change trends, so the simulation deviation is acceptable.

Based on this simulation platform, the following tests are performed under different navigation environments:

1. Calm water, engine speed from 400 to 700 r/min;
2. Ordinary water (wind speed: 4 m/s; wave height: 1 m; water speed: 1 m/s), engine speed from 400 to 700 r/min;
3. Critical water (wind speed: 15 m/s; wave height: 2.5 m; water speed: −2 m/s), engine speed from 400 to 700 r/min;
4. Under the engine speed of 460 r/min, wind speed from 0 to 20 m/s;
5. Under the engine speed of 550 r/min, wave height from 0 to 3 m/s;
6. Under the engine speed of 650 r/min, water current speed from −2.5 to 2.5 m/s;

The simulation results are shown in Figure 3.

## 5 CONCLUSION

The objective of this work is to demonstrate a model and its simulation method for the operational energy efficiency of inland river ships. From the simulation results, the following conclusions can be drawn:

1. As container ship, the EEOI of inland river one (around $15*10^{-6}$) is lower than that of seagoing ship (around $200*10^{-6}$) (Ni & Zhao, 2010).
2. EEOI varied with the engine speed distinctly. Navigation with lower speeds cannot only save fuel, but also improve the energy efficiency.
3. EEOI is influenced by navigation environment. Water current changed the navigation speed to ground directly, and brought about a distinct effect. Wave and wind reduced EEOI by adding ship resistance.

These conclusions would provide a foundation to engage the public, industry players and policy makers for inland river shipping industry. However, it should be recognized that the simplified methods employed in this work introduced deviation to some extent. On the other hand, according to the influence from the navigation environment, it would be necessary to solve the optimization problem of EEOI, that is develop energy improvement methods under different navigation conditions.

## ACKNOWLEDGEMENT

This work is supported by the National Natural Science Foundation of China (51279149) and the China Scholarship Council (File No. 201306950035).

## REFERENCES

Feng, P.Y., Ma, N., et al. 2012. Analysis of the ship speed loss coefficient based on hull-engine-propeller matching and wave statistics. *Journal of Shanghai Jiaotong University* 46(8): 1248–1253.

Georgakaki, A. & Sorenson, S.C. 2004. *Report on collected data and resulting methodology for inland shipping.* Report No. MEK-ET-2004-2.

Lindstad, H., Jullumstro, E., et al. 2013. Reductions in cost and greenhouse gas emissions with new bulk ship designs enabled by the Panama Canal expansion. *Energy Policy* 59: 341–349.

Ni, J.K. & Zhao, Y.F. 2010. Study on energy efficiency operational indicator of container ships. *Ship & Ocean Engineering* 39(5): 140–143.

Shi, W. 2008. Simulation of the influence of ship voyage profiles on exhaust emissions. *ASME International Mechanical Engineering Congress and Exposition. 31 October-11 December 2008*. Boston: ASME.

Sun, X., Yan, X.P., et al. 2013. Analysis of the operation energy efficiency for inland river ships. *Transportation Research Part D: Transport and Environment* 22: 34–39.

Tzannatos, E. & Papadimitrou, S. 2013. The energy efficiency of domestic passenger shipping in Greece. *Maritime Policy & Management* 40(6): 574–587.

Yan, Z.Z., Yan, X.P., et al. 2010. Green Yangtze River intelligent shipping information system and its key technologies. *Journal of Transportation Information and Safety* 29(6): 76–81.

Ying, Y.J. (ed.), 2007. *Ship speed and resistance*. Beijing: China Communications Press.

*Maritime-Port Technology and Development – Ehlers et al. (Eds)*
*© 2015 Taylor & Francis Group, London, ISBN 978-1-138-02726-8*

# Reinforcing existing port facilities to withstand natural disasters: An analysis of increased construction costs

K. Takahashi, Y. Kasugai & I. Fukuda
*Port and Airport Research Institute, Yokosuka, Kanagawa, Japan*

ABSTRACT: The importance of preparing for unexpected phenomena is being seen in a new light after the Great East Japan Earthquake of 2011. In order to manage container terminals soundly, it is essential that stakeholders clarify for investors the trends and future outlook of construction costs, which are important indicators of business management. The authors have analyzed the factors that increase these costs. As a result, the costs are likely to increase further because of natural factors such as earthquakes and softening ground. Based on the above analysis, the authors propose effective measures for unexpected phenomena.

## 1 INTRODUCTION

The major shipping companies of the world are increasing the sizes of their containerships in order to reduce the construction and transportation costs per Twenty-foot Equivalent Unit (TEU) of containerships. Accordingly, anchorage sites for such large containerships need to prepare deep water mooring facilities, and the construction costs of the container terminals is bound to increase.

Further, the 2011 off the Pacific Coast of Tohoku Earthquake (so-called "the Great East Japan Earthquake of 2011") and Hurricane Sandy of 2012 showed port authorities of the container terminals around the world that some large-scale natural disasters might exceed the conventional maximum disaster prevention level.

As mentioned earlier, the expansion in containership sizes and preparation for preventive countermeasures against large-scale natural disasters is causing an increase in the construction costs of container terminals, and port authorities all over the world share a common concern, namely limiting this increase in the construction costs.

Therefore, we first compared construction costs across major global ports, to clarify the gap between internalized and externalized costs within the construction costs and also to identify the social and natural factors that would increase the construction costs. Then, we introduced efforts to reduce the construction costs in Japan and verified how the costs for improving the existing facilities at the Port of Nagoya, Japan, would rise in response to the upgradation of its disaster prevention level for expected large-scale natural disasters.

By verifying the Japanese cases, we clarified that it is possible to suppress an increase in the rate of the construction costs to around 10% of its total. Note that we could not estimate the total sum of money required for upgrading the disaster prevention level for future large-scale natural disasters, because the gaps between the internalized and externalized costs within the construction costs were large among countries.

Given current progress in the horizontal and vertical specializations of the global economy, any stoppage in the port functions caused by a large-scale natural disaster in one country may seriously damage the economic activities of many countries. Therefore, we hope this study will highlight the importance of reinforcing existing port facilities against large-scale natural disasters.

## 2 GAP BETWEEN THE INTERNALIZED AND EXTERNALIZED COSTS WITHIN THE PORT CONSTRUCTION COSTS

As mentioned in section 1, the port construction costs of countries increases owing to social and natural factors. However, the extent of this increase varies according to social and natural conditions of each country and site. In this section, we compared the construction costs of each country and verified the gap between the internalized and externalized costs within the port construction costs.

### 2.1 *Prerequisite for verifying the gap between internalized and externalized costs within the port construction costs*

Port construction costs vary largely according to the social and natural conditions of the construction site. We focused on the gap between the internalized and externalized costs within the port construction costs for Europe, South Korea, and Japan, and we compared them under the following prerequisites in this study.

<Prerequisites>

Water depth: Quay around 17 m deep

Quay length: Quay around 400 m long (Converted to construction cost permeter)

Seismic load: Seismic load outside Japan is indicated via a design seismic coefficient. Since Japanese quays are designed considering seismic motion, seismic load is not indicated (if using the conventional design seismic coefficient, this value corresponds to 0.25 g)

Construction cost: Direct construction costs (Port cost estimation standards) (Outside Japan, the construction costs may include project costs other than the construction costs of the quays)

### 2.2 *Comparison between port construction costs inside and outside Japan*

Table 1 shows the result of the comparison in quay construction costs among countries for a high-standard container terminal of depth 16 m or more. Generally, the structural form varies according to the design and conditions during construction, standards such as ground conditions and earthquake resistance standards, and the ordering system. In this study, we estimated that the ports for large containerships in all countries include mooring structures, thus lowering the construction costs to the extent possible. In particular, we compared the construction costs in terms of secure mooring functions for large container ships. Since the port construction costs for outside Japan might include the construction costs for other port facilities except quays (for example, freight handling area), there was a possibility that the actual gap between the internalized and externalized costs may be larger than that indicated in the comparison. However, when comparing costs for quays of 16 m depth or more, we found that the construction costs of the Ports of Rotterdam and Antwerp, which are not typically subject to earthquakes, were overwhelmingly lower than those of Japanese ports. Further, when comparing quays of 18 m depth, we found that the construction costs of the Port of Busan, an area subject to earthquakes albeit at lower seismic loads than those experienced at the Tokyo and Osaka Ports, was lower.

Thus, it is clear that the port construction costs vary widely across countries. In other words, we cannot compare construction costs simply on the basis of the large gap between external and internal prices.

## 3 FACTORS INCREASING PORT CONSTRUCTION COSTS

Among the factors increasing port construction costs around the world, we selected offshore deployment to support larger ships as a social factor and countermeasures against earthquakes and soft ground as a natural factor.

### 3.1 *Offshore deployment to support larger ships*

Worldwide, value-added expensive products are carried by containerships. By utilizing economies of scale, shipping companies are rapidly increasing the sizes of their containerships (see Fig. 1), which go in service in the global liner routes, in order to reduce the transportation cost per TEU of the marine container. Until recently, the companies

Table 1. Comparison of container quay construction costs.

| Port | Project cost ($ million/m) | Maximum water depth (m) | Structure form |
|---|---|---|---|
| Rotterdam | 16.7 | 0.060 | Reinforced concrete sheet pile type with relieving platform |
| Antwerp | 17 | 0.052 | Gravity type (Reinforced concrete L-shaped retaining wall) |
| Busan | 18 | 0.24 | Gravity type (Vertical wave dissipating caisson) |
| Tokyo | 18 | 0.43 | Piled quay type |
| Yokohama | 18 | 0.32 | Cellular-bulkhead type |
| Osaka | 16 | 0.28 | Piled quay type |
| Kobe | 16 | 0.25 | Piled quay type |

*The project costs outside Japan may include project costs other than the construction costs of the quays. **Port project costs outside Japan are calculated as construction costs according to Koizumi et al. (2011). ***Japanese port project costs are based on the hearing data of the Ministry of Land, Infrastructure, Transport and Tourism of Japan.

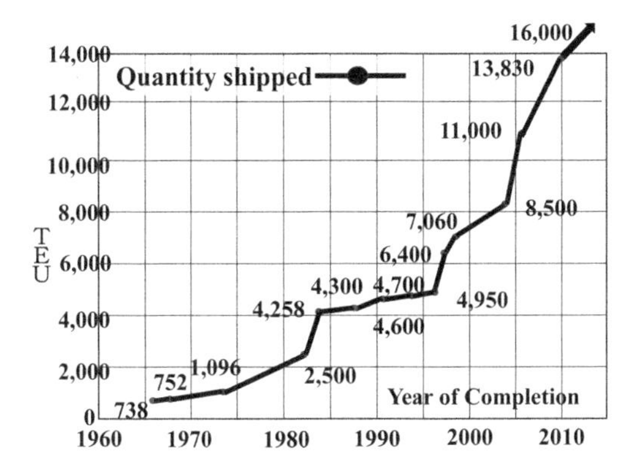


Figure 1. Transition in the largest containership size.

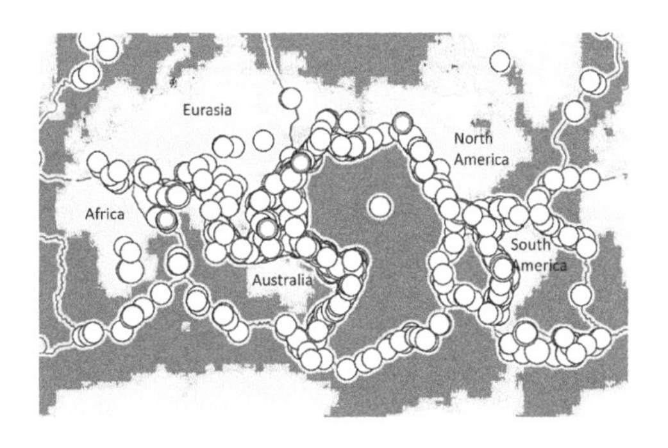


Figure 2. Distribution map of earthquakes. (Magnitude 6 or higher at depths of 100 km or less from 2004.07.14 00:00:00 UTC to 2014.07.21 23:23:59 UTC), Source: United States Geological Survey (USGS).

operated large container ships of total lengths of up to 400 m, drawing 16 m, and of 16,000 TEU capacity. They just let larger container ships go into service with loadable capacities of 18,000 TEU. Although some believe that very large container ships cancel out economies of scale and that the restriction at the Strait of Malacca will not allow passage of larger ships, globally, shipping companies continue to promote the construction of ever-larger container ships; this has been evidenced since the 1960s in three prime shipping areas: Europe, Asia, and North America. The mooring facilities of the ports in these regions have also been enlarged.

Enlarging ports requires not just deepening the water depth at the front of the quay, where ships come alongside, to match the draft line of the large ships but also enlarging the seaway, anchorage site, and quay site areas and deepening the seaway at the anchorage site.

Enlarging ports in areas without a large ground area requires offshore deployment of the construction site and the construction of port structures in deep sea areas.

Enlarging port structures increases the construction costs considerably. For example, for a caisson type quay (a port structure) in deeper water, the landside ground pressure increases proportionally with depth, and the resultant force of the ground pressure increases proportionally to the square of the depth. Generally, construction costs rapidly grow exponentially with the increase in water depth.

### 3.2 *Countermeasures against earthquakes*

Figure 2 shows the world earthquake distribution map issued by the United States Geological Survey (USGS), indicating the distribution of earthquakes of magnitude 6 or higher at depths of 100 km or less. Earthquakes originating at depths of 100 km or deeper have their hypocenters in specific regions, such as Japan and the west coast of North America (but not in Europe). Although the Japanese archipelago constitutes only 0.1% of the land area on earth, 10% of magnitude 6 or higher earthquakes have occurred in Japan. Some researchers estimate that this rate increased to 21% in 1994 and later years.

Figure 3 shows the estimated financial loss for gravity-type quays at each earthquake motion level indicated by Ichii (2002). This figure helps us understand that as the seismic load increases, the estimated loss increases rapidly. A facility with a structure receiving ground pressure consistently from one direction, such as a quay or seawall, is designed so that it remains stable on account of the horizontal force resisting the seismic load. To resist a larger seismic load, the cross section must be enlarged horizontally to match the increased resisting force. To improve safety against seismic loads, quays constructed in earthquake zones like Japan and the west coast of North America entail higher construction costs than their counterparts in East Asian counties, such as South Korea, China, and Vietnam, and European countries.

### 3.3 *Countermeasures against soft ground*

Flat lands along coastal areas called "alluvial plains" were formed on the drowned valleys, which were eroded when the sea level fell during the Ice Age, by the accumulation of soft soil from the rivers. Therefore, the ground in alluvial plains comprises soft soil deposit.

Constructing a port facility on such thick and soft ground entails additional cost for ground improvement and settlement of long piles into the foundation layer, which increases the construction costs tremendously. Compared to the ground condition outside Japan where the foundation layer is shallower, the ground in alluvial plains is

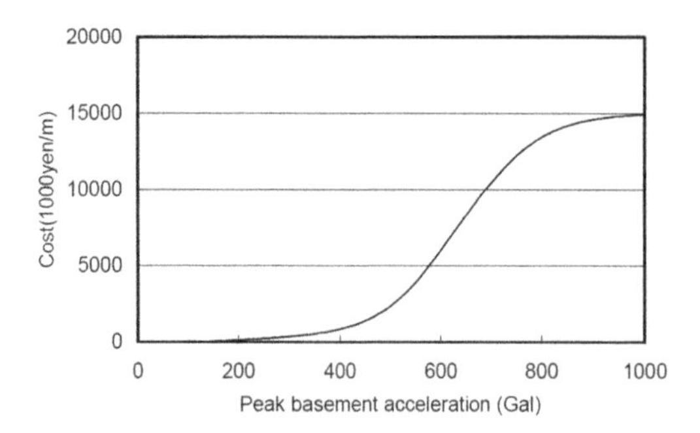


Figure 3. An example of estimated loss for each input excitation level, Source: Ichii (2002).
Note: Unit 'Gal' means gravitational acceleration, Gal = $cm/s^2$.

too soft and necessitates the introduction of new technologies, special work barges, and skillful workers in every design and construction phase. Therefore, it is not possible to reduce the construction costs of a port sited on soft ground.

Typical examples of structures constructed on thick and soft ground are the Sakishima Tunnel of the Port of Osaka and Kansai International Airport in Japan. Located on the sea bottom of Osaka Bay, resting on alluvial and diluvial deposits several hundred meters thick, they were designed and constructed assuming that a large-scale consolidation settlement would occur. The construction costs for these projects increased drastically compared to the costs for cases without consolidation settlement.

### 3.4 *Increase in administrative and maintenance expenses*

The Ministry of Land, Infrastructure, Transport and Tourism (Japan) estimated future public works expenditures, upgradation costs, and administrative and maintenance expenses for existing facilities in Japan. Since the facilities were mainly constructed in the 1990s, they are in need of upgradation. The Ministry of Land, Infrastructure, Transport and Tourism estimates that these costs and the associated administrative and maintenance expenses will exceed the total investible funds.

Since Asian countries have relatively newer port facilities compared to those in Japan, we can estimate that their upgradation costs and administrative and maintenance expenses will be less.

## 4 EFFORTS TO REDUCE PORT CONSTRUCTION COSTS

How do countries reduce port construction costs? They are making efforts to suppress increases in port construction costs. In this study, we introduced the case of Japan, wherein efforts for reducing port construction costs are divided into two periods: from 1950 to 1999, when the design and analysis methods were improved through technological development, and post-1999, when reductions in the construction costs and administrative and maintenance expenses were practically realized.

### 4.1 *Approach to technological development in Japan*

Table 2 shows the history of technical standards applied for port construction works.

Constructing ports in Japan according to the specification requirements has necessitated moving from the conventional experience-oriented engineering approach to the knowledge-based approach. The first design standard in the postwar period, "Design Specification Guideline for Port and Harbor Construction," was issued in 1950.

In 1973, the Japanese government incorporated the "Technical Standards for Port and Harbor Facilities in Japan" (hereafter, "Technical Standards") in the Port and Harbor Act, thus extending them legal sanction. These technical standards include the structural functions and safety procedures to be followed for all port-related construction works. In doing so, the Japanese government employed the Technical Standards not just to ensure safety at under-construction port facilities but also to serve as the standards that port authorities would use to approve constructions within the port zone.

At the same time, the Japanese government urged the port authorities and private business operators to improve their technological skills by improving the technical guidance and information required for port construction. These measures contributed to improving the functions of and the safety required at the port facilities of port authorities and private business operators.

These Technical Standards included the findings of conventional theoretical studies, indoor experiences, and field observations, which contributed to secure the functions and safety of port structures. It defines criteria regarding the deformation and strength of the facilities as well as the materials to be used. The original technological developments mainly helped to refine the methods used to calculate the external forces the port facilities have to withstand and improve the methods for port design, thus aiming at reliable and safe construction.

In 1999, the Technical Standards reached a major turning point. In responding to public opinions about the construction costs being higher in Japan compared to costs in other countries, many

Table 2. Revision history of the technical standards.

Technical developments to improve accuracy and safety
Technological developments for cost reduction

| | |
|---|---|
| 1950 | Setting the earth pressure calculation formula and frictional coefficient of soil |
| | Setting the design seismic coefficient |
| | Setting the safety factors for sliding and falling |
| | Clarification of Hiroi's formula and Sainflou's formula as wave pressure formulae |
| 1959 | Introduction of the wave prediction method (SMB, etc.) |
| | Safety consideration of slope and slip circle analysis |
| 1967 | Specification design methods for various facilities |
| | Accuracy improvements in the design seismic coefficient considering the local seismic activity |
| | Indication of standard ship size |
| 1973 | Including safety requirements in the Port and Harbor Act |
| 1979 | Adoption of Goda's wave pressure formula as the standard wave pressure formula |
| | Inclusion of design methods for various facilities |
| | Introduction of rules pertaining to anchoring site calmness |
| 1989 | Setting the estimation method for liquefaction |
| | Introduction of bearing capacity analysis with the Bishop method |
| | Introduction of the soil improvement construction method |
| | Lowering the end bearing capacity coefficient for pile foundations |

The technological development can be charted through the history of changes in the external load setting method and the analytical and design methods used for the rational planning and design of safe facilities.
These developments also include steps taken toward reducing construction costs. However, cost increases were inevitable in some cases where it was necessary to secure a certain level of safety.

| | |
|---|---|
| 1999 | Introduction of Level 1 and Level 2 earthquake motions |
| | Introduction of the reliability design method with the expected sliding volume |
| | Improvement in flexibility introduced by notifications in the standards |
| 2006 | Definition of performance and transition to the reliability design method |
| | Performance-based design methodology was introduced. |
| | Introduction of Life Cycle Management (LCM) by preparation of maintenance and management standards |

researchers tried devising ways to reduce the cost while maintaining safety. The Japanese government included these results into the Technical Standards by revising them in 1999.

However, besides the Technical Standards in 1999, the conventional Technical Standards were specification-oriented standards, which defined the standardized materials and design methods for port construction works. This type of standard was convenient and reliable for the government and port authorities, who were responsible for checking conformity with functions and safety. However, since these standards adopted methods that secured safety using overdesigned structural reinforcement, they tended to ignore cost issues.

### 4.2 *Reducing construction costs using performance definitions in the technical standards*

The conventional revisions of the Technical Standards show the history of safety-related improvements. Although the government revised the Technical Standards in 1999 in response to its policy of reducing public works expenditures, it did not review the fundamental system employed therein, and hence, it could not actually reduce the construction costs.

On the other hand, in view of the international trend to define performance standards, the government revealed plans to define performance in the Technical Standards through a three-year deregulation program. As a part of its Public Works Cost Structural Reform Program (March 2003), it decided to change the Technical Standards for port facilities from specification-oriented standards to performance-oriented standards.

In 2006, the government revised the Port and Harbor Act to change the Technical Standards for port facilities, from the conventional specification-oriented standards to performance-oriented standards, and at the same time, hand over the responsibility of conformance judgments to the Technical Standards to the government or a third party. It is possible that the government defined the technical standards as being related to the required safety issues and commissioned the popular belief that specifications and designs must be performance-based instead of relying on safety certifications for port facilities by checking their adaptability to the standardized national Technical Standards. Accordingly, new design methods and special structures based on private inventive approaches could be adopted; notably, it was possible for private corporations to design the facility's strength and durability to fit its intended importance and lifecycle. It was expected that these measures would help reduce not just the construction costs but the upgradation costs and administrative and maintenance expenses also.

Figure 4 shows an example of a breakwater with a reduced construction costs, designed according to the performance-oriented Technical Standards. This breakwater was designed using the expected sliding volume method, allowing approximately 10% reduction in the construction costs.

### 4.3 *Reduction in administrative and maintenance expenses by introducing lifecycle management*

Matsubuchi and Yokota (1999) released their "Basic Study on Lifecycle Cost Generation of a Mooring Facility and Establishment of a Maintenance Management Decision-making Support System.

The results described how ideal mooring facilities could be designed alongside their lifecycle costs. Several other studies about cost reduction through maintenance management are ongoing.

The Japanese government defined performance-oriented Technical Standards by revising the Port and Harbor Act in 2006, thus allowing the introduction of new design methods based on novel ideas. One particular revision was quite significant; the Technical Standards also stipulated the preparation of a maintenance management plan for all port facilities so as to allow long-term cost reduction by maintenance management across the lifecycle.

With regard to maintenance management, although the Japanese government stated in the Port and Harbor Act that "port facilities should be constructed, upgraded, or maintained so that they meet the Technical Standards," the conventional Technical Standards focused on securing a certain level of safety within the facilities, and definitions about the lifecycle cost, including maintenance management costs, were insufficient. It is undeniable that, until then, the authorities had overlooked the significance of maintenance and management. The new Technical Standards, however, clarified the importance of maintenance and management, stipulating the preparation of maintenance and management plans and the minimization of lifecycle cost (and thus, total of construction costs), upgradation costs, and administrative and maintenance expenses. The government also released the "Port Facility Maintenance and Management Technical Manual," which introduced preventive maintenance and management procedures such as the Checkup/Diagnosis Plan and the Maintenance/Repair Plan.

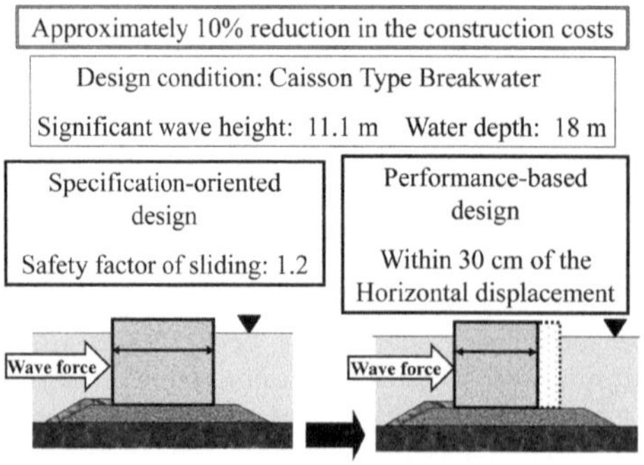

## 5 ELICITATION OF NEW RISK

In spite of the aforementioned efforts to reduce port construction costs in Japan, the Great East Japan Earthquake of 2011 indicated new risks that would increase these costs. We now describe these risks.

L1 events and L2 events are defined as Table 3 (Government of Japan (2011)).

### 5.1 *New risks revealed by the great east Japan earthquake*

The Great East Japan Earthquake of 2011 underscored the importance of preparing for unexpected events. In the past, the possible occurrence of events exceeding the design conditions was considered highly improbable. In addition, arguments about

Approximately 10% reduction in the construction costs
Design condition: Caisson Type Breakwater
Significant wave height: 11.1 m Water depth: 18 m
Specification-oriented design
Safety factor of sliding: 1.2
Performance-based design
Within 30 cm of the Horizontal displacement
Wave force
Wave force

Figure 4. Design example of a breakwater with expected sliding volume.

Table 3. Definition of L1 events and L2 events.

| |
|---|
| L1 ground motion: Appropriately as a stochastic time history with considerations of source, path and site effects, based on the results of earthquake observation |
| L1 tsunami: Occurring more frequently than the largest-possible tsunamis and causing major damage despite their relatively lower tsunami heights |
| L2 ground motion: Appropriately as a time history with considerations of source, path and site effects, based on the results of earthquake observation and the source parameters of the scenario earthquake |
| L2 tsunami: Envisaged on the basis of developing comprehensive disaster management measures, which focus on the evacuation of local residents as the main pillar |

preparing for Level 2 events were unacceptable to port authorities as no budgets were allocated for taking measures against such events. Therefore, arguments in favor of preparing for Level 2 events had been inactive for a long time.

A post-earthquake analysis revealed the following problems: lack of preliminary discussions about damage reduction using a combination of hard and soft measures for preventing and mitigating Level 2 disasters, and insufficient information dissemination to the public although some concerns in this regard had been voiced previously.

Therefore, in the future, it would be prudent to analyze all events corresponding to Level 2 disasters and foreplan countermeasures against them on national and regional scales, in order to clarify the estimated risks.

Possible Level 2 disasters include earthquakes and tsunamis. Other presumable events, such as unexpected storm surges and tidal waves caused by strong typhoons generated as a result of global warming, should be targets for such discussions.

However, given the extreme improbability of certain events, such as the giant meteorite strike that killed off the dinosaurs, it is inappropriate to include such events in the discussion about Level 2 disasters in port operations.

Further, many locations are yet to establish facilities to help withstand Level 1 events, in terms of hard measures. Therefore, besides soft measures, port facilities must establish and maintain reliable hard measures to withstand Level 1 disasters. Moreover, they must also prepare for Level 2 disasters by including soft measures such as establishing refuge instruction methods and constructing refuge facilities for employees and inhabitants, especially in cases where hard measures such as construction and reinforcement of existing port facilities are unavailable.

In light of this discussion, the estimated emerging risks for port facilities are as follows.

a *Increase in the probability of earthquakes*
Toward the end of 2012, the Headquarters for Earthquake Research Promotion of the Ministry of Education, Culture, Sports, Science and Technology (Japan) created the "Probabilistic Seismic Hazard Map". It shows that the possibility of earthquakes with seismic intensities of lower 6 or more as per the Japan Meteorological Agency's (JMA's) seismic scale, increased significantly within 30 years after 2012 mainly around the Kanto area. The possibilities of large-scale earthquakes in all three major harbors, which are the focal points of the Japanese economy, are very high, and thus, port authorities must undertake immediate and urgent countermeasures.

b *Breakwater sinking caused by liquefaction after an earthquake*
In addition to the quays and seawalls, breakwaters founded on sandy soil are likely to sink because the soil will undergo liquefaction during an earthquake. One example of this phenomenon is the sinking of the breakwaters of the Port of Kobe, which sunk by 2 m owing to the Hanshin Awaji Great Earthquake (formally 'the Southern Hyogo Earthquake') in 1995. In order to ensure that they continue to function as required in the event of a tsunami, a certain level of breakwater raising and reinforcement are necessary.

c *Partially functional breakwater after a tsunami*
Damage reduction and (at least) partial functionality have been hotly debated topics after the earthquake. Clearly, measures and policies allowing the construction of robust breakwaters (Fig. 5) are the need of the hour.

### 5.2 *Lateral flow of landfilled areas caused by liquefaction*

The 2004 Chuetsu Earthquake caused land deformation severe enough to create lateral spreading across a wide landfilled area owing to the liquefaction. However, the danger of this phenomenon was not taken into consideration in the design measures stated in the Technical Standards. Figure 6 shows a model of such a phenomenon. Hamada (2012) had pointed out the danger posed by such an event; in

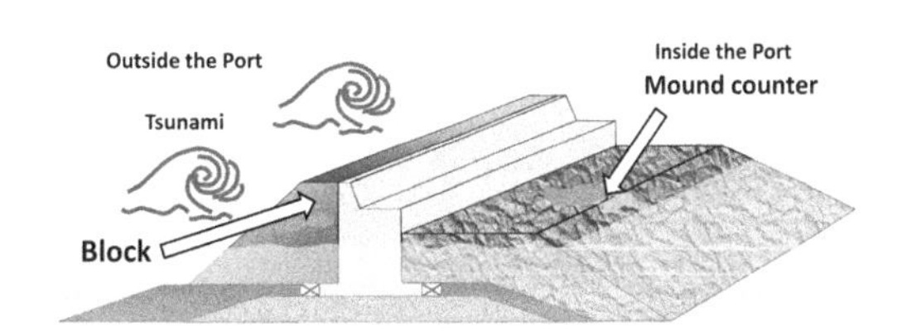


Figure 5. Reinforcing the breakwater.

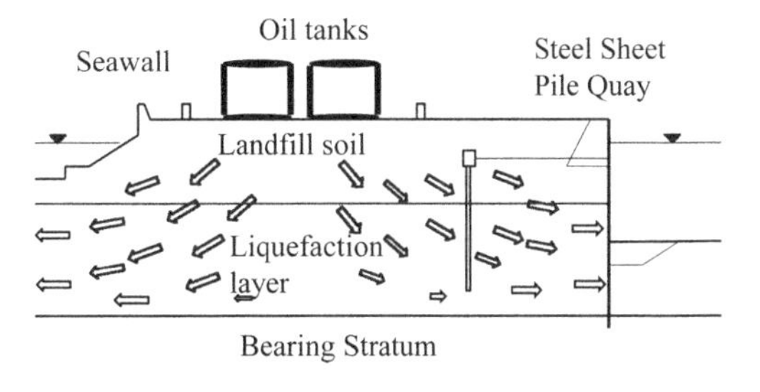


Figure 6. Horizontal soil displacement caused by liquefaction and lateral flow of landfilled areas.

particular, should such an event occur, he indicated the risks posed by the large oil tanks located in the coastal areas of the three major harbors. Since it is difficult to secure safety with breakwater design alone, it is important to consider the safety of the overall landfilled areas and take appropriate countermeasures.

## 6 VERIFYING INCREASE IN THE CONSTRUCTION COSTS: THE CASE OF THE STORM SURGE BREAKWATER OF THE PORT OF NAGOYA

Although efforts to decrease port construction costs are already underway in Japan, it is expected that the costs will increase in the future owing to social and natural factors, such as the emergence of new risks as shown by Table 4. In this section, we provide estimates of the increase in the construction costs using the case of the breakwater of the Port of Nagoya as an example.

Located in the Chubu district (midland of Japan), the Port of Nagoya is the largest and busiest trading port in Japan. It is the largest exporter for several global manufacturing industries, such as automobiles and the aerospace industry. We provide the estimated increase in cost owing to the construction of storm surge breakwaters to reduce the threat from storm surges. To do so, we referred to the damage caused by Typhoon Vera (Isewan Typhoon) at the Port of Nagoya.

The main objective of building these storm surge protection breakwaters was to raise the breakwater crown so as to protect the port from a tsunami that would be generated after a large earthquake; the breakwaters would presumably sink owing to the liquefaction caused by the earthquake. Further, since they were already very (50 years) old, measures to upgrade them in preparation for large earthquakes were also necessary. Therefore, the Ministry of Land, Infrastructure, Transport and Tourism based their estimates considering that a tsunami likely to occur during the design life of the structure (the so-called "L1 tsunami") and the maximum predicted tsunami that may attack the structure (the so-called "L2 tsunami") would eventually occur. Accordingly, the Bureau estimated the amount by which the storm surge protection breakwaters would sink after the earthquake and checked the protection function against each tsunami and the storm surges that would attack the structure after the sinking of the storm surge protection breakwaters.

The Central Disaster Prevention Council of the Japanese government predicted that if a megaquake (Mw 9.1) occurs in the Nankai Trough off the Pacific coast as shown by Figure 7, the quake and its ensuing tsunami will cause serious economic damage, and accordingly, the Ministry of Land, Infrastructure, Transport and Tourism was urged to consider taking the following appropriate protective measures.

### 6.1 *Checking the breakwater crown by analyzing the seismic response to L1 and L2 earthquakes*

The calculation clearly showed that the storm surge protection breakwaters of the Port of Nagoya would sink by 3.4 m at most after an L2 earthquake, and almost all cross sections would fail to meet the required breakwater crown (N.P. + 5.4 m). To take measures against L1 earthquakes and to secure calm after L2 earthquakes, the Ministry of Land, Infrastructure, Transport and Tourism

Table 4. Assumed megaquakes and megaquakes of the past.

| Assumed megaquakes | Assumed magnitude |
|---|---|
| Nankai Trough Earthquake | 9.1 (Mw) |
| Tokai/Tonankai/Nankai consolidated Earthquake | 8.7 (Mj) |
| Earthquake that directly hits the Tokyo area | 7~8 (Mj) |
| **Megaquakes of the past** | **Mw (year)** |
| 1946 Nankai Earthquake | 8.4 (1946) |
| Great East Japan Earthquake | 9.0 (2011) |
| Hanshin Awaji Great Earthquake | 6.8 (1995) |

*'Mw' is the Moment Magnitude Scale, which is synonymous with the Richter Scale and 'Mj' is the Magnitude Scale used by the Japan Meteorological Agency (JMA).

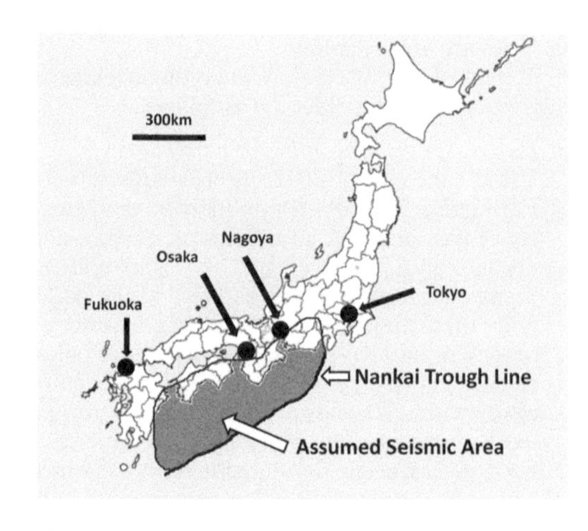


Figure 7. One of new risks: Nankai Trough Earthquake (Assumption Mw 9.1).

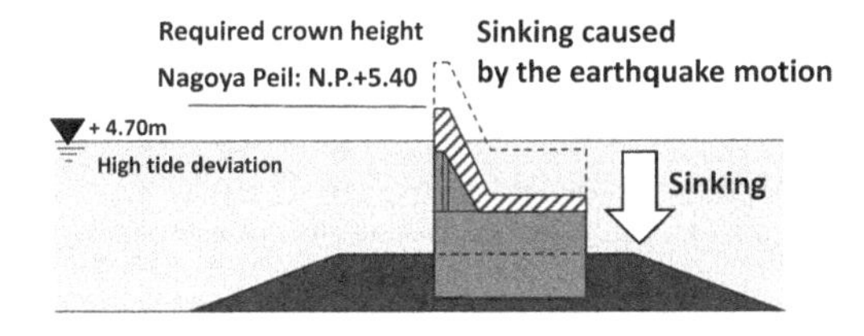


Figure 8. Concept of preventing breakwater sinking.

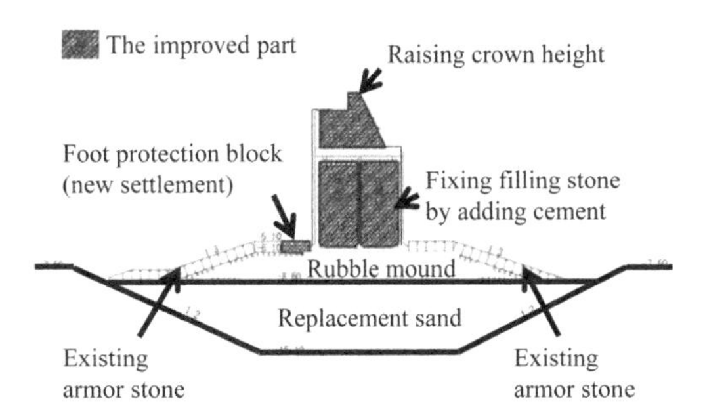


Figure 9. Cross section of the upgraded storm surge protection breakwater of the Port of Nagoya.

plans to raise the breakwaters to the required level as shown in Figure 8. It also plans to replace the packing sands inside the caisson with mortar for reinforcement, which are of questionable strength owing to aging.

### 6.2 *Stability of the storm surge protection breakwaters against L1 and L2 tsunamis*

The results of this verification indicated that all breakwater cross sections would be stable against both L1 tsunami and L2 tsunamis. We also confirmed that they were sufficiently robust.

As mentioned earlier, the Ministry of Land, Infrastructure, Transport and Tourism discussed the robustness of the storm surge protection breakwaters of the Port of Nagoya to protect the port against L1 and L2 tsunamis, such that they can operate as intended even if they sink owing to liquefaction after earthquakes. Accordingly, concrete reinforcement measures have been in place since 2012. At the same time, measures are also being taken to reinforce the aging breakwaters, which were built nearly 50 years ago.

Figure 9 shows the cross section of the upgraded storm surge protection breakwater of the Port of Nagoya, wherein the crown has been raised (among other measures).

The total amount of investment (construction costs) for these measures amounted to 6 billion yen, including the cost for raising the breakwaters, fixing the caisson fillings, and reinforcing the foot protection blocks according to the breakwater cross sections. Thus, assuming that the same measures apply to a new storm surge protection breakwater, the increase in the construction costs derived from reinforcing the breakwater to withstand L1 and L2 earthquakes corresponds to 10% of the total project cost of a new construction, that is, 60 billion yen. In other words, it is clear that to secure measures against the emerging risks posed by L1 and L2 tsunamis and earthquakes, the construction costs would increase by about 10% over the conventional costs.

## 7 CONCLUSION

Considering the current progress in the horizontal and vertical specializations of the global economy, when a port of one country becomes functionally paralyzed, many countries suffer economic damage. However, countries charge either private companies or local public authorities with the operation of ports, and these port authorities cannot meet the construction costs for reinforcing port infrastructure, mainly owing to poor fundraising capacity. In order to manage the container terminal soundly, it is essential that stakeholders clarify for investors the trends and future outlook of construction costs, which are important indicators of business management. To eliminate such an eventuality, in this study, we verified the increase in the rate of the construction costs in the Japanese context. Note that we could not compare the absolute increase in the cost of port construction across countries, because of the large gaps between the external and internal prices within the construction costs. Our results showed that port construction costs increased by around 10% even though there were gaps between the external and internal prices within the construction costs for each country. The operators of the container terminals that increase only 10% of conventional facilities investment can reduce the risk of the large-scale natural disaster. This 10% is the cost that they can take in in the management of their container terminals. To ascertain the total increase in construction costs, we would need to verify the amount of increase in cases other than the port facilities mentioned in this document; however, our result provides an approximated trend applicable to all ports in Japan.

Operators of the container terminals need to reinforce existing facilities for natural disaster. In addition, they need to construct refuge facilities for employees and inhabitants and establish refuge instruction methods for the time when a large-scale natural disaster strikes.

To sum up, this study underscores the importance of reinforcing existing port facilities, to prepare them to withstand large-scale natural disasters.

## REFERENCES


Akakura, Y. 2011. Dimensions of Mega Container ship and berth Dimensions/Container Terminal Area Compatible, *RESEARCH REPORT of National Institute for Land and Infrastructure management, Vol 45*, Yokosuka, Japan.

Goda, Y. 2006. Changing engineering standards and the revision of ports, Changing! Port Construction and Usage, *KOWAN, Vol. 8,* The Ports and Harbours Association of Japan, Tokyo, Japan.

Government of Japan. 2011. Report of the Committee for Technical Investigation on Countermeasures for Earthquake and Tsunamis Based on the Lessons Learned from the "2011 off the Pacific coast of Tohoku Earthquake", Central Disaster Management Council, Cabinet Office, Tokyo, Japan, http://www.bousai.go.jp/kaigirep/chousakai/tohokukyokun/pdf/Report.pdf#page=1.

Hamada, M. 2012. Risk of Liquefaction of landfilled areas, *Iwanami Shoten,* Tokyo, Japan, ISBN978-4-00-028523-0 C0336.

Ichii, K. 2002. A seismic risk assessment procedure for gravity type quay wall: Structural Engineering/Earthquake Eng., *JSCE, Vol. 19, No. 2*:131–140, Tokyo, Japan.

Koizumi, T., Watanabe, T., Suzuki, K. 2011. An Investigation Analysis on the Overseas Supersized Container Terminals, *TECHNICAL NOTE of National Institute for Land and Infrastructure Management No. 628,* Ministry of Land, Infrastructure, Transport and Tourism, ISSN 1346-7328, Yokosuka, Japan.

Matsubuchi, S., Yokota H. 1999. Life-Cycle Cost Analysis of Berthing Facilities and Development of A Decision Support System during their Maintenance Work, *REPORT OF THE PORT AND HARBOUR RESEARCH INSTUTUTE, Vol. 8, No. 2,* MINISTRY OF TRANSPORT, Yokosuka, Japan.

Ministry of Land, Infrastructure, Transport and Tourism. 2011. Japan http://www.nagoya.pa.cbr.mlit.go.jp/topics/111011/index_files/data030.pdf.

Takahashi, K., Urabe, S., Umeno, S., Kozawa, K., Fukuda, I., Kond, T. 2013. PORT LOGISTICS POLICY OF JAPANESE GOVERNMENT FOR STRENGTHENING GLOBAL COMPETITIVENESS OF INDUSTRY IN CASE OF OCEAN SPACE UTILIZATION, *Proceedings of the ASME 2013 32nd International Conference on Ocean, Offshore and Arctic Engineering OMAE2013*, Nantes, France.

United States Geological Survey (USGS), Earthquake Hazards Program, Earthquake Archive Search & URL Builder, USA http://earthquake.usgs.gov/earthquakes/search/.

*Maritime-Port Technology and Development – Ehlers et al. (Eds)*
*© 2015 Taylor & Francis Group, London, ISBN 978-1-138-02726-8*

# Designing an autonomous collision avoidance controller respecting COLREG

H.-C. Burmeister & W. Bruhn
*Fraunhofer Center for Maritime Logistics and Services CML, Hamburg, Germany*

ABSTRACT: This paper aims to show how a ship controller is implemented within the MUNIN project complying with the rules for steering and sailing as per COLREG Part B. After a short introduction into the project and a review of current approaches with regards to collision avoidance at sea, this paper will introduce the current process of collision avoidance and COLREGs. Afterwards, the different functionalities and solution methods of the collision avoidance controller are described, before the paper concludes with the depiction of further research needs.

## 1 INTRODUCTION

Autonomous vehicles are already state-of-the-art in e.g. aviation, public transportation and the automotive sector. In the maritime domain autonomous vessels are also frequently used, especially for military or underwater research purposes. Additionally, due to sustainability reasons, it is reasonable to aim for autonomous freight carriers, too. Within the MUNIN project (Maritime Unmanned Navigation through Intelligence in Networks) a concept for an unmanned dry bulk carrier sailing in deep-sea is developed and exemplarily tested. Two of the key design requirements for this future bulker are that it must be applicable on an ocean, where manned vessels still exist and that it needs to be at least as safe as current vessels. Ship safety is basically concerned with operations in harsh weather as well as collision avoidance. For the latter, the International Convention on the International Regulations for Preventing Collisions at Sea (COLREG) is currently the baseline for determining the right of way between manned vessels on the oceans. However, COLREG partly relies on fuzzy formulations as well as good seamanship customs and practices. Thus it is not a straight forward task to implement them in an autonomous controller.

Based on an introduction into the MUNIN project in section 2 and the necessity to conduct lookout on an unmanned vessel, section 3 will give a short overview about COLREGs and actual approaches to maritime collision avoidance routeing. Based on that, the methodology of the MUNIN Collision Avoidance Module is outlined in section 4, while section 5 concludes the paper and gives an outlook for the next steps in the project.

## 2 THE MUNIN PROJECT

### 2.1 *Objective and rationale*

The objective of the MUNIN project is to develop a technical concept for an autonomous vessel and to validate its feasibility with regards to technical, commercial and legal requirements (Rødseth & Burmeister 2012). By describing an autonomous vessel, MUNIN refers to a "next generation modular control systems and communications technology [that] will enable wireless monitoring and control functions both on and off board. These will include advanced decision support systems to provide a capability to operate ships remotely under semi or fully autonomous control" as described by Waterborne TP 2011. Hereby, MUNIN aims for a vessel that is at least completely unmanned during the deep-sea voyage.

Maritime shipping is taking place in a global competition putting high pressure on cost efficiency, while at the same time environmental issues become more and more important. MUNIN can contribute to improve the sustainability of shipping by (Rødseth & Burmeister 2012):

- Increasing cost efficiency by reduced on board crew requirements and less fuel utilisation,
- Improve work-life-balance for seafarers as jobs will be shifted ashore with positive effects on job attractiveness and
- Facilitating slow steaming and thus $CO_2$ reductions, that might not have been reachable due to an expected shortage of seafarers (BIMCO 2010).

### 2.2 *MUNIN concept and operational modes*

Reaching the goal of a completely unmanned ship during the deep-sea voyage requires a redesign of processes and implementation of new technical systems on board and ashore. From a system architecture point of view, MUNIN foresees three additional entities in its technical concept for the navigational process (Burmeister et al. 2014b):

- An Advanced Sensor Module that conducts lookout duties,
- An Autonomous Navigation System that monitors and controls the voyage of the unmanned vessel within a certain operational envelope and
- A Shore Control Centre, where skilled nautical officers and engineers continuously monitor the vessel, with the possibility to intervene in extraordinary circumstances.

Thereby, the Advanced Sensor Module uses radar, AIS and camera data to detect, classify and identify objects in the vicinity of the vessel with the help of sensor and feature fusion (Bruhn et al. 2014). At the Shore Control Centre instead, the data transferred to shore must be grouped and displayed in a way to allow for sufficiently good situational awareness of the onshore operators. The aim is for them to actively perform his monitoring tasks, but not to be occupied with too many active interventions to allow him monitoring several vessels at the same time (Porathe 2014). These operational tasks shall be performed by the Autonomous Navigation System instead, that basically covers the functions of "collision avoidance" and "weather routeing" (Walther et al. 2014).

Thereby, the Autonomous Navigation System shall operate in the four different modes depicted in Figure 1. The normal operation comprises the Autonomous Execution mode, where it just constantly assesses the surrounding, while it is actively taken action during the Autonomous Problem Solving mode. The extraordinary modes are represented by the Remote Control case, where the Shore Control Centre is overwriting the decisions made by the Autonomous Navigation System and the Fail-to-Safe mode. Here, the Autonomous Navigation System is still operating the vessel, but the goal is no longer to fulfil the transport mission but only to maintain safety of the vessel on the seas.

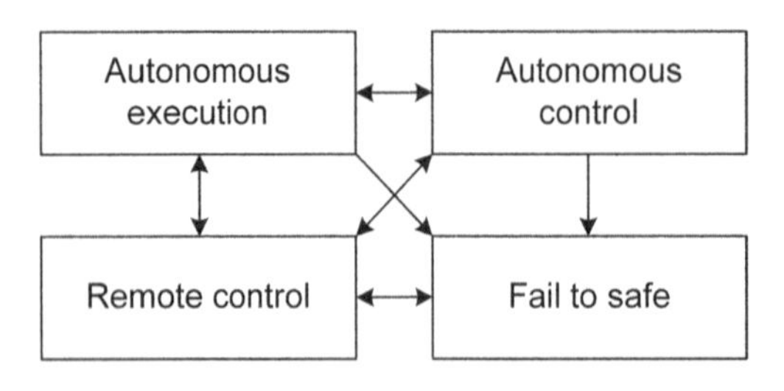


Figure 1. Operational modes of the MUNIN vessel in unmanned operation (Reference: Rødseth et al. 2013).

### 2.3 *Collision avoidance in the context of MUNIN*

Most of the collision avoidance functionalities on the MUNIN autonomous vessel will be handled by the Autonomous Navigation System. As the MUNIN vessel is mainly operating unmanned and in open waters, navigation is mainly constrained by the rules for steering and sailing imposed by COLREGs (Burmeister et al. 2014a). COLREGs must of course be respected, to enable operations of the MUNIN vessels within the existing legal framework. However, enabling a system to respect these regulations requires, amongst others, a defined, machine-understandable model of the collision avoidance regulations.

## 3 COLREG

### 3.1 *COLREGs in a nutshell*

The International Regulations for Preventing Collisions at Sea aims to prevent a high level of safety at sea by establishing a common, worldwide regulation on collision avoidance (IMO 2002a). It consists of five different parts covering 38 rules. Part B covers the steering and sailing rules that also define the main measure to determine whether a risk of collision exists: A constant bearing towards an approaching vessel.

After a risk of collision has been determined, COLREG foresees to identify the obligation of the vessels in open waters primarily on different navigational statuses that are displayed according to Part C, or, in case of the same status, based on the kind of encounter situation given non-restricted visibility:

- Head-on encounter,
- Overtaking or
- Crossing situation.

The outcome of this process is the identification of the ship's obligations being either give-way or stand-on vessel (Fig. 2). Based on this classification, COLREGs describe certain actions that shall be fulfilled until the other ship has been clearly passed. However, COLREG does not relieve the stand-on vessel from its collision avoidance obligation, as it is still required to conduct the manoeuver-of-the-last-second in case of close-quarter-situations.

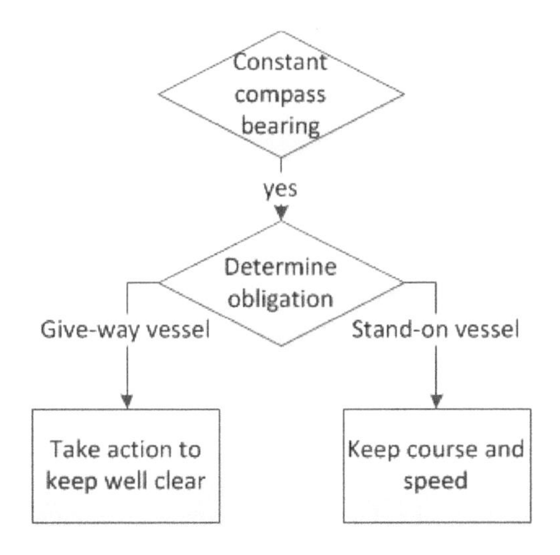


Figure 2. Generic description of COLREG decisions.

### 3.2 *State-of-the-art in COLREG systems*

Automated evaluation of COLREG situations and counter measures are not a new scientific domain, but have been addressed by different approached over the years. Kreutzmann et al. 2013 exemplarily developed a spatio-temporal logic to formalise COLREG rule 12 and thus to enable a machine to determine a COLREG situation correctly based on ideal information conditions. Based on ARPA-data Zeng 2000 tested and implemented a genetic algorithms approach for a simple head-on collision situation. Instead, Perera et al. 2009 proposed a fuzzy logic based approach to reason which of two encountering power-driven vessels is give-way vessel using AIS and ARPA data. Liu et al. 2006 used a fuzzy-neural inference network to classify encounter situations and suggest respective counter measures for two-ship encounters. A rather far-fetched model has also been proposed by Xue et al. 2008, where multi power-driven ship encounter problems can be solved respecting COLREG restrictions with regards to overtaking, head-on and crossing situations.

However, there are a few restrictions that the investigated models live with. Beyond the fact, that most algorithms are only covering certain parts of COLREGs, they often assume perfect information conditions during the determination process. Furthermore, most methods do not cover a procedure for COLREG rule 19 that regulates the behaviour of vessels in restricted visibility. Additionally, Hyundai lately released that they will provide a commercial anti-collision system for maritime applications in 2016, even though the exact functionalities are not yet known (Hyundai 2014).

## 4 COLLISION AVOIDANCE IN MUNIN

For designing a controller capable to operate a vessel in the real world, uncertainty in the input data as well as real world conditions cannot be neglected. Thus, the Collision Avoidance Module within MUNIN's autonomous ship controller must be capable to deal with these situations. In general, collision avoidance can be split up into the two tasks (Froese & Mathes 1995):

- Analyse actual traffic situation and
- Determine COLREG-conform counter measures.

While the former belongs to the autonomous execution mode of the MUNIN vessel, the latter represents the problem-solving phase. Further obligations, like the execution of the manoeuvre of the last second, are implemented within the Fail-to-Safe functionalities.

### 4.1 *Analysing actual traffic situations*

The data input to the collision avoidance process is ensured by the Advanced Sensor Module (Bruhn et al. 2014). Based on analysing AIS-, radar and camera data, this module provides the Autonomous Ship Controller with the necessary data for determining its obligations according to COLREG. Thereby, it will be distinguished between three different kinds of object categories:

- Detected objects (includes positions and speed data)
- Classified objects (includes type of object; like ship or solid flotsam) and
- Identified ships (includes name, call sign and navigational status).

The detailed data available for the different kind of objects are given in Table 1. In addition, the Advanced Sensor Module also provides the Colli-

Table 1. Available object data per category.

| Available data | Detected | Classified | Identified |
|---|---|---|---|
| Position | x | x | x |
| Speed over ground | x | x | x |
| Course over ground | x | x | x |
| Heading | x | x | x |
| Bearing | x | x | x |
| Rate of turn | x | x | x |
| CPA | x | x | x |
| TCPA | x | x | x |
| Object type | – | x | x |
| MMSI-number | – | – | x |
| Ship type | – | – | x |
| Navigational status | – | – | x |

sion Avoidance Module with an indication of visibility restrictions.

Based on the kind of object category, Table 2 lists the general manoeuvre responses. While objects not being classified as a ship must simply be evaded, but with no COLREG restrictions towards the evasive manoeuvre, it is still aspired to avoid detected objects according to COLREG suggestions, in case their status might change during the operations. Reasonably, the unmanned ship's obligations towards identified vessels as well as the restrictions towards the evasive manoeuvre routeing are determined based on COLREG, as all necessary information is available.

It is already clear from the different object categories, that a full appraisal of the traffic situation according to COLREG is not possible for "detected" and "classified" objects, as navigational status of the object is missing in all situations. Only in the case of "identified ship", this data relevant for determining the ship's obligations according to rule 18 can be provided analogously to Table 3, even though it will only be obtained from AIS-data and not from visual detection of lights and shapes. This is still critical as even though IMO Resolution A.917(22) requires the OOW (Officer of the Watch) to manually input this data Harati-Mokhtari et al. 2007 reports that

Table 2. Response strategies per object category.

| Object categories | Own-ship | Evasive manoeuvre |
|---|---|---|
| Detected | Give-way | Preferably COLREG* |
| Classified—ship | Acc. COLREG* | COLREG |
| Classified—no ship | Give way | No restrictions |
| Identified ship | Acc. COLREG | COLREG |

*Assuming normal power-driven vessel.

Table 3. COLREG-rule base depending on AIS-Status.

| Code | AIS-status | COLREG rule |
|---|---|---|
| 0 | Underway using engine | §13–15 |
| 1 | At anchor | – |
| 2 | Not under command | § 18a |
| 3 | Restricted manoeuvrability | § 18a |
| 4 | Constrained by her draught | §9/§18d |
| 5 | Moored | – |
| 6 | Aground | – |
| 7 | Engaged in fishing | §18a |
| 8 | Under way sailing | §18a |
| 9–15 | *Special or empty fields* | – |

up to 30% of all vessels display an incorrect navigational status.

While it can be argued that the implementation of the unmanned vessel might require a revision of COLREG's Part C to make AIS equivalent to displaying lights and shapes, the actual layout tries to use this data as seldom as possible to reduce the likelihood for false decisions.

However, to cope with the uncertainty of the navigational status the following rule set can be applied on the basis of the relative movements between the vessels and under the assumptions that the own-ship has the lowest navigational status "power-driven vessel" and operates in open waters:

$$\chi = -\text{sgn}(\sin\theta\cos\omega - \cos\theta\sin\omega) \cdot \text{arcos}(\sin\theta\sin\omega + \cos\theta\cos\omega) \quad (1)$$

where $\chi$ is the angle of attack between own-ship and object, $\omega$ is the true heading of the own-ship and $\theta$ the one of the object. As several situations result in the own-ship being the give-way vessel regardless of the navigational status, the result of this assessment is in certain conditions over-determined. Table 4 depicts these over determined conditions.

Thus, the occurrence of a situation which is still attached with some uncertainty is limited down to an 110°-wide band from $67.5° < \chi < 177°$. In that area, a normal "power-driven vessel" is assumed for classified objects and in case of identified ships, their AIS-information is taken for granted. Additionally, a method is implemented that continuously monitors the behaviour of the object with regards to its stand-on or give-way obligations respectively, which then involves the Shore Control Centre in case of detected misalignments. This corresponds to an operational mode change from autonomous to remote control.

In case of restricted visibility, $\chi$ cannot be used as the main decision criterion, but instead the relative bearing $\varphi$ of the object from the own-ship will be used. Recommended evasive manoeuvres described in Rule 19 can thus be assigned to a certain value of $\varphi$ as shown in Table 5.

Table 4. Over determination possible (unrestricted visibility).

| Relative bearing | Give way | Remarks |
|---|---|---|
| $\lvert\phi\rvert <= 67.5°$ | Rule 13 | Nav status independent* |
| $\lvert\phi\rvert >= 177°$ | Rule 14 | Both are give-way** |
| $-177 < \phi < -67.5°$ | Rule 15 | |

*Provided own-ship overtakes.
**Design evasive manoeuvre independent of object.

Table 5. Give-way obligations in restricted visibility.

| Bearing | Recommendation |
|---|---|
| $\varphi >= 90°$ | Alternate course away from object |
| $\varphi <= -90°$ | Alternate course away from object |
| $-90° < \varphi < 90°$ | Alternate course |
| and $\lvert\chi\rvert <= 67.5°$ | But not towards port |

### 4.2 *Determine COLREG-conform counter measures*

After the analyses, all own-ships have an assigned COLREG obligation towards this object and possibly a generic evasive manoeuvre strategy. Determining an evasive manoeuvre now requires to maintain a certain safety level but also to keep economic impacts as low as possible. Hereby, course instead of speed alterations are preferred.

As the MUNIN vessel will not be operated in high traffic density areas at the beginning, a simple heuristics based on prioritised strategies, like e.g. in Froese & Mathes 1995, provides a robust control strategy. This will be combined with the progress achieved in fast time manoeuvring simulation, like e.g. by Benedict et al. 2014, to be capable to take hydrodynamic limitations better into account. In addition, the calculated routes will be checked by the MUNIN harsh weather operation system to avoid evasive manoeuvres that pose a risk to the own-ship due to harsh weather (Walther et al. 2014). The proposed procedure is given in Table 6.

### 4.3 *Manoeuvre of the last second*

Besides the planning functionalities, the Collision Avoidance Module also includes Fail-to-Safe-functionalities that execute the manoeuvre of the last second. Primarily, this will be executed in case a certain threshold defined by the Shore Control Centre, e.g. a minimum CPA or TCPA value, is reached, but also more farfetched solutions are possible due to the inclusion of the fast time manoeuvring simulation. This provides the autonomous ship controller constantly with the following values:

- Emergency stopping distance,
- Time and distance to get off-track (starboard and portside),
- Turning circles for a given rudder value (starboard and portside)

This information can be used to dynamically and precisely define no-go-areas around the vessel and can assist in robustly determining the last moment, in which an action must be conducted to avoid a collision, analogously to crash avoidance features in the automobile industry.

Table 6. Pseudocode prioritised collision avoidance manoeuvre.

```
If New give-way obligation identified Then
  CollisionRisk = True
  EncounterType = RuleCausingGiveWayObligation()
  StrategyNo = 1

  While(CollisionRisk)
    NewTrack = CalculateEvasiveManoevre(StrategyNo,
               Encounter type)
    NewTrack = HydrodynamicSmooth(NewTrack)

    ObjectAlert = CollisionCheck(NewTrack)*
    WeatherAlert = IMOSafetyCheck(NewTrack)*

    If ObjectAlert Or WeatherAlert Then
      StrategyNo = StrategyNo + 1
    Else
      CollisionRisk = False
    If StrategyNo > MaxNo(EncounterType) Then
      RequestSCCInput()
      Abort
Loop

UpdateAutopilot(NewTrack)
```

*False if no alert occurred.

## 5 CONCLUSION AND OUTLOOK

This paper proposes an integrated approach to explain which information about collision avoidance is available in a real-world application, how the uncertainty within this information can be taken into account and which methods can be used to easily identify a COLREG-compliant track for the unmanned vessel. As the MUNIN vessel shall primarily operate in deep-sea and unrestricted waters, encounter situations are rather unlikely and the focus of this paper is not to identify the best evasive manoeuvre strategy with regards to fuel efficiency, but to quickly find a safe strategy with the help of simple heuristics based on a robust classification of COLREG obligations. However, to improve the data quality and reliability for automated collision avoidance systems, it is suggested to consider an amendment to COLREG Part C in order to allow for an alignment of obligations with those described in SOLAS V Regulation 19 in relation to nautical data sent by AIS. Of course, the data may still need to be manually entered into the AIS transceiver, thus human errors might occur, but it provides the unmanned vessel as well as masters and nautical officer of manned vessels with confidence, that if a navigational status is displayed by

AIS, then it also has a legal meaning for determining its COLREG-obligations.

The described approach is currently being implemented in the MUNIN unmanned ship simulation environment and shall then be validated by prerecorded AIS-data and usability assessment through nautical officers. In a next step, the approach can of course be extended by more advanced optimisation techniques for fuel-optimal evasive manoeuvres. Furthermore, rules 9 and 10 shall also be included into the obligation identification process to make the system applicable to all sea areas.

## ACKNOWLEDGEMENTS

The research leading to these results has received funding from the European Union Seventh Framework Programme under the agreement SCP2-GA-2012-314286.

## REFERENCES

Benedict, K., Kirchhoff, M., Gluch, M., Fischer, S., Schaub, M., Baldauf, M. & Klaes, S. 2014, Simulation Augmented Manoeuvring Design and Monitoring—a New Method for Advanced Ship Handling. *TransNav* 8 (1): 131–141.

BIMCO 2010, *BIMCO ISF Manpower 2010 Update: The Worldwide demand for and supply of seafarers*, http://www.marisec.org/Manpower%20Study.pdf, last accessed June 2012.

Bruhn, W.C., Burmeister, H.-C., Long, M.T. & Moræus, J.A. 2014, Conducting look-out on an unmanned vessel: Introduction to the advanced sensor module for MUNIN's autonomous dry bulk carrier, *Proceedings of International Symposium Information on Ships—ISIS 2014*, Hamburg, Germany, September 04–05, 2015, in press.

Burmeister, H.-C., Bruhn, W., Rødseth, Ø.J. & Porathe T. 2014a, Can unmanned ships improve navigational safety?, *TRA 2014—Proceedings*, http://www.traconference.eu/pa-pers/pdfs/TRA2014_Fpaper_17909.pdf, last accessed July 2014.

Burmeister, H.-C., Bruhn, W., Rødseth, Ø.J. & Porathe T. 2014b, Autonomous unmanned merchant vessel and its contribution towards the e-Navigation implementation: The MUNIN perspective, *Proceedings of International Symposium on Advanced Intelligent Maritime Safety and Technology* 1: 53–61.

Froese, J. & Mathes, S. 1995, *RKV Computer Assisted Collision Avoidance Final Report.*

Harati-Mokhtari, A., Wall, A., Brooks, P. & Wang, J. 2007, Automatic Identification System (AIS): data reliability and human error implications, *Journal of Navigation* 60 (3): 373–389.

Hyundai 2014, *Hyundai Heavy Industries Developed HiCASS*, http://english.hhi.co.kr/news/view?idx=517, last accessed July 2014.

IMO 2002a, *COLREG Convention on the International Regulations for Preventing Collisions at Sea, 1972, Consolidated Edition 2002*, London: IMO.

IMO 2002b, Guidelines for the onboard operational use of shipbourne Automatic Identification System, Resolution A. 917(22).

Kreutzmann, A., Wolter, D., Dylla, F. & Lee, J.H. 2013, Towards Safe Navigation by Formalizing Navigation Rules, *TransNav* 7 (2): 161–168.

Liu, Y.-H., Du, X.-M. & Yang, S.-H. 2006, The Design of a Fuzzy-Neural Network for Ship Collision Avoidance, In D.S. Yeung, Z.-Q. Liu, X.-Z. Wang & H. Yan (eds), *Advances in Machine Learning and Cybernetics*: 804–812, Germany: Springer.

Perera, L.P., Carvalho, J.P. & Guedes Soares, C. 2010, Bayesian Network based sequential collision avoidance action execution for an Ocean Navigational System, *Proceedings of the 8th IFAC Conference on Control Applications in Marine Systems*: 301–306.

Porathe, T. 2014, Remote Monitoring and Control of Unmanned Vessels—The MUNIN Shore Control Centre, In V. Bertram (ed), *13th International Conference on Computer and IT Applications in the Maritime Industries, Redworth, 12–14 May 2014*: 460–467. Hamburg, Technische Universität Hamburg-Harburg.

Rødseth, Ø.J. & Burmeister, H.-C. 2012, Developments toward the unmanned ship, *Proceedings of International Symposium Information on Ships—ISIS 2012*, Hamburg, Germany, August 30–31, 2012.

Rødseth, Ø.J., Kvamstad, B., Porathe, T. & Burmeister, H.-C 2013, Communication Architecture for an Unmanned Merchant Ship. *OCEANS—Bergen, 2013 MTS/IEEE*, http://dx.doi.org/10.1109/OCEANS-Bergen.2013.6608075, last accessed July 2014.

Walther, L., Burmeister, H.-C. & Bruhn, W. 2014, Safe and Efficient Autonomous Navigation with Regards to Weather, In V. Bertram (ed), *13th International Conference on Computer and IT Applications in the Maritime Industries, Redworth, 12–14 May 2014*: 303–317. Hamburg: Technische Universität Hamburg-Harburg.

Waterborne T.P. 2011, *Waterborne Implementation Plan: Issue May 2011*, http://www.waterbornetp.org/index.php/documents, last accessed September 2013.

Xue, Y., Lee, B.S. & Han, D. 2009, Automatic collision avoidance of ships, *Journal of Engineering for the Maritime Environment* 223 (1): 33–46.

Zeng, X., Ito, M. & Shimizu, E. 2000, Collision avoidance of moving obstacles for ship with genetic algorithm, *Proceedings on the 6th International Workshop on Advanced Motion Control*: 513–518, http://dx.doi.org/10.1109/AMC.2000.862927, last accessed July 2014.

*Maritime-Port Technology and Development – Ehlers et al. (Eds)*
*© 2015 Taylor & Francis Group, London, ISBN 978-1-138-02726-8*

# A generic modelling approach for heavy lifting marine operations

J. Xu & K.H. Halse
*Aalesund University College, Aalesund, Norway*

ABSTRACT: This paper introduces a generic modelling approach for heavy lifting marine operation based on modelling and simulation software 20-sim (Controllab 2014). The model is a multi-body dynamic system which can be divided into vessel, crane, cable, load, and control system. Physical entities are modelled either in bond graph or directly using 3D Mechanics toolbox and connected by interactive power port. All control schemes are modelled as signal flows separated from physical entities. The vessel is modelled as six Degrees of Freedom (DOF) bond graph and connected to the crane model inside 3D Mechanics unit. Crane model is controlled by outside manual/auto control scheme. Cable and load are modelled inside 3D Mechanics with hydrodynamic behaviour represented by actuators. The performance of each system is evaluated respectively by regulations and analysis. Examples of different combinations of sub-systems are given at the end. The project is aiming at developing a generic modelling method to serve as a multi-user training and design platform.

## 1 INTRODUCTION

### 1.1 *Complexity of marine operations*

Marine operations are usually multi-system involved activities with interaction and coordination behind. Heavy lifting operation is a typical example often used for remotely operated underwater vehicle deployment and subsea installation/demolition, which normally involves vessel, crane, cable, load and the manual/auto control system. Because of the complexity and diverse research focus of its nature, people tend to isolate the issue by neglecting the insignificant and simplifying the complicated, e.g. the researchers who intend to study the sea-keeping feature of the platform supply vessel normally regard the dynamic behavior of the crane and the load as negligible (Lloyd 1989), while people who study the Active Heave Compensation control (AHC) treat vessel motion as unaffected by the crane hence an independent variable outside the equation. Moreover, people who study the hydrodynamic behavior of a submerged load usually only apply simple motion to the load in Computational Fluid Dynamics (CFD) calculation (Fackrell 2011) or experiment.

### 1.2 *The present modelling approach*

This paper introduces a generic modelling approach for heavy lifting operation by using bond graph technique, signal blocks and 3D Mechanics toolbox in simulation software 20-sim. The input data resources are from ShipX (MARINTEK 2014), DNV rules and CFD calculation, processed in MATLAB (MathWorks 2014) panel and communicated with 20-sim. A fully interactive system of vessel, crane, cable, and control system is realized and able for easy parameter selection. If real-time simulation can be reached, manual control can also be added as well as automatic control.

The work presented in this paper is a part of the ongoing research in the Ship Operation Lab at Aalesund University College. The activities herein support the ongoing development of the activities at the Offshore Simulator Centre (OSC 2014). The model consists of several modelling contributions from other researchers, yet with necessary modification for a compatible system. The general second order differential equation of vessel hydro-dynamics is from Fossen & Fjellstad (2011). MATLAB files for ShipX data transformation are found in Fossen & Perez (2014). Bond graph model is based on Pedersen (2008). Crane hydraulics and control model are developed by Chu (2013). Cable model is inspired by Johansson (1976). Load model inspired by Halse et al. (2014).

## 2 METHODOLOGY

The main method used in this paper is virtual simulation and the primary target of this paper is to build a realistic real-time simulation model for marine operation. As for economic and safety reasons, a full size experiment of marine operation is practically impossible for the industry. An alternative is to develop a virtual environment where maritime operations can be carried out without any danger for the personnel or equipment. Physical objects are being represented as components with different functionality in the system. The system can accept input from the user (e.g. wave spectrum, vessel inertia matrix, hydrodynamic damping parameters, crane model, CFD data of the load) and produce output to the user (e.g. vessel motion during the operation, crane hydraulics performance, load motion in the water, power output, and mechanical behavior). The model is built based on studies about realistic feature of the vessel and hydrodynamic statistics, and coded according to general physical principles with simplification for faster performance. The selected parameters depend on field measurements but are also tuned for better approximation to the reality. Each sub-system is modelled separately. Accuracy and app-licability are tested before being integrated into bigger system. During the simulation, variables are controlled to test the influence of each. Although virtual simulations have the flexibility and compatibility to adjust and to replace input with minimum cost and maximum fidelity, it will always rely on certain assumptions which may appear questionable in specific situations. Thus, simulation results are recorded and analyzed both theoretically and statistically. DNV rule-checking will be conducted to make sure its correspondence with legitimate requirements.

## 3 SYSTEMS MODELLING

### 3.1 *Vessel model*

Fossen presented the hydrodynamic behavior of a vessel in traditional quadratic linear differential equations by utilizing the Kirchhoff equations (Fossen & Fjellstad 2011). The Kirchhoff equations, derived from generalized Lagrangian equations, describe the motion of a rigid body in an ideal fluid (Kirchoff 1877).

$$\frac{1}{2}(\boldsymbol{\omega}^T \boldsymbol{I}\boldsymbol{\omega} + mv^2) = T \tag{1}$$

$$\frac{d}{dt}\frac{\partial T}{\partial \boldsymbol{\omega}} + \boldsymbol{\omega} \times \frac{\partial T}{\partial \boldsymbol{\omega}} + \upsilon \times \frac{\partial T}{\partial \boldsymbol{v}} = \boldsymbol{Q}_h + \boldsymbol{Q} \tag{2}$$

$$\frac{d}{dt}\frac{\partial T}{\partial \boldsymbol{v}} + \boldsymbol{\omega} \times \frac{\partial T}{\partial \boldsymbol{v}} = \boldsymbol{F}_h + \boldsymbol{F} \tag{3}$$

$$-\int p\boldsymbol{x} \times \boldsymbol{n} d\sigma = \boldsymbol{Q}_h \tag{4}$$

$$-\int p\boldsymbol{n} d\sigma = \boldsymbol{F}_h \tag{5}$$

where $\boldsymbol{\omega}$ and $\boldsymbol{v}$ are angular and linear velocity vectors at point $\boldsymbol{x}$ respectively; $\boldsymbol{I}$ is the moment of inertia tensor, $m$ is body's mass; $\boldsymbol{n}$ is a unit normal to the surface of the body at point $x$; $p$ is a pressure at the point $x$; $\boldsymbol{Q}_h$ and $\boldsymbol{F}_h$ are the hydrodynamic torque and force acting on the body respectively; $\boldsymbol{Q}$ and $\boldsymbol{F}$ likewise denote all other torques and forces acting on the body and the integration is performed over the fluid-exposed portion of the body surface. Additional with two equations from Momentum Conservation Principle

$$\frac{\partial T}{\partial \boldsymbol{v}} = m \cdot \boldsymbol{v} - (m \times \boldsymbol{r}_G) \cdot \boldsymbol{\omega} \tag{6}$$

$$\frac{\partial T}{\partial \boldsymbol{\omega}} = (m \times \boldsymbol{r}_G) \cdot \boldsymbol{v} + \mathrm{I} \cdot \boldsymbol{\omega} \tag{7}$$

The equations can be restated into standard form of quadratic linear differential equations of motion.

$$\boldsymbol{M} \cdot \ddot{\boldsymbol{\eta}} + \boldsymbol{C}' \cdot \dot{\boldsymbol{\eta}} = \boldsymbol{\tau}_h + \boldsymbol{\tau} \tag{8}$$

where $\dot{\boldsymbol{\eta}} = [\boldsymbol{v}, \boldsymbol{\omega}]^T$, $\boldsymbol{\tau}_h = [\boldsymbol{F}_h, \boldsymbol{Q}_h]^T$ and $\tau = [\boldsymbol{F}, \boldsymbol{Q}]^T$. The inertia matrix $\boldsymbol{M}$ is recognized to be

$$\boldsymbol{M} = \begin{bmatrix} m & -m \times r_G \\ m \times \boldsymbol{r}_G & \boldsymbol{I} \end{bmatrix} \tag{9}$$

and the remaining terms which are known to contribute to the Coriolis force can be stated as

$$\boldsymbol{C}' = -\begin{bmatrix} 0 & 0 & 0 & 0 & -a_3 & a_2 \\ 0 & 0 & 0 & a_3 & 0 & -a_1 \\ 0 & 0 & 0 & -a_2 & a_1 & 0 \\ 0 & -a_3 & a_2 & 0 & -b_3 & b_2 \\ a_3 & 0 & -a_1 & b_3 & 0 & -b_1 \\ -a_2 & a_1 & 0 & -b_2 & b_1 & 0 \end{bmatrix}$$

$$[a_1, a_2, a_3, b_1, b_2, b_3]^T = \boldsymbol{M} \cdot \dot{\boldsymbol{\eta}} \tag{10}$$

The hydrodynamic force can also be separated into two components

$$\boldsymbol{\tau}_h = \boldsymbol{\tau}_{rad} + \boldsymbol{\tau}_{exc} \tag{11}$$

where $\boldsymbol{\tau}_{rad}$ is the hydrodynamic radiation forces and $\boldsymbol{\tau}_{exc}$ is environmental excitation forces from wind and waves. The radiation forces can be expressed in frequency domain as

$$\boldsymbol{\tau}_{rad} = -\boldsymbol{A}(\omega)\ddot{\boldsymbol{\eta}} - \boldsymbol{B}(\omega)\dot{\boldsymbol{\eta}} - \boldsymbol{C}\boldsymbol{\eta} \tag{12}$$

where $\boldsymbol{A}(\omega)$ and $\boldsymbol{B}(\omega)$ denote the frequency—dependent added mass and damping matrices. The restoring forces are assumed to be a linear formulation $\boldsymbol{C\eta}$. The resulting motion can be stated as $\boldsymbol{\eta} = \hat{\boldsymbol{\eta}} e^{jwt}$ and the dynamic equation in frequency domain becomes

$$-\omega^2[\boldsymbol{M} + \boldsymbol{A}(\omega)] \cdot \boldsymbol{\eta}(j\omega) + j\omega \boldsymbol{B}(\omega) \cdot \boldsymbol{\eta}(j\omega) + \boldsymbol{C} \cdot \boldsymbol{\eta}(j\omega) = \boldsymbol{\tau}_{exc} + \boldsymbol{\tau} \tag{13}$$

Naval architects usually write the equation in a mixed frequency—time domain formulation

$$(\boldsymbol{M}_{RB} + \boldsymbol{M}_A(\boldsymbol{\omega})) \cdot \ddot{\boldsymbol{\eta}}(t) + (\boldsymbol{C}_{RB} + \boldsymbol{C}_A(\omega)) \cdot \dot{\boldsymbol{\eta}}(t) + \boldsymbol{B}(\boldsymbol{\omega}) \cdot \dot{\boldsymbol{\eta}}(t) + \boldsymbol{C} \cdot \boldsymbol{\eta}(t) = \boldsymbol{\tau}_{exc} + \boldsymbol{\tau} \tag{14}$$

Strictly speaking Eq. (14) is not correct because the real vessel motion is not purely harmonic, but it is tolerated in a dominated frequency motion at the moment. The excitation forces including forces from wave, current and wind can be derived from force RAO in ShipX. Other forces from control units such as rudder, propeller, foil and fin stabilizer can be either assumed or estimated through classical calculation. However, without real time CFD calculation, most of the force representations are still restricted for only uniform flow condition. The motion equation is established in body-fixed coordinate system $B(x_b, y_b, z_b)$ in order to keep the rigid body inertia and added mass as constants. The restoring force and damping force are treated with respect to the hydrodynamic frame $H(x_h, y_h, z_h)$. Therefore a coordinate transformation is required before those two elements are being put into the equation. The overall 6DOF kinematic equation between $H(x_h, y_h, z_h)$ and $B(x_b, y_b, z_b)$ is

$$\dot{\boldsymbol{\eta}}_h = \boldsymbol{J}_b^h \cdot \dot{\boldsymbol{\eta}}_b \tag{15}$$

$$\boldsymbol{J}_b^h = \begin{bmatrix} \boldsymbol{R}_b^h & \boldsymbol{0}_{3\times3} \\ \boldsymbol{0}_{3\times3} & \boldsymbol{T}_b^h \end{bmatrix} \tag{16}$$

$$\boldsymbol{T}_b^h = \begin{bmatrix} 1 & sin\phi \cdot tan\theta & cos\phi \cdot tan\theta \\ 0 & cos\phi & -sin\phi \\ 0 & sin\phi/cos\theta & cos\phi/cos\theta \end{bmatrix} \tag{17}$$

$$\boldsymbol{R}_b^h = \boldsymbol{R}_{x,\phi} \cdot \boldsymbol{R}_{y,\theta} \cdot \boldsymbol{R}_{z,\psi} \tag{18}$$

$$\boldsymbol{R}_{x,\phi} = \begin{bmatrix} 1 & 0 & 0 \\ 0 & cos\phi & -sin\phi \\ 0 & sin\phi & cos\phi \end{bmatrix} \tag{19}$$

$$\boldsymbol{R}_{y,\theta} = \begin{bmatrix} 1 & 0 & 0 \\ 0 & cos\theta & -sin\theta \\ 0 & sin\theta & cos\theta \end{bmatrix} \tag{20}$$

$$\boldsymbol{R}_{z,\psi} = \begin{bmatrix} 1 & 0 & 0 \\ 0 & cos\psi & -sin\psi \\ 0 & sin\psi & cos\psi \end{bmatrix} \tag{21}$$

The final equation then becomes

$$(\boldsymbol{M}_{RB} + \boldsymbol{M}_A(\omega)) \cdot \ddot{\boldsymbol{\eta}}(t) + (\boldsymbol{C}_{RB} + \boldsymbol{C}_A(\omega)) \cdot \dot{\eta}(\boldsymbol{t}) + \boldsymbol{J}^{-1}{}_b^h \cdot \boldsymbol{B}(\omega) \cdot \boldsymbol{J}_b^h \cdot \dot{\eta}(t) + \boldsymbol{J}^{-1}{}_b^h \cdot \boldsymbol{C} \cdot \int \boldsymbol{J}_b^h \cdot \boldsymbol{\eta}(t) = \boldsymbol{J}^{-1}{}_b^h \cdot \boldsymbol{\tau}_{exc} + \boldsymbol{\tau} \tag{22}$$

This equation is used as a basis to establish a bond graph model of the whole system (Pedersen 2008). Figure 1 shows the established bond graph model where the vessel is modelled as a 3D

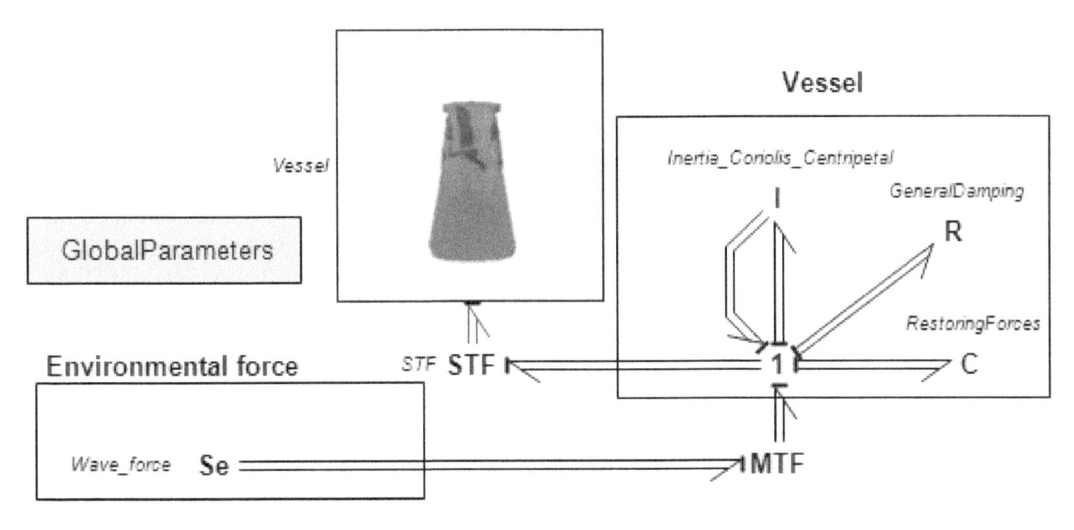


Figure 1. Bond graph of a simple vessel.

Mechanics module. The $\boldsymbol{I}$ element in the model represents the Inertia force and Coriolis force.

$$(\boldsymbol{M}_{RB} + \boldsymbol{M}_A(\omega)) \cdot \ddot{\boldsymbol{\eta}}(t) + (\boldsymbol{C}_{RB} + \boldsymbol{C}_A(\omega)) \cdot \dot{\eta}(t) \tag{23}$$

To solve the rigid body causality problem, the causality of two ports in $\boldsymbol{I}$ element shall be both changed to 'indifferent'.

The $\boldsymbol{R}$ element represents the damping force

$$\boldsymbol{J}^{-1h}_{b} \cdot \boldsymbol{B}(\omega) \cdot \boldsymbol{J}^h_b \cdot \dot{\boldsymbol{\eta}}(t) \tag{24}$$

The Effort Source **Se** and a **MTF** acting as coordinate transformation Jacobian matrix. The **C** element represents the restoring force with Jacobian matrix $\boldsymbol{J}$ in the code.

$$\boldsymbol{J}^{-1h}_{b} \cdot \boldsymbol{C} \cdot \int \boldsymbol{J}^h_b \cdot \eta(t) \tag{25}$$

The Jacobian matrix is set as a global variable $\boldsymbol{J}$. Only wave force is included in this model, but other hydrodynamic forces can be added by the same modelling approach.

$$\boldsymbol{J}^{-1h}_{b} \cdot \boldsymbol{\tau}_{exc} \tag{26}$$

The vessel in the 3D Mechanics block has a negligible inertia, thus it can be considered as a shadow vessel. By connecting the bond graph to an actuator located on the origin of body-fixed coordinate system, the outside bond graph is acting equivalent to a Flow Source to determine the vessel motion inside the 3D Mechanics block. However, in 3D Mechanics block, an actuator puts torque before translational force, as a result, to connect bond graph model to 3D Mechanics block, a transformer **STF** is added to shift the translational force above torque.

$$\begin{bmatrix} Torque \\ Force \end{bmatrix} \leftarrow \boldsymbol{STF} \leftarrow \begin{bmatrix} Force \\ Torque \end{bmatrix}$$

### 3.2 *Crane model*

The 3D Mechanics toolbox in 20-sim provides 3D animation environment for rigid body dynamics modelling, where the crane can be modelled as well as the vessel. The 3D modelling of crane was firstly done in Solidworks and then transferred into 3D Mechanics by using a converter called "COLLADAto20-sim" (Chu 2013), see Figure 2. "COLLADAto20-sim" is developed by the Controllab group (Controllab Product BV 2013) using C++ and COLLADA, an inter application exchange file

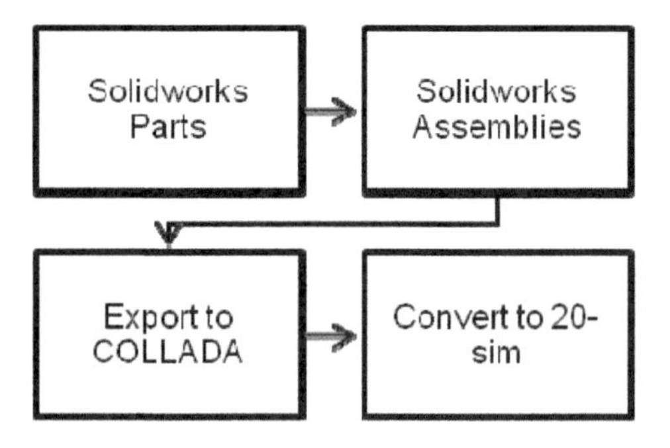


Figure 2. Steps from Solidworks to 20-sim.

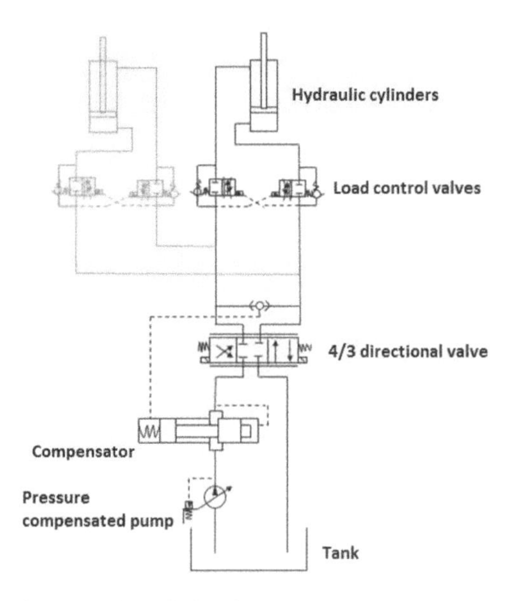


Figure 3. Simplified main derrick hydraulic sketch.

format (Janssen 2013). There are special requirement for this kind of conversion however.

20-sim also provides hydraulic components library for modelling and simulation hydraulic systems. In Chu's work, the model is designed as close as possible to the Modelica hydraulic library. However the library does not provide all components that are required. So some components are specially designed and added into the library later. Figure 3 is an example of hydraulic sketch of the main derrick of the crane in Chu's work.

Manual/auto control algorithms are modelled as signal flow model or codes in Jacobian (Control Box), where also an AHC system can be implemented. An example of an AHC algorithm using crane kinematics will be given in Figure 16. The crane tip motion will be compensated by a velocity signal equal to the vertical velocity at the position of the crane tip (induced by the ship motion). The physical entities, hydraulics and dynamics, are modelled as power flow model (hydraulic library, 3D Mechanics, bond graph) which remains intact from the variations of control system.

### 3.3 *Cable model*

Offshore cable is usually treated as flexible body with infinite degrees of freedom. It has physical features of elasticity and plasticity in both axial direction and radial direction. In practice, the external forces offshore cable may experience includes concentrated force on both ends, gravity force along the entire length, buoyancy force on the submerged part and damping forces caused by relative motion between cable and the fluid around. Those external forces can cause axial tension, axial torque, radial shear and in extreme cases bending moment in local area. Because of the compound material inside the cable, some may have non-linear elastic modulus and unevenly distributed stress level in local area.

In this paper, the cable is modelled as rigid bars with free rotational joint and 1-DOF translational joint on two ends respectively, see Figure 4. The rigid bars are represented by an inertia matrix which has both mass and moment of inertia. The free rotational joint has no spring or damping but constrained translational motions between bars. The 1-DOF translational joint has spring and damping property which represent the elasticity and structural damping of the cable respectively. The joint connecting the crane tip is free rotational and the joint connecting to the load can be either free rotational or translational depending on the structure of the connecting point. If the number of discretized parts is $n$, the number of rigid bars, free rotational joint and 1-DOF translational joint shall be $2n$, $n$, $n$ respectively, not including the joint on crane tip and load.

By this approach, the elasticity, mass, moment of inertia and structural damping of the cable are all realized in this modelling method. Using the same approach as in vessel motion input, the external forces such as current force and buoyancy force can be applied as bond graph model on each rigid bar as a constant or the function of motion and velocity. The port power should be defined in world coordinates, because of the coordinate where current force is defined.

Outside the 3D Mechanics, each actuator is linked with a bond graph effort source where current force and buoyancy force can be defined by simple Morison equation and Archimedes law or complex expressions. The current speed is defined in a signal block by an array with length of $n$ or $2n$ depending on if two bars connected by translational joint shall have the same current speed, see Figure 9.

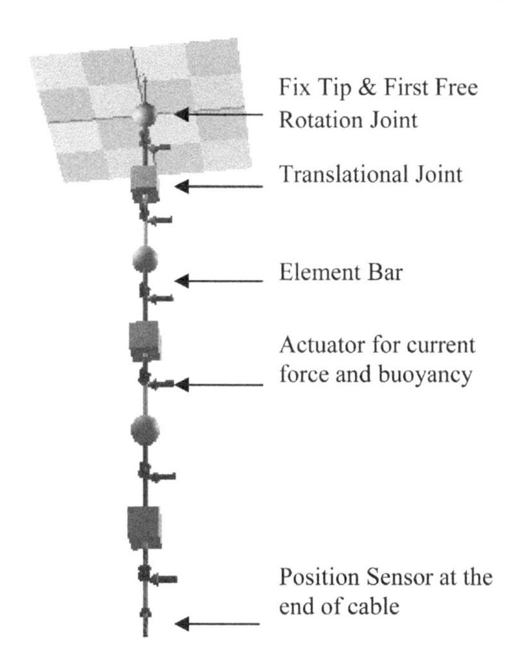


Figure 4. Discretized cable model in 3D mechanics.

### 3.4 *Load model*

When an object is deeply submerged, it experiences a concentrated/scattered lifting force from the cable, an evenly distributed gravity force downwards and buoyancy force upwards under the premise of a constant submerged volume. It also experiences hydrodynamic forces including the inertia force and the damping force by the definition in the Morison equation, see Figure 5.

Similar to the ship motion equation, for a harmonically oscillating object, the motion function of a submerged object in its own body-fixed coordinate can be written as

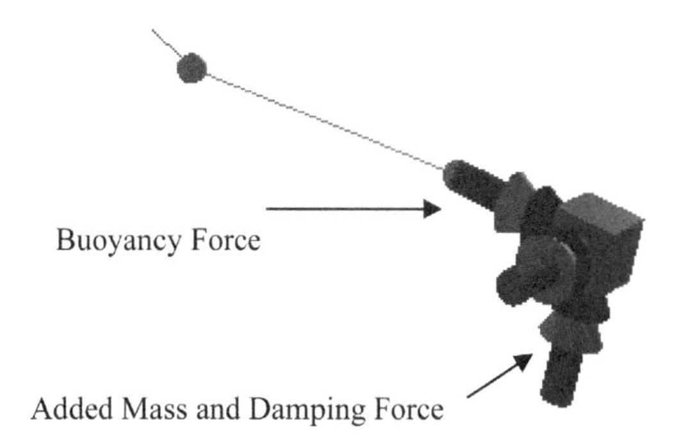


Figure 5. Load model.

$$
\begin{aligned}
&M_{RB}\cdot\ddot{\eta}(t)+M_A(\omega,\zeta)\cdot\ddot{\eta}_r(t)+C_{RB}\cdot\dot{\eta}(t)\\
&\quad+C_A(\omega,\zeta)\cdot\dot{\eta}_r(t)+B(\omega,\zeta)\cdot\dot{\eta}_r(t)\,|\,\dot{\eta}_r(t)\,|\\
&\quad=\tau_l+\tau_g+\tau_b
\end{aligned}
\tag{27}
$$

where

$M_{RB}$ rigid body inertia matrix of the object
$\omega$ oscillation frequency of the object
$\zeta$ oscillation amplitude of the object
$M_A(\omega, \zeta)$ added mass matrix for a specific $\omega$ & $\zeta$
$C_{RB}$ Coriolis force matrix of the object
$C_A(\omega, \zeta)$ added mass Coriolis for a specific $\omega$ & $\zeta$
$\eta(t)$ object motion
$\eta_r(t)$ object relative motion to the fluid
$B(\omega, \zeta)$ damping matrix for a specific $\omega$ & $\zeta$
$\tau_l$ lifting force from the cable
$\tau_g$ gravity force of the object
$\tau_b$ buoyancy force of the object

In 20-sim 3D Mechanics, the loading object can simply be modelled as an object attached to the end of the rope with a $6 \times 1$ actuator in body fixed coordinate system representing extra added mass inertia force, damping force and another $1 \times 1$ actuator in world coordinate system representing the buoyancy force.

The reason transformation Jacobian $J_b^h$ is not used here is because added mass and damping coefficient for a submerged object is constant with respect to the body fixed coordinate, while floating object has a constant added mass and damping coefficient with respect to the steadily translating coordinate. The overall model with cable and load in 20-sim is shown in Figure 9.

## 4 SIMULATION EXAMPLE

### 4.1 *Vessel*

To assess the accuracy and authenticity of the bond graph model, an evaluation between 20-sim simulation results and ShipX motion RAO is conducted. The default ship 's175' from ShipX database is used. The time domain 6 DOF motion are compared in amplitude. The results show good matching between 20-sim and ShipX in most wave periods and headings yet poor accuracy of both heave and roll in vessel's natural rolling period (18 sec), (Figs. 6, 7 and 8).

The reason is ShipX will add extra viscous damping coefficient in its calculation when the vessel reaches its natural rolling frequency.

One way to improve the model quality is to add linear viscous damping coefficient for natural rolling frequency, and to tune the motion RAO in 20-sim closer to that in ShipX or user defined value since ShipX itself also has poor accuracy in natural rolling frequency. The user can tune

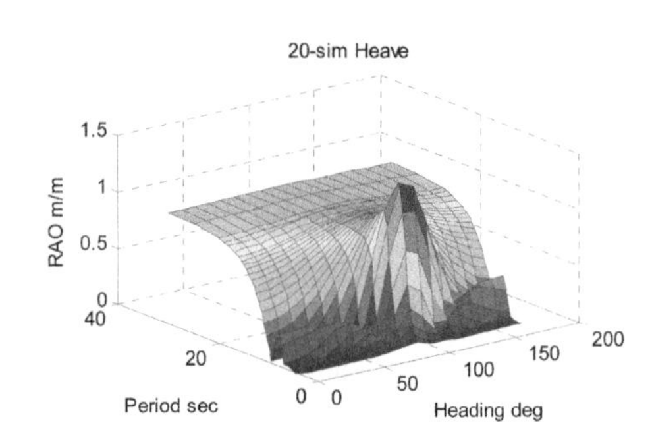


Figure 6. Heave motion in 20-sim.

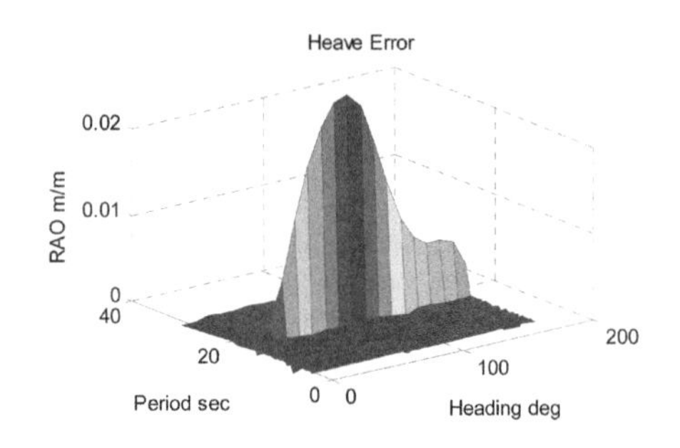


Figure 7. Heave error between 20-sim & ShipX.

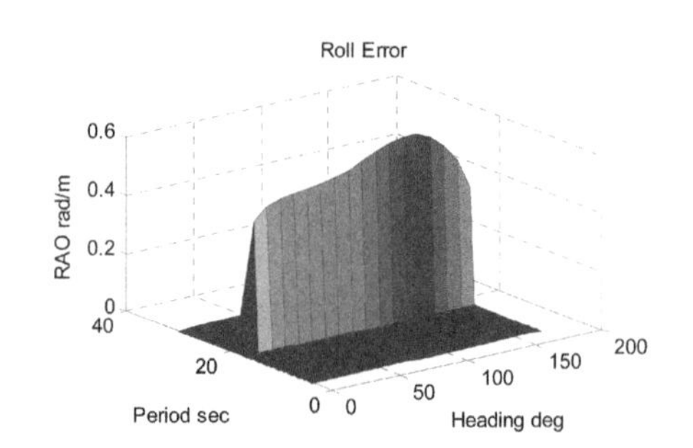


Figure 8. Roll error between 20-sim & ShipX.

the damping coefficient for each wave period and heading to have a better approximation towards ShipX motion RAO.

### 4.2 *Cable and load*

The sample is Lankhorst Ropes' $6 \times 36$ WS+IWRC standard wire rope with higher breaking strength, the simulation result of cable and load model will be compared with DNV-RP-H103 5.2 (DNV 2010), with all identical parameters for both DNV rules and 3D Mechanics model. The horizontal assessment was conducted with variables of parts number, current speed and linearity of the shape. The parameter variations are shown in Table 1.

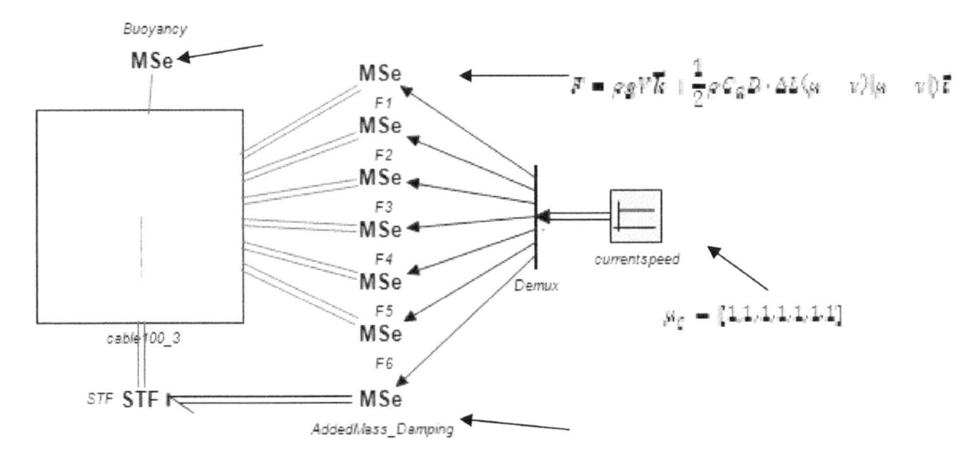


Figure 9. Cable and load model.

Table 1. Vertical offset with lifted load [m].

| | L = 10 m | L = 100 m | L = 1000 m |
|---|---|---|---|
| DNV | 0.00387 | 0.0403 | 0.562 |
| n = 1 | 0.00387 | 0.0403 | 0.562 |
| n = 2 | 0.00387 | 0.0403 | 0.562 |
| n = 3 | 0.00387 | 0.0403 | 0.562 |

The result shows that the vertical offsets of the 3D Mechanics cable have perfect compliance with DNV rules no matter how much discretized parts were set because both DNV rules calculation and 3D Mechanics cable have analytical results of the problem. Thus for cases where people's only interest is in vertical offset without current force, a single straight line with one translational connector is sufficient enough.

Figures 10, 11 and 12 show the horizontal offset as a function of water depth for the various parameter variations. For horizontal offset, under small horizontal offset condition ($\zeta/L < 1{:}5$), the offset at the bottom end of the cable has good compliance with DNV rules calculation results, while under large horizontal offset condition ($\zeta/L \geq 1{:}5$), the bottom end of the cable has significant deviation from DNV rules calculation result. This is because by the definition of DNV-RP-H103 5.2.

For an axially stiff cable with negligible bending stiffness the offset of a vertical cable with a heavy weight at the end of the cable in an arbitrary current with unidirectional (x-direction) velocity profile is

$$\zeta(z) = \int_z^0 \left[ \frac{F_{D0} + 0.5\rho \int_{-L}^{z1} C_{Dn} D_c [U_c(z_2)]^2 dz_2}{W + w(z_1 + L)} \right] dz_1 \tag{28}$$

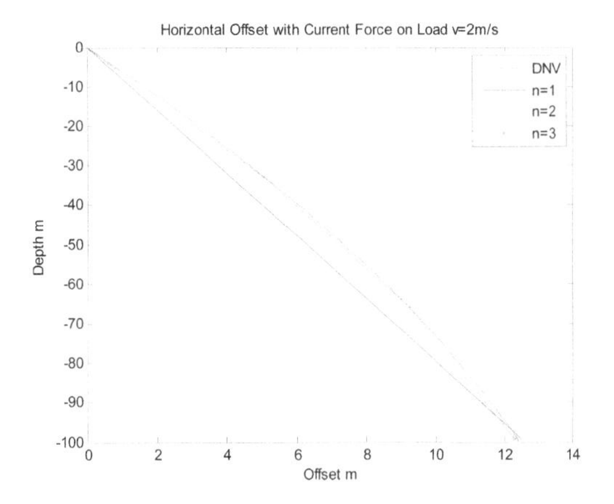


Figure 10. Horizontal offset with current force on load V = 2 m/s.

The formula is only valid under small-angle approximation where the vertical depth of the cable equals to the axial length of the cable, yet large horizontal offset cases ($\zeta/L \geq 1{:}5$) cannot be treated under small-angle approximation anymore, where the resultant cable length under large horizontal offset is significantly longer than the depth.

The large horizontal offset condition also contradicts the premise of *'a vertical cable with a heavy weight at the end of the cable'* from the description of the formula.

However, the 3D Mechanics model is built according to a fixed cable length. Large horizontal offset will cause the end of the cable to be lifted hence a smaller depth. From the small horizontal offset cases the validity of the subject model has been proved.

For large horizontal offset condition the subject model is still applicable for simulation purpose but

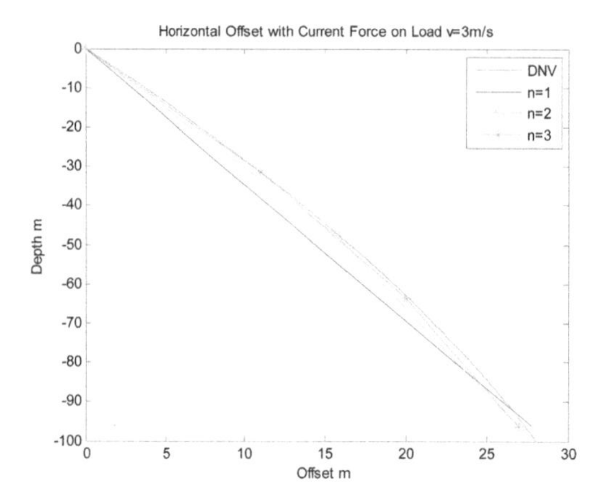


Figure 11. Horizontal offset with current force on load V = 3 m/s.

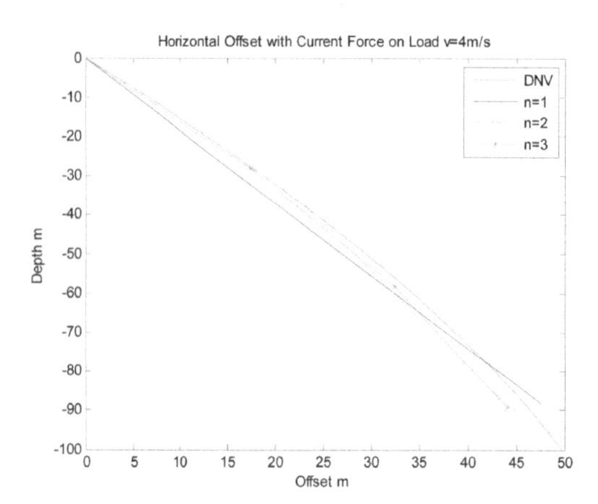


Figure 12. Horizontal offset with current force on load V = 4 m/s.

requires extra evidence to support. Since under small horizontal offset condition, the subject model shows good compliance regardless of the parts number, the choice of the part number only depends on the distribution of current force and nonlinearity of the cable. Higher parts number will have better approximation of the shape, finer distribution of the current force and more accurate dynamic simulation result. But it will also increase the computation amount and time. The decision shall be made under different practical conditions.

### 4.3 *Overall assembly*

All the various sub-models can be assembled into one big model selectively. The interactive physical entities are connected with power port whereas

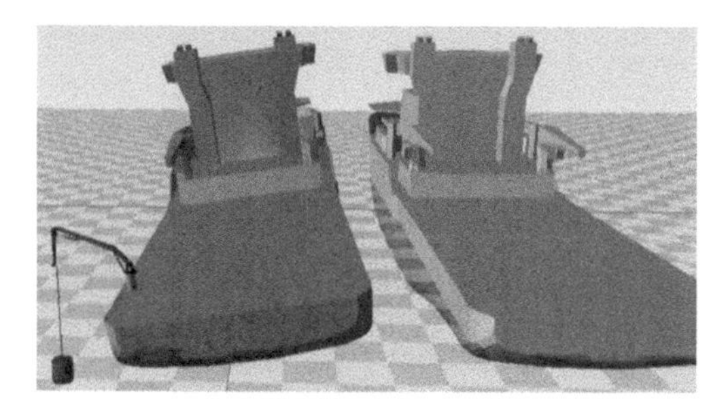

Figure 13. Animation: Vessel + Crane + Cable + Load.

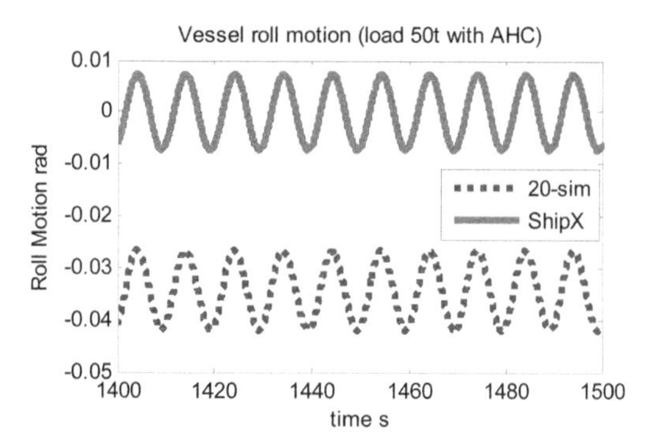


Figure 14. Vessel roll motion.

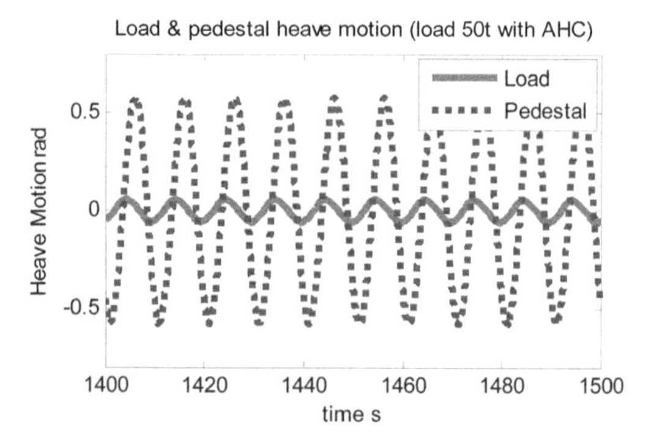


Figure 15. Load & pedestal heave motion.

control loop are connected with signal port. An example of vessel, crane, cable, load and AHC control model is shown in Figure 16. Because of the computation limits on PC, hydraulic sub-model was replaced with direct PID control. Also as for now, added mass and damping of load model is set as constant, e.g. ($C_a = 1$; $C_d = 1.2$). For most cases, the cable can be represented as a straight bar. With the help of AHC system and joystick, the crane cannot only be manually but also automatically controlled.

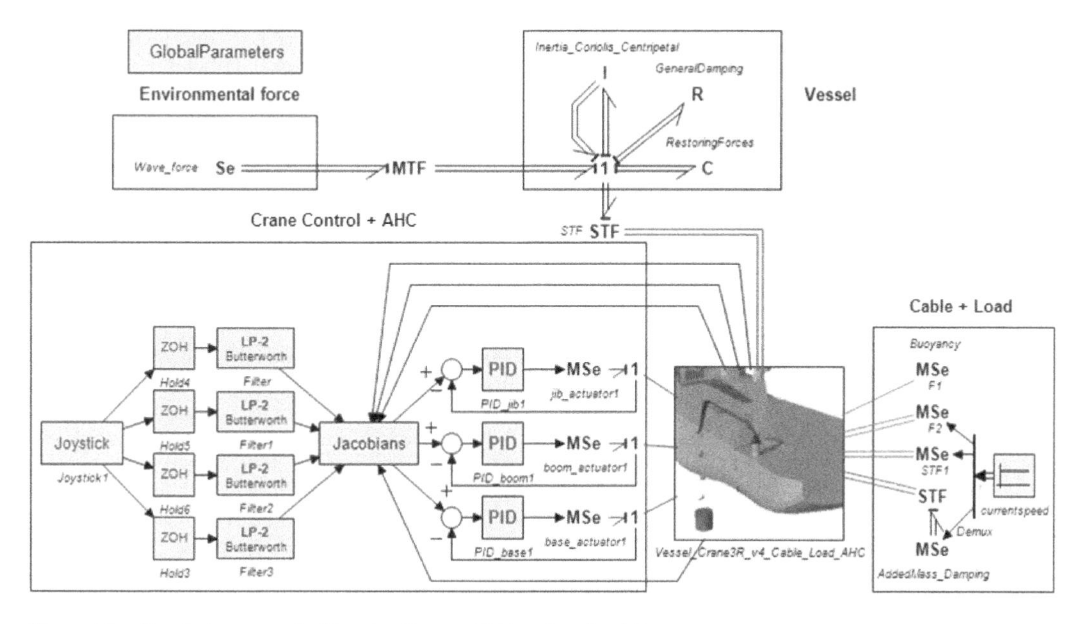


Figure 16. Simulation example: Vessel + Crane + Cable + Load + AH.

Figure 13 shows a snapshot of an animation of two offshore vessels moving in waves. The vessel on the right hand side shows the motion of an offshore vessel moving in waves according to the motion RAO defined by ShipX. The vessel on the left hand side includes a crane, a cable and a load in addition to the original vessel. From the snapshot, we see clearly that the vessel to the left has a port side inclination due to the weight of the load and cable. This will be found in the simulations as a dynamic roll motion around a mean angle different from zero (a mean static inclination angle). The dynamic char acteristics of the two vessels are the same (no update is carried out due to the mean inclination).

Figure 14 shows the time history of the roll motion with and without the lifted object. Figure 15 shows the vertical heave motion at the foot of the crane pedestal and for the lifted object, when the AHC is active.

The AHC system is based on crane kinematics for now (Chu 2013). Winch based AHC system may have different performance in this case, which can be investigated in the future.

## 5 FUTURE WORK

As a part of the ongoing research, there are many things yet to be done. In the load model part, the motion of the vessel model is assumed to be only periodic with constant added mass and damping matrices for each frequency and amplitude. But the load could be lowered, lifted, dragged and rotated without periodic feature. As the simulation is on a time-domain basis, the hydrodynamic coefficients of the load should be transformed from frequency-amplitude dependent into velocity-acceleration dependent $\boldsymbol{M}_A(\dot{\boldsymbol{\eta}},\ddot{\boldsymbol{\eta}})$ and $\boldsymbol{B}(\dot{\boldsymbol{\eta}},\ddot{\boldsymbol{\eta}})$. Although the function is not strictly correct because the fluid has memory effect, i.e. the historical motion of the fluid will affect the presence. But if a good correlation can be shown by CFD calculation, the trajectory of the load can then be removed from the variables.

To investigate the feasibility of velocity-acceleration dependent coefficients, a CFD experiment can be conducted to compare the difference of coefficients under same velocity and acceleration condition but with and without periodic movement, Figure 17 illustrates the idea. If valid, the velocity-acceleration dependent coefficients can then be fitted and input into 20-sim model as functions of velocity and acceleration.

So far the modelling method is still based on 20-sim, which has many restrictions, e.g. only diagonal matrix of inertia tensor is allowed inside 3D Mechanics; only linear spring/damping is allowed inside 3D Mechanics; the function of real-time interaction with Matlab has not been realized. In future developments, a better platform or direct programming of physics engine may be required to expand the functionality of the model. Also, in 3D Mechanics, the model is eventually being solved as

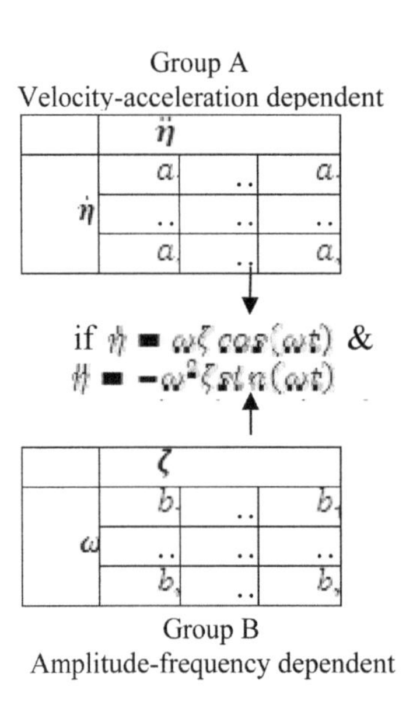


Figure 17. Velocity-acceleration dependent coefficient.

a big matrix yet only occupying single CPU thread, which is time consuming and inefficient. Multi-threads solution using GPU and parallel computation should be applied in the future. The model also requires collision property and event detector to control the different state of marine operation, e.g. crossing splash zone will have a different model which requires splash model and detectors to detect the start and the end of the phase.

The ideal model shall have the function of 'Plug and Play' which minimize the modelling procedure and maximize the flexibility for different models. Thus, a standard interface between sub-models is also required. Different parties shall have the same User Interface (UI) to develop their own sub-model, which in 3D gaming industry often called as modification (MOD). Many sub-models can then be added into a product library let users and customers to choose and test, e.g. a winch, a propeller or a drilling tower can also be modelled by following the same approach. Remote control and local PC experience is also to be expected in the final package. There are still so much to be done.

## 6 CONCLUSION

This paper demonstrated a generic modelling approach for marine operation. The aim is to propose a standard application of modelling each sub-system in the marine operation whatever the design is. The model consists of several modelling methods from other researchers yet with necessary modification to make a compatible system. The general second order differential equation of vessel hydrodynamics is from (Fossen and Fjellstad 2011). MATLAB files for ShipX data transformation is from (Fossen and Perez 2014). Bond graph modelling technique is from (Pedersen 2008). Crane hydraulics and control model is from (Chu 2013). The cable model is inspired by (Johansson 1976). The load model is inspired by (Halse, et al. 2014). Combining the knowledge of previous researches, the modelling method proposed in this paper enables the user to assemble all sub-models into a complete operation model. The complete model can then serve as a virtual reality for training purpose. In the complete model, most of the 3D mechanics modelling are done inside the 3D Mechanics, a toolbox of 20-sim, but with connectors, actuators and sensors, the 3D Mechanics model receives information and interacts with bond graph and control scheme outside. The bond graph and control scheme expand the function of 3D Mechanics model and give user the freedom to modify the design. The realistic physical entities interact as power flow and information interacts as signal flow, which gives a clear idea to separate the different level of modelling.

## REFERENCES


Controllab Product BV. 2013. *20-sim Reference 4.3.*

Chu, Yingguang. 2013. "20 sim-based Simulation of Offshore Hyddraulic Crane Systems with Active Heave Compensation and Anti-sway Control."

DNV. 2010. DNV-RP-H103 Modelling and analysis of marine operations, April 2010. DNV.

Fackrell, Shelagh. 2011. *Study of the Added Mass of Cylinders and Spheres.* University of Windsor.

Fossen, Thor I., and Ola-Erik Fjellstad. 2011. *Handbook of Marine Craft Hydrodynamics and Motion Control.* Trondheim: John Wiley & Sons, Ltd.

Fossen, Thor I., and Tristan Perez. 2014. *MSS. Marine Systems Simulator (2010).* 16 3. http://www.marine-control.org.

Halse, Karl H., Vilmar Æsøy, Dimitry Ponkratov, Yingguang Chu, Jiafeng Xu, and Eilif Pedersen. 2014. Lifting Operations For Subsea Installations Using Small Construction Vessels and Active Heave Compensation System—A Simulation Approach. OMAE.

Janssen, Sander. 2013. *A Conversion from SolidWorks to 20-sim through COLLADA.* Controllab Product BV.

Johansson, Per I. 1976. *A Finite Element Model for Dynamic Analysis of Mooring Cables.*

Kirchhoff, G.R. 1877. *Vorlesungen ueber Mathematische Physik, Mechanik.* Leipzig.

Lloyd, A.R.J.M. 1989. *Seakeeping: Ship Behaviour in rough weather.* Ellis Horwood Limited.

MARINTEK. 2014. http://www.sintef.no/home/MARINTEK/ Software/Maritime/.

MathWorks. 2014. http://www.mathworks.se/products/matlab/.

Offshore Simulator Centre (2014). Company web page, www.offsim.no/eng, last visited 25.06.2014.

Pedersen, Eilif. 2008. «Bond Graph Modeling of Marine Vehicle Dynamics.» Trondheim.

*Maritime-Port Technology and Development – Ehlers et al. (Eds)*
 *ISBN 978-1-138-02726-8*

# Model Predictive Control of a waterborne AGV at the operational level

H. Zheng, R.R. Negenborn & G. Lodewijks
*Department of Maritime and Transport Technology, Delft University of Technology, Delft, The Netherlands*

ABSTRACT: Increasing container transport volume has been seen not only inside container terminals, but also among terminals in the port area, known as Inter Terminal Transport (ITT). While conventional Automated Guided Vehicles (AGVs) have been put into practice to enhance efficiency inside Automated Container Terminals, this paper proposes the novel concept of waterborne Automated Guided Vessels (w-AGVs) for the application of ITT. Given route and timing information, a controller guaranteeing both smooth path convergence and Required Time of Arrival (RTA) at the operational level is proposed. Path convergence is achieved by introducing a path parameter, adding one extra degree of freedom to the original optimization problem. RTA is approximately guaranteed considering both the distance-to-go and time-to-go in terms of the current position. Model Predictive Control (MPC) is proposed for solving the control problem formulated for its advantages of being optimization-oriented and able to explicitly handle various constraints. The possible online computational burden of MPC is eased by a successive linearization around a seed trajectory from the previous step. Furthermore, an iterative framework is implemented to account for the linearization errors. Simulation tests are run based on a small-scale vessel model to demonstrate the effectiveness of the proposed scheme in terms of solving the control problem of a w-AGV system at the operational level.

## 1 INTRODUCTION

Large ports like the Port of Rotterdam have seen a rapid increase of transport volume in the port area since 1970s. More than 30 million TEUs (Twenty-foot Equivalent Unit) per year are expected to be handled towards 2035 (Port of Rotterdam, 2014). It thus can be foreseen that container handling efficiencies will be critical both inside and inter terminals. AGVs have been adopted in automated container terminals to ease problems like long operation times and high personnel expenses, etc. Outside terminals, increasing transportation demands on punctual (neither early nor late) collection and delivery of containers also happen among various terminals through various modalities (rail, road, sea etc.), known as ITT. The most important performance criterion of ITT is "Non-performance" which happens when the completion time of ITT tasks is later than the permitted latest arrival time (Duinkerken, *et al.*, 2007). Therefore, we propose waterborne AGVs with both path following and guaranteed arrival time capabilities in the scenario of ITT.

The above application scenario of w-AGVs differs them from Unmanned Surface Vessels (USVs) which have been studied for a wide range of purposes, including environmental survey, rescue, military or pure research platforms (Zheng, *et al.*, 2013). The criterion of delivering/picking-up the right amount in the right place at the right time in an economical way is considered priority for w-AGVs. Typically, development of an entire w-AGV system in the port area requires planning and control at three levels, i.e. strategic level, tactical level and operational level, as shown in Figure 1. At the strategic level, types, numbers and the way w-AGVs are operated etc. are decided considering factors in a wide area in the long term. Scheduling decisions

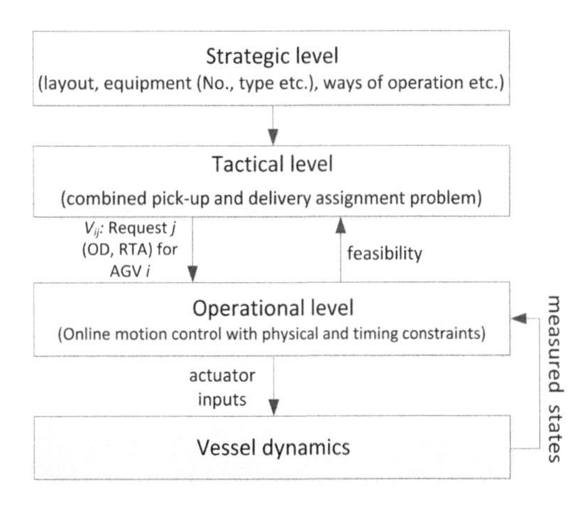


Figure 1. Planning and control levels of w-AGVs. OD stands for origin/destination.

are often made at the tactical level for multiple requests assignment given multiple w-AGVs available. Single or multiple w-AGVs motion control problems are solved at the operational level to fulfil the tasks scheduled at the tactical level. The outputs from the operational level are then applied to the w-AGV dynamics. System responses are measured and feedback to the operational level for real-time decision-making. This paper, as a starting point, will first formulate and solve the control problem of a single w-AGV at the operational level based on given transportation tasks with specific route and timing requirements.

Applications of system and control theory have a long history in vessel automation. The first recognized and most widely implemented controller until now still favors the classical PID (Proportional-Integral-Derivative) control theory (Minorski, 1922), because of its simplicity both in theoretical analysis and implementation. However, this simplicity is at the cost of heavy tuning work which largely depends on personal experiences, and no optimality can be guaranteed. Still, system nonlinearities and constraints are often hard to be incorporated in the controller design in a systematic way. Another large family of vessel motion control problems in the literature are solved using Lyapunov based design methods (Jiang, 2002; Do, Pan, 2006). The idea of backstepping with theoretical stability analysis and experimental control performances has been illustrated by (Fossen, *et al.*, 2003; Skjetne, *et al.*, 2005) based on a small-scale model vessel—Cybership II. Their Lyapunov based backstepping based controller design technique and the vessel model have seen a large amount of references in the field. Nonetheless, system constraints are still not considered and optimality still cannot be guaranteed. Neglecting constraints in controller design might result in infeasibility which might further result in instability or even damages to the system; neglecting optimality, intuitively, means that the system will be driven to achieve the desired performance regardless the energy required, namely not in an economical way. In reality, constraints do exist due to limited engine power, mechanical maximum deflections/revolutions, maximum sailing speed or spatial no-sailing zone, etc.; and for w-AGVs, as mentioned before, economical operation is a design requirement.

Model Predictive Control (MPC) solves an optimal constrained problem online in a receding horizon way, then naturally offers an alternative. MPC has become arguably the most widely implemented advanced control methodology currently in industry (Camacho and Bordons, 2013). Its theoretical basis as well as the stability, optimality, and robustness properties are well understood (Mayne, *et al.*, 2000). The main reason for its popularity is that it can handle constraints on inputs and states in a systematic way. In vessel motion control, limitations on actuators and states arise from physical, economical, or safety considerations. The first application of MPC for vessels in the literature is (Wahl and Gilles, 1998). More recent applications are made by researchers from the University of Michigan, who have taken the advantage of MPC in conjunction with line-of-sight guidance for vessel path following control (Oh and Sun, 2010) and a disturbance compensating scheme for vessel heading control (Li and Sun, 2012). In their work, linear prediction models are used in MPC, but showed satisfactory control results nonetheless. However, none of the controllers for automated vessels are explicitly designed for transportation purposes where vessel behaviors are influenced by transportation requests from a scheduling level. Spatial route and timing requirements are expected to be fulfilled in an economical way, which poses challenges to the control problem.

In this paper, we propose a novel concept—waterborne Automated Guided Vessels (w-AGVs), which are dedicated to inter terminal transport in the port area. Together with the individual automated container terminals and the automated equipment, including AGVs, inside of them, these w-AGVs are expected to contribute to a fully automated transport system. The control problem at the operational level is formulated for a single w-AGV considering both smooth path following and RTA by using a parameterized path and an online adjustable speed reference. MPC is proposed to solve the above control problem; trade-offs between the online computational burden and control performances are obtained by linearization and multiple iterations.

This paper is organized as follows. Since MPC is a model-based control method, we first describe the vessel dynamical model used for MPC controller design and simulation in Section 2. A predictive path following problem with required time of arrival for w-AGVs at the operational level is formulated in Section 3. Then in Section 4, test scenarios, simulation results and discussions are given, followed by concluding remarks and future work in Section 5.

## 2 VESSEL MODEL

A vessel model that can capture the main dynamics at the operational level is required for MPC controller design. Marine surface vessels experience motions along 6 Degrees of Freedom (DOF) which are normally described in two coordinate frames: body-fixed frame $\{b\}$ and inertial coordinate system $\{n\}$.

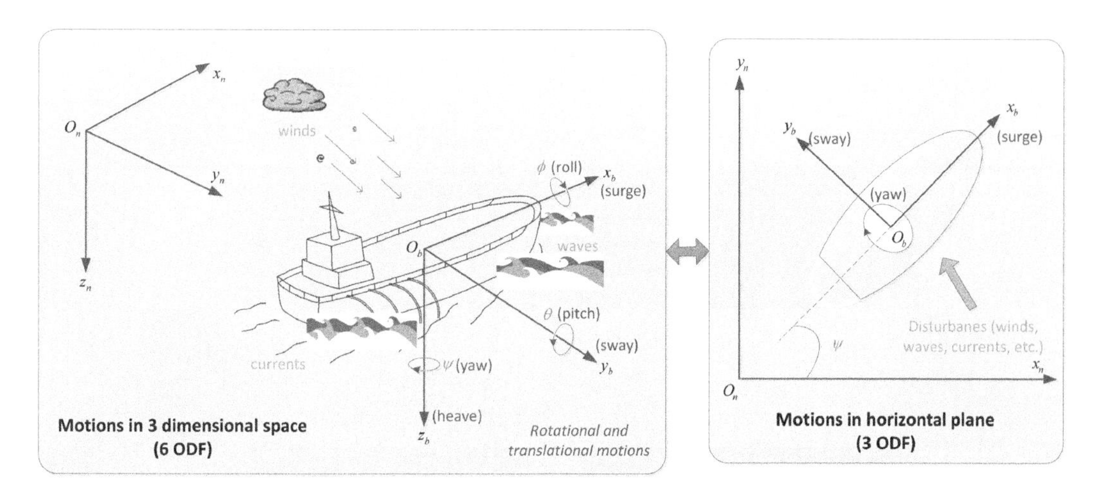


Figure 2. Vessel motions in 6 DOF and 3 DOF.

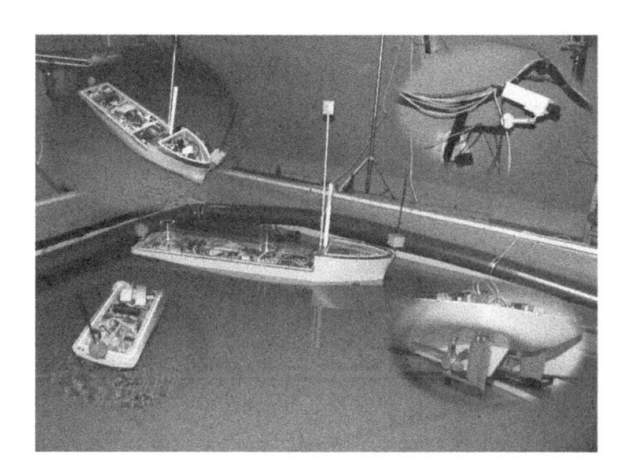

Figure 3. Cybership II at Marine Cybernetics Laboratory of NTNU (Fossen, 2008).

Motions in the horizontal plane are referred to as surge (longitudinal motion), sway (sideways motion) and yaw (rotation around the vertical axis). The other three DOF are roll (rotation about the longitudinal axis), pitch (rotation about the transverse axis), and heave (vertical motion). For vessel maneuvering control, it is common to formulate a 3 DOF vessel model as a coupled *surge-sway-yaw* model and neglect heave, roll and pitch motions (Fossen, 2011). The degrees of motions of a surface vessel are shown in Figure 2.

Next, the 3 DOF nonlinear dynamic model of Cybership II (as shown in Fig. 3) is described, because its hydrodynamic parameters have been identified and published in the literature (Skjetne, *et al.*, 2005). The kinematics and kinetic model of the vessel (neglecting forces caused by environmental disturbances at this stage) can be written as:

$$\begin{aligned} \boldsymbol{\eta}(t) &= \boldsymbol{T}(\boldsymbol{\eta}(t))\boldsymbol{\nu}(t) \\ \boldsymbol{M}\dot{\boldsymbol{\nu}}(t) + \boldsymbol{C}(\boldsymbol{\nu}(t))\boldsymbol{\nu}(t) + \boldsymbol{D}\boldsymbol{\nu}(t) &= \boldsymbol{\tau}(t) \end{aligned} \tag{1}$$

where $\boldsymbol{\eta}(t)$ is the pose (position and orientation) vector in the inertial frame, $\boldsymbol{\nu}(t)$ is the body-fixed velocity vector and $\boldsymbol{\tau}(t)$ is the control vector. These vectors are given by:

$$\boldsymbol{\eta}(t) := \begin{bmatrix} x_n(t) \\ y_n(t) \\ \psi(t) \end{bmatrix}, \boldsymbol{\nu}(t) := \begin{bmatrix} u(t) \\ v(t) \\ r(t) \end{bmatrix}, \boldsymbol{\tau}(t) := \begin{bmatrix} f_u(t) \\ f_v(t) \\ t_r(t) \end{bmatrix}, \tag{2}$$

where $x_n(t)$(m), $y_n(t)$(m) are the positions along axis $x_n$, $y_n$, respectively, and $\psi(t)$(rad) is heading angle in the inertial frame; $u(t)$(m/s), $v(t)$(m/s) and r($t$) (rad/s) are the surge, sway velocities and yaw rate in the body-fixed frame, respectively; $f_u(t)$(N), $f_v(t)$ (N) and $t_r(t)$(Nm) are the surge, sway forces and yaw moment produced by the vessel actuators. The system matrices $\boldsymbol{M}$, $\boldsymbol{C}(\boldsymbol{\nu}(t))$ and $\boldsymbol{D}$ are the inertial mass matrix (invertible), Coriolis and centrifugal matrix, and damping matrix, respectively, which are defined as:

$$\boldsymbol{M} := \begin{bmatrix} m_{11} & 0 & 0 \\ 0 & m_{22} & m_{23} \\ 0 & m_{32} & m_{33} \end{bmatrix}, \boldsymbol{D} := \begin{bmatrix} d_{11} & 0 & 0 \\ 0 & d_{22} & d_{23} \\ 0 & d_{32} & d_{33} \end{bmatrix},$$

$$\boldsymbol{C}(\boldsymbol{\nu}(t)) := \begin{bmatrix} 0 & 0 & -c_{31}(\boldsymbol{\nu}(t)) \\ 0 & 0 & c_{23}(\boldsymbol{\nu}(t)) \\ c_{31}(\boldsymbol{\nu}(t)) & -c_{23}(\boldsymbol{\nu}(t)) & 0 \end{bmatrix}, \tag{3}$$

where $c_{23} = m_{11}u(t)$ and $c_{31} = m_{22}v(t) + 0.5(m_{23} + m_{32})$ $r(t)$. The Jacobian matrix $\boldsymbol{T}(\boldsymbol{\eta}(t))$ transforms the body-fixed velocities $\boldsymbol{\nu}(t)$ into the inertial velocities $\boldsymbol{\eta}(t)$ and is given by

$$\boldsymbol{T}(\boldsymbol{\eta}(t)) = \begin{bmatrix} \cos(\psi(t)) & -\sin(\psi(t)) & 0 \\ \sin(\psi(t)) & \cos(\psi(t)) & 0 \\ 0 & 0 & 1 \end{bmatrix}. \quad (4)$$

## 3 PREDICTIVE PATH FOLLOWING CONTROL WITH REQUIRED TIME OF ARRIVAL

The control objectives of motion control can be roughly classified into three categories: 1) set-point regulation, which requires a target point, e.g. a constant speed, target pose, to be provided to the regulator, 2) trajectory tracking, where a time-varying/parameterized reference trajectory needs to be tracked, 3) path following, where the system is forced to follow a varying geometric reference path with no temporal constraints.

For the operational level control problem of a w-AGV, the outputs from a scheduling level (see Fig. 1) are usually assigned tasks with routing and timing requirements. The exact time-dependent trajectory, arrive where at when, is not known a priori. In reality, we do observe that captains always have in mind a coarse estimation about how much more time they still need before arriving at the destination so that they will roughly be on schedule to decrease non-performance rate. This is done by looking ahead the distance-to-go and time-to-go and then adjusting the desired vessel speed within system constraints. Once this timing prerequisite is satisfied, the vessel will then be operated in a way as economical as possible. Next, we mathematically formulate a predictive path following problem with required time of arrival to mimic this human intelligence.

### 3.1 *Model predictive control*

MPC is a control strategy based on online repetitive numerical optimization. At each time step, a new optimization problem is formulated based on the current new measurements and predicted system responses; a sequence of optimal control inputs is calculated with respect to a performance index in such a way that the prediction of the system output is driven close to the reference, see Figure 4. The first element of this optimal control sequence is the output of the controller and applied to the real system.

For MPC controller design, it is convenient to rewrite model (1)~(2) into a state space format as Ordinary Differential Equations (ODEs):

$$\dot{\boldsymbol{x}}(t) = f(\boldsymbol{x}(t), \boldsymbol{u}(t)), \quad (5)$$

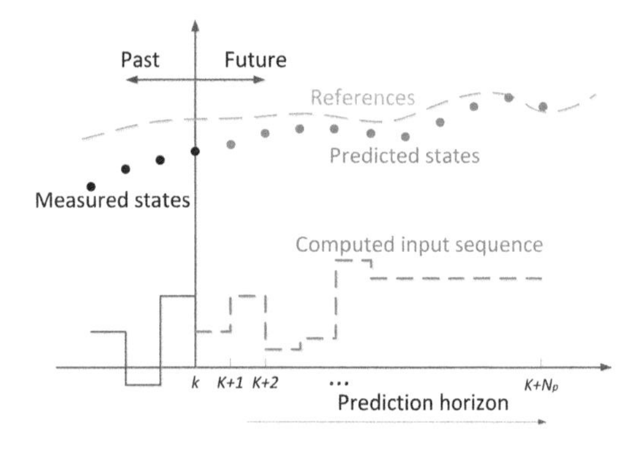


Figure 4. Model predictive control at time step $k$.

where $\boldsymbol{x}(t) = [\boldsymbol{\eta}(t)^{\mathrm{T}}\ \boldsymbol{\nu}(t)^{\mathrm{T}}]^{\mathrm{T}}$, $\boldsymbol{u}(t) = \tau(t)$ and $f$: $\Re^6 \times \Re^3 \rightarrow \Re^6$ is a nonlinear smooth function, given as:

$$f(\boldsymbol{x}(t), \boldsymbol{u}(t)) = \begin{bmatrix} \boldsymbol{T}(\boldsymbol{p}_1\boldsymbol{x}(t))(\boldsymbol{p}_2\boldsymbol{x}(t)) \\ \boldsymbol{M}^{-1}(-\boldsymbol{C}(\boldsymbol{p}_2\boldsymbol{x}(t))(\boldsymbol{p}_2x(t)) - \boldsymbol{D}(\boldsymbol{p}_2\boldsymbol{x}(t)) + \boldsymbol{u}(t)) \end{bmatrix}, \quad (6)$$

where $\boldsymbol{p}_1 = [0\ 0\ 1\ 0\ 0\ 0]$ and $\boldsymbol{p}_2 = \begin{bmatrix} 0 & 0 & 0 & 1 & 0 & 0 \\ 0 & 0 & 0 & 0 & 1 & 0 \\ 0 & 0 & 0 & 0 & 0 & 1 \end{bmatrix}$.

In numerical simulations, a zero order hold is usually applied by assuming a constant control input during the interval $[(k-1)T_s, kT_s]$, $(k = 1,2, \ldots, N_p)$, where $N_p$ is the prediction horizon and $T_s$ is the sampling time. Next, we will denote continuous time $t = kT_s$ as discretized time step $k$ for simplification.

Nonlinear MPC that predicts future system responses from nonlinear system dynamics directly can be much more computationally complex than MPC utilizing a linear prediction model (Zheng, *et al.*, 2014). Therefore, for fast dynamics like vessel systems, we take advantage of linearized models for prediction. In particular, we make full use of the calculated optimal control input sequence $\{\boldsymbol{u}^0(k+i|k), I = 0,1, \ldots, N_p - 1\}$ from the previous step $k$ (conventionally implement the first one and disregard the rest) to obtain a seed trajectory (Kouvaritakis, *et al.*, 1999) $\{\boldsymbol{x}^0(k+i|k), I = 0,1, \ldots, N_p\}$ about which the nonlinear dynamics are approximated by linear incremental models. Hereby, $k + i|k$ stands for the *ith* element at step $k$ and the superscript $^0$ denotes a seed trajectory that satisfies:

$$\boldsymbol{x}^0(k+i+1\,|\,k) = \boldsymbol{f}(\boldsymbol{x}^0(k+i\,|\,k), \boldsymbol{u}^0(k+i\,|\,k)) \qquad i = 0, 1, \ldots, N_p - 1, \quad (7)$$

with $\boldsymbol{x}^0(k|k) = \boldsymbol{x}(k)$ as the current measurement. ODE solvers (e.g. ode45 in MATLAB) are available to provide solutions at specified sampling time within certain error tolerances and can be used to obtain the numerical results in (7). We also denote

$$x(k+i\,|\,k) = x^0(k+i\,|\,k) + \Delta x(k+i\,|\,k)$$
$$u(k+i\,|\,k) = u^0(k+i\,|\,k) + \Delta u(k+i\,|\,k), \tag{8}$$

with $\Delta x(k+i|k)$ and $\Delta u(k+i|k)$ to be state and input deviations from seed trajectories. By Taylor's theorem and neglecting the higher order terms than first order, we get the approximated incremental models around the seed trajectory as:

$$\begin{aligned} f(x(k+i\,|\,k), u(k+i\,|\,k)) &= f(x^0(k+i\,|\,k), \\ &\quad u^0(k+i\,|\,k)) + A(k+i\,|\,k)\Delta x(k+i\,|\,k) \\ &\quad + B(k+i\,|\,k)\Delta u(k+i\,|\,k) \end{aligned} \tag{9}$$

where

$$A(k+i\,|\,k) = \partial f / \partial x(x^0(k+i\,|\,k), u^0(k+i\,|\,k))$$
$$B(k+i\,|\,k) = \partial f / \partial u(x^0(k+i\,|\,k), u^0(k+i\,|\,k)),$$

with $\Delta x(k|k) = 0$.

### 3.2 *Path following and required time of arrival*

In order to achieve smooth path convergence, the geometric reference path $r(k)\Re^3$ is parameterized by a scalar path parameter $s(k)$ instead of time step $k$. Considering the shortest path between two points is a straight line, we define a straight line reference path between two terminals:

$$r(k) = p(s(k)) = \begin{bmatrix} s(k)\cos(\theta) \\ s(k)\sin(\theta) \\ \theta \end{bmatrix}, \tag{10}$$

where $p$: $\Re^1 \rightarrow \Re^3$ and $\theta$ is the angle between the straight path and x axis. The time evolution of $s(k)$ provides an extra degree of freedom to the original system dynamics which is controlled by an extra input $v_p(k)$ as:

$$\dot{s}(t) = v_p(t),\ v_p \in \Re^1 \tag{11}$$

in continuous time form or

$$s(k+1) = s(k) + v_p(k)T_s,\ v_p \in \Re^1 \tag{12}$$

in discrete time form. Path parameter $s(k)$ can be seen as a virtual mass point (Ghabcheloo, *et al.*, 2009), acting the role of a tractable reference point on the path. If the path following error between the w-AGV and route is large, the mass point will slow down to '*wait for*' the w-AGV to achieve route convergence. To avoid stationary behavior of the mass point and achieve continuous forward motion, usually a specified constant along-path speed is set for the path parameter, as have been done in the maneuvering problem proposed in (Skjetne, *et al.*, 2004). However, their formulation cannot guarantee a time convergence to fulfill the task of a w-AGV.

We propose, in this paper, an online adjustable desired path velocity based on the current distance-to-go and time-to-go of the mass point:

$$v_d(k) = s_1(k)/t_1(k), \tag{13}$$

where $s_1(k)$ is the distance left to be travelled and $t_1(k)$ is the time left in terms of the required time of arrival on schedule:

$$s_l(k) = L - s(k)$$
$$t_l(k) = T - kT_s, \tag{14}$$

with $L$ being the total path length and $T$ being the total time scheduled to fulfill the task.

Remark 1: $v_d(k)$ is an average speed over the distance to go, so (13) only approximately calculates a desired speed. However, we argue that since the actual motion pattern, e.g. constant speed, accelerating or decelerating, is unknown and since we update this desired speed every prediction step in the predictive path following problem, this approximation is sufficient to guarantee a time convergence.

Remark 2: Since the path parameter is just a scalar, the approximate calculation of the average desired speed will not bring too much extra computational burden to the online optimization problem, which is another advantage of introducing the path parameter $s(k)$.

Therefore, in general, path convergence is achieved by tracking the reference signals provided by (10); the above convergence is guaranteed in a smooth way by assigning and optimizing the extra dynamics of the path parameter as (11) or (12); an approximate required time of arrival is then realized by penalizing the difference between $v_p(t)$ and $v_d(t)$ as specified by (13) and (14). Whenever off the path or behind the schedule, large control efforts (but within the boundaries) will be made to steer the w-AGV to track the path and avoid being '*lagged behind*'; once on track, the vessel is operated in an economical way by solving online optimization problems. If an emergency occurs, e.g. an obstacle on the path, the speed of the virtual mass point is limited during obstacle avoidance and accelerated after that. In this way, the w-AGV is expected to maneuver smoothly and safely in this hazardous region while still stay as close as possible to the path; still, the time delay can be compensated after passing this region by accelerating. The above characteristics thus resemble the captain's intelligence elaborated on at the beginning of this section.

### 3.3 *Predictive path following with required time of arrival*

Now, we can formulate the optimization problem for predictive path following with time convergence in the framework of MPC. At each time step $k$, the following optimization problem is solved online to fulfill both smooth path following and required time of arrival of a w-AGV at the operational level:

$$\Delta\mathbf{u}^*(k), \mathbf{v}^*(k) = \arg\min_{\Delta\mathbf{u},\mathbf{v}} \mathbf{J}(k), \tag{15}$$

with

$$\begin{aligned}\boldsymbol{J}(k) &= \sum_{i=1}^{N_p}\Bigg(\underbrace{[\boldsymbol{\eta}(k+i\,|\,k)-\boldsymbol{r}(k+i\,|\,k)]^{\mathrm{T}}\boldsymbol{Q}[\boldsymbol{\eta}(k+i\,|\,k) - \boldsymbol{r}(k+i\,|\,k)]}_{1} \\ &+ \underbrace{\Delta\boldsymbol{u}(k+i\,|\,k)^T\boldsymbol{R}\Delta\boldsymbol{u}(k+i\,|\,k)}_{2} \\ &+ \underbrace{F\frac{1}{2}\boldsymbol{v}(k+i\,|\,k)^T\boldsymbol{M}\boldsymbol{v}(k+i\,|\,k)}_{3} \\ &+ \underbrace{T\left[v_{\mathrm{p}}(k+i\,|\,k)-v_{\mathrm{d}}(k+i\,|\,k)\right]^2}_{4}\Bigg)\end{aligned} \tag{16}$$

subject to $\forall\, i = 0, 1, \ldots, N_p - 1$

$$\begin{aligned}\Delta\boldsymbol{x}(k+i+1\,|\,k) &= \boldsymbol{A}(k+i\,|\,k)\Delta\boldsymbol{x}(k+i\,|\,k) \\ &\quad + \boldsymbol{B}(k+i\,|\,k)\Delta\boldsymbol{u}(k+i\,|\,k), \\ \boldsymbol{x}(k+i\,|\,k) &= \boldsymbol{x}^0(k+i\,|\,k)+\Delta\boldsymbol{x}(k+i\,|\,k), \\ \boldsymbol{\eta}(k+i\,|\,k) &= \boldsymbol{C}\boldsymbol{x}(k+i\,|\,k), \\ s(k+i+1\,|\,k) &= s(k+i\,|\,k)+v_p(k+i\,|\,k)T_s,\end{aligned} \tag{17}$$

$$\boldsymbol{r}(k+i\,|\,k) = \begin{bmatrix} s(k+i\,|\,k)\cos(\theta) \\ s(k+i\,|\,k)\sin(\theta) \\ \theta \end{bmatrix}, \tag{18}$$

$$\begin{aligned}s_1(k+i\,|\,k) &= L - s(k+i\,|\,k), \\ t_1(k+i\,|\,k) &= T-(k+i)T_s, \\ v_d(k+i\,|\,k) &= s_1(k+i\,|\,k)/t_1(k+i\,|\,k),\end{aligned} \tag{19}$$

$$\Delta\boldsymbol{x}(k\,|\,k) = 0, \tag{20}$$

$$|\,\boldsymbol{u}(k+i\,|\,k)\,| \le \boldsymbol{u}_{max}, \tag{21}$$

$$\boldsymbol{v}_{\min} \le \boldsymbol{v}(k+i\,|\,k) \le \boldsymbol{v}_{\max}, \tag{22}$$

$$x(k+i\,|\,k) \leqslant obs_{x,\min} + Ma_1, \tag{23}$$

$$-x(k+i\,|\,k) \leqslant -obs_{x,\max} + Ma_2, \tag{24}$$

$$y(k+i\,|\,k) \leqslant obs_{y,\min} + Ma_3, \tag{25}$$

$$-y(k+i\,|\,k) \leqslant -obs_{y,\max} + Ma_4, \tag{26}$$

$$\sum_{j=1}^{4} a_j \le 3 \text{ and } a_j \in \{0,1\}. \tag{27}$$

In objective function (16), we have four terms to penalize:

- Term 1 is the path tracking error over the prediction horizon which guarantees path convergence;
- Term 2 penalizes too larges changes in control inputs;
- Term 3 stands for the energy consumption where $\boldsymbol{M}$ is the inertial mass matrix as defined in Section 2 and
- Term 4 penalizes the error between the speed of the virtual mass point and the desired average speed that is a function of the distance-to-go and time-to-go and thus guarantees time convergence.

Constraints (17)-(19) are prediction dynamics; (20) is the initial constraint for the incremental model in (17); (21) and (22) are constraints due to system physical limits on maximum actuator forces/moment and maximum speed, respectively; and (23)-(27) introduce integer variables and poses constraints on positions to realize obstacle avoidance (obstacles are handled as rectangles and safe margins are implemented to avoid crossings in corners (Richards, *et al.*, 2002)).

The above optimization problem is solved in a receding horizon style as introduced in Section 3.1. By Jacobian linearization about a seed trajectory calculated from the previous step optimal control input sequence, the possible computational burden from online non-convex optimization problems is alleviated by approximated convex ones. Even though we linearize at each prediction step instead of only at the start of the prediction horizon so that higher linearization precision is expected, linearization errors due to neglecting the remainder of the Taylor series expansion are inevitable. In order to reduce the possible model mismatch brought by linearization errors, we implement an iterative structure which can be terminated after a predefined maximum number of iterations (denoted as $I_{max}$), or when the input perturbation vector falls below a given tolerance (denoted by $\varepsilon$). The entire algorithm for predictive path following with time convergence in given in Algorithm 1.

**Algorithm 1.** Predictive path following with required time of arrival.

---

Initialization at $k = 0$: $\mathbf{x}(k|k) = \mathbf{x}_0$ and $\mathbf{u}(k) = \mathbf{u}_0$;
**while** $s_f(k) \geqslant 0.01 * l$ (vessel length) **do**
  Measure current states $\mathbf{x}(k|k)$;
  Set $n = 0$ (iteration loop);
  **while** $\|\Delta\mathbf{u}_k\| > \varepsilon$ and $n < I_{max}$ **do**
    Set $\mathbf{x}^0(k|k) = \mathbf{x}(k|k)$ and calculate $\mathbf{x}^0(k)$ according to (7);
    Linearize nonlinear model (5) about the seed trajectory $(\mathbf{x}^0(k), \mathbf{u}^0(k))$ and obtain $\mathbf{A}(k+i|k)$ and $\mathbf{B}(k+i|k)$
    for $i = 0, 1, \ldots, N_p - 1$;
    Solve the above optimization problem to determine $\Delta\mathbf{u}^*(k)$, $\mathbf{v}_p{}^*(k)$
    Determine $\mathbf{u}(k)$ by (11);
    Go to next iteration $n = n + 1$
  **end while**
  Apply the first input $\mathbf{u}^*(k|k)$ to process and $\mathbf{u}^0(k+1) = \mathbf{u}(k+1|k), \ldots, \mathbf{u}(k+N_p-1|k), \mathbf{u}(k+N_p-1|k)$;
  $k = k + 1$;
**end while**

---

## 4 SIMULATION RESULTS AND DISCUSSIONS

### 4.1 *Test scenario*

Being developed to become an innovative and sustainable container terminal cluster with complex spatial layout, Maasvlakte 2 (Maasvlakte 2, 2014), the new port area in the Port of Rotterdam, is one of the possible application domains for w-AGVs. W-AGVs are especially advantageous between terminals with longer waterborne than land distances and with heavy transportation demands. A pair of origin/destination terminals has been selected as the APM terminal (latitude: 51.959334, longitude: 4.060373) and Rhenus logistics deep-sea terminal (latitude: 51.961501, longitude: 4.049278), as shown in Figure 5. The RTA at Rhenus logistics deep-sea terminal is 10 minutes later after a departure from APM terminal. Since we use the small-scale (1:70) vessel model Cybership II for simulation, these quantities also need to be scaled properly according to Froude scaling law (1:70 for length (m) and 1:√70 for time (s)), so we get the following data for simulation.

Table 1. Test scenario data after scaling.

| Origin [x y] (m) | Destination [x y] (m) | RTA t (s) |
|---|---|---|
| [0 0] | [−10.8957 3.4443] | 71.71 |

Figure 5. Maasvlakte 2 and APM terminal and Rhenus logistics deep-sea terminal from Google Earth.

### 4.2 *Model and controller parameters*

In terms of model parameters, the following specifications of Cybership II are implemented for simulation:

- System matrices:

$$\boldsymbol{M} = \begin{bmatrix} 25.8 & 0 & 0 \\ 0 & 33.8 & 1 \\ 0 & 1 & 2.8 \end{bmatrix}, \boldsymbol{D} = \begin{bmatrix} 0.72 & 0 & 0 \\ 0 & 0.89 & 0.03 \\ 0 & 0.03 & 1.9 \end{bmatrix}.$$

- Maximum speeds: 0.2 m/s for surge velocity, 0.1 m/s for sway velocity and 0.5236 rad/s for yaw rate (which corresponds to 1.7 m/s or 3.3 knots of the corresponding full scale vessel).
- Maximum actuator forces ($f_u$ and $f_v$) and yaw moment ($t_r$): 2 N, 2 N and 1.5 Nm, respectively (which corresponds to 686 kN, 686 kN and 36015 kNm of the corresponding full scale).

In terms of parameters for the MPC controller design, weight matrices in (16) are as follows:

$$\boldsymbol{Q} = \begin{bmatrix} 1000 & 0 & 0 \\ 0 & 1000 & 0 \\ 0 & 0 & 100 \end{bmatrix}, \boldsymbol{R} = \boldsymbol{I}_{3\times3}, F = 1, T = 100,$$

Prediction horizon $N_p$ is set to 10; $I_{max}$ is set to 10 and $\varepsilon$ is 0.1. The value of big M in (23) is $10^6$. Initial states of the system $x_0 = [-1\ 0\ 0\ 0\ 0\ 0]^T$ and $\boldsymbol{u}_0 = [0\ 0\ 0]^T$.

### 4.3 *Results and discussions*

In this paper, the optimization problems and constraints are formulated using *YALMIP* (Lofberg, 2004) and solved with *Gurobi* (Gurobi Optimization, Inc., 2014). All the simulations are run in MATLAB 2011b at a platform with Intel (R) Core (TM) i5–3470 CPU @3.20 GHz.

For comparison, three numerical experiments are carried out, each with a different controller: controller 1 uses the algorithm proposed in this paper with one extra degree of freedom for optimization and an online flexible desired average speed, as described in Section 3; controller 2 still takes the advantage of the extra degree of freedom of the path parameter but the desired average speed is constant ($L/T$ = 0.1593 m/s as the scenario set above); controller 3 does not optimize the path speed but uses the constant average speed as calculated in the second controller to obtain the desired trajectory directly. Therefore, the third controller, essentially, has become a trajectory tracking controller. For a controller designed for a w-AGV with the application for ITT, three performance indices are considered critical, namely smooth path following, required time of arrival and satisfaction of the above two capabilities in a way as economical as possible.

Path following performances of the three controllers during experiments are shown as Figure 6(a). Generally, satisfactory path convergences are observed for all of them. Figure 6(b) and (c) show details at the initial point and obstacle area, respectively. All of the trajectories converge to the route and avoid the obstacle area successfully. While trajectories from controller 1 and controller 2 demonstrate similar and smoother convergence to the desired route, controller 3 with no extra optimization degree, showed larger cross-track errors (distance of the vessel to the route), which may result in aggressive control inputs. Figure 7 with the control inputs confirms this.

The control inputs of controller 3 fluctuate more intensively and last longer when the vessel is around the initial point or the obstacle region. If we sum the absolute values of the control inputs of each controller, then we get 307.40, 320.04 and 400.42 for controller 1, controller 2 and controller 3, respectively. A 23.23% decrease of controller 1 is observed compared with controller 3. Figure 8 further shows the cross track errors of the 3 controllers. Controller 3 has much larger cross tracking errors than the other two, with controller 1's

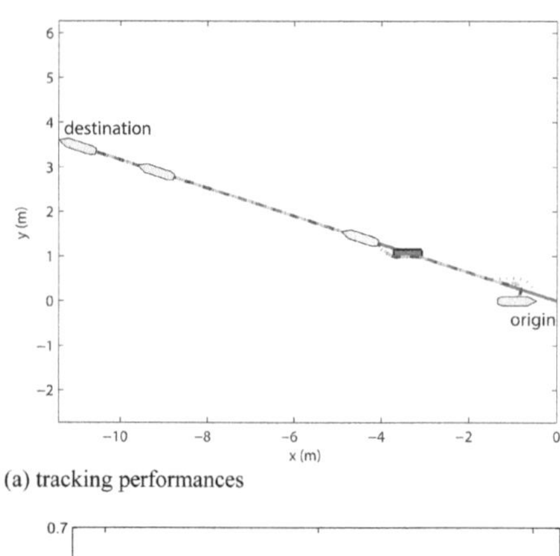

(a) tracking performances

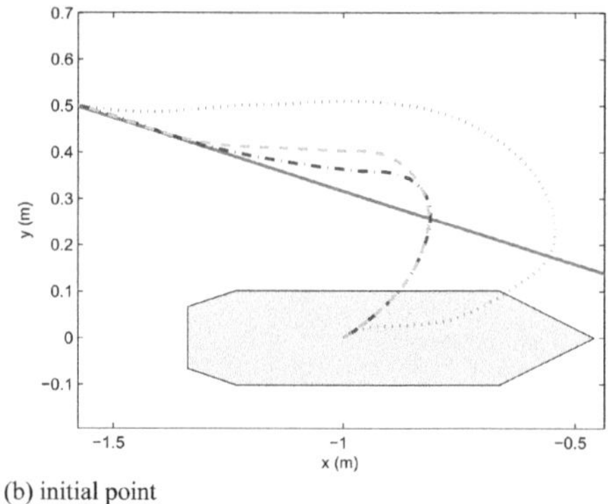

(b) initial point

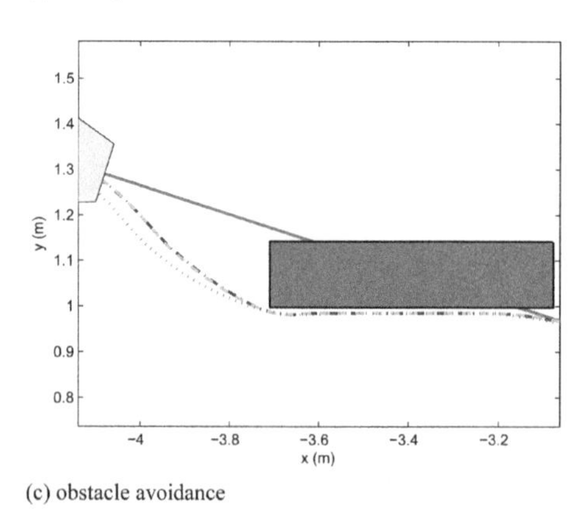

(c) obstacle avoidance

Figure 6. Tracking performances of the three controllers: solid line-route between two terminals; dash dot line-controller 1 path; dashed line-controller 2 path; dotted line-controller 3 path.

slightly smaller than controller 2's. The mean values of the tracking errors of the 3 controllers are 0.0232 m, 0.0236 m and 0.0337 m, respectively, which are corresponding to 1.624 m, 1.652 m and 2.359 m for full scale vessels in reality.

Another important criterion for the path following problem for a w-AGV is the required time of

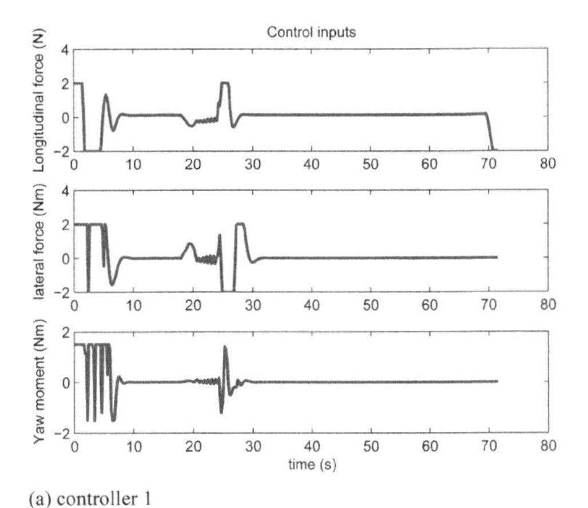


(a) controller 1

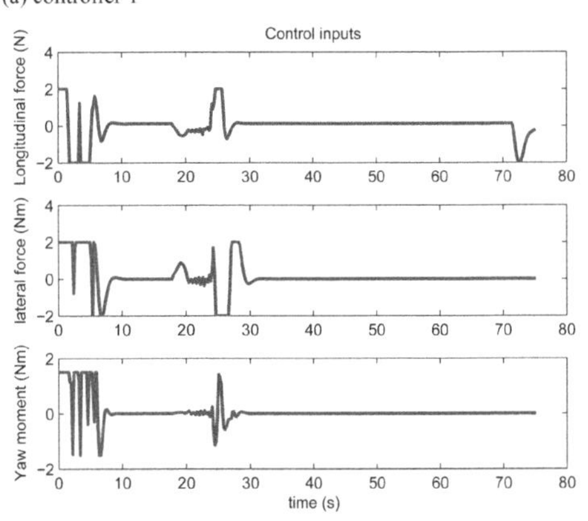


(b) controller 2

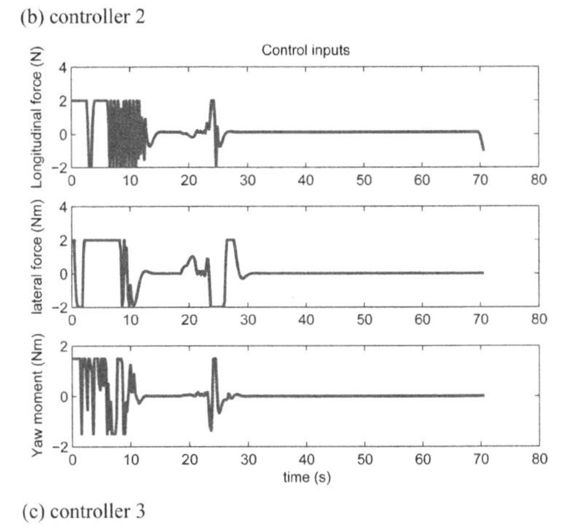


(c) controller 3

Figure 7. Control inputs from 3 controllers.

arrival, which means the vessel not only needs to stay close to the route, but also arrive punctually as scheduled. In the scenario between APM terminal and Rhenus logistics deep-sea terminal, we have set the required time as 10 minutes, which is 71.71 s for the vessel model we used in the experiments.

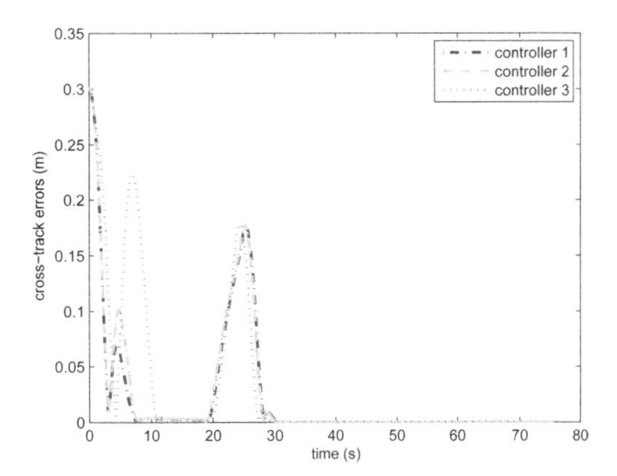


Figure 8. Cross track errors.

A total of 71.60 s, 75.20 s and 70.60 s are recorded for controller 1, controller 2 and controller 3, respectively. Since the third controller is just a trajectory tracking controller, the time it takes should be equal to the set time 71.71 s. The slightly smaller actual value is because we terminate the simulation once the distance to destination is smaller than 0.01 times vessel length, see Algorithm 1. Controller 1 shows good punctuality while controller 2 has been behind the schedule about 3.5 s, which is about half a minute for a real vessel. We argue that since the total voyage time is only 10 minutes, this delay has been significant.

## 5 CONCLUSIONS AND FUTURE RESEARCH

A novel concept, waterborne Automated Guided Vessels (w-AGVs), has been proposed in this paper for the application in Inter Terminal Transport (ITT). W-AGVs are transportation oriented and thus need to be operated to fulfill scheduled tasks with both route and time requirements in an economical way. A control problem at the operational level for a w-AGV has been formulated and solved to meet those requirements. By introducing one extra degree of freedom of a path parameter dynamics and an online adjustable speed profile, both route and time convergences have been achieved. Online optimization problems have been solved in the framework of MPC so that constraints are systematically handled and optimal system performances are obtained. The performance of the algorithm is compared with another two approaches to illustrate its advantages in solving the problem in a case study. Extension to multiple w-AGVs in coordination and integration of the operational level with a higher scheduling level problem for ITT are interesting directions for future research.

## ACKNOWLEDGEMENTS

This research is supported by the China Scholarship Council under Grant 201206950021 and the VENI project "Intelligent multi-agent control for flexible coordination of transport hubs" (project 11210) of the Dutch Technology Foundation STW, a subdivision of the Netherlands Organization for Scientific Research (NWO).

## REFERENCES

Camacho, E.F. & Bordons, C. A. (ed.) 2013. *Model predictive control.* Springer.

Do, K.D. & Pan, J. 2006. Robust path-following of underactuated ships: Theory and experiments on a model ship. *Ocean Engineering* 33(10): 1354–1372.

Duinkerken, M.B., Dekker, R., Kurstjens, S.T., Ottjes, J.A., & Dellaert, N.P. 2007. Comparing transportation systems for inter-terminal transport at the Maasvlakte container terminals. In *Container Terminals and Cargo Systems* (ed.) 37–61. Berlin Heidelberg: Springer.

Fossen, T.I., Breivik, M. & Skjetne, R. 2003. Line-of-sight path following of underactuated marine craft. *In Proceedings of the 6th IFAC on Maneuvering and Control of Marine Craft*, Girona, Spain: 244–249.

Fossen, T.I. 2008. *Cybership II*, URL: http://www.itk.ntnu.no/ansatte/{Fossen_Thor}/GNC/cybership2.htm. Accessed on May 21, 2014.

Fossen, T.I. (ed.) 2011. *Handbook of marine craft hydrodynamics and motion control.* West Sussex, U.K.: John Wiley and Sons Ltd.

Ghabcheloo, R., Aguiar, A.P., Pascoal, A., Silvestre, C., Kaminer, I. and Hespanha, J. 2009. Coordinated path-following in the presence of communication losses and time delays[J]. SIAM *Journal on Control and Optimization*, 48(1): 234–265.

Gurobi Optimization Inc. 2012. *Gurobi optimizer reference manual.* URL: http://www.gurobi.com/. Accessed on May 21, 2014.

Jiang, Z.P. 2002. Global tracking control of underactuated ships by Lyapunov's direct method. *Automatica* 38(2): 301–309.

Kouvaritakis, B., Cannon, M. & Rossiter, J.A. 1999. Non-linear model based predictive control. *International Journal of Control*, 72(10): 919–928.

Li, Z. & Sun, J. 2012. Disturbance compensating model predictive control with application to ship heading control. *IEEE Transactions on Control Systems Technology*, 20(1): 257–265.

Lofberg, J. 2004. YALMIP: A toolbox for modeling and optimization in MATLAB. In *Proceedings of 2004 IEEE International Symposium on Computer Aided Control Systems Design*, Taiwan, Taipei: 284–289.

Mayne, D.Q., Rawlings, J.B., Rao, C.V. & Scokaert, P.O. 2000. Constrained model predictive control: Stability and optimality. *Automatica* 36(6): 789–814.

Minorsky, N. 1922. Directional stability of automatically steered bodies. *Journal of the American Society of Naval Engineers* 42(2): 280–309.

Oh, S.R. & Sun, J. 2010. Path following of underactuated marine surface vessels using line-of-sight based model predictive control. *Ocean Engineering*, 37(2): 289–295.

Port of Rotterdam. 2014. *Projects*, URL: http://www.portofrotterdam.com/en/Business/Containers/Pages/projects.aspx. Accessed on May 21, 2014.

Project Organization Maasvlakte 2. 2014. Maasvlakte 2. URL: https://www.maasvlakte2.com/en/index/. Accessed on May 21, 2014.

Richards, A., Schouwenaars, T., How, J.P. & Feron, E. 2002. Spacecraft trajectory planning with avoidance constraints using mixed-integer linear programming. *Journal of Guidance, Control, and Dynamics*, 25(4): 755–764.

Skjetne, R., Fossen, T.I. & Kokotović, P.V. 2005. Adaptive maneuvering, with experiments, for a model ship in a marine control laboratory. *Automatica,* 41(2): 289–298.

Wahl, A. & Gilles, E. 1998. Track-keeping on waterways using model predictive control. *In Proceedings of the IFAC Conference on Control Applications in Marine Systems*, Fukuoka, Japan: 149–154.

Zheng, H., Negenborn, R.R. & Lodewijks, G. 2013. Survey of approaches for improving the intelligence of marine surface vehicles. *In Proceedings of the 16th International IEEE Conference on Intelligent Transportation*, The Hague, The Netherlands: 1217–1223.

Zheng, H., Negenborn, R.R. & Lodewijks, G. 2014. Trajectory tracking of autonomous vessels using model predictive control. *Accepted for the 19th IFAC World Congress (IFAC WC'14),* Cape Town, South Africa.

*Maritime-Port Technology and Development – Ehlers et al. (Eds)*
*© 2015 Taylor & Francis Group, London, ISBN 978-1-138-02726-8*

# Estimation of Dynamic Positioning performance by time-domain simulations—a step toward safer operations

Dong Trong Nguyen, Luca Pivano & Øyvind Smogeli
*Marine Cybernetics, Trondheim, Norway*

ABSTRACT: The importance of Dynamic Positioning (DP) capability is steadily increasing as the industry is moving into harsher environments, and focus on risk management and Health, Safety and Environment (HSE) is increasing. It is thus essential that the vessel operators can rely on realistic estimates of the vessel DP capability to determine the operational weather window. The current industry standard for computing the DP capability has been proven non-conservative. It has been also shown that by including the complete vessel dynamics the results are much closer to reality. This paper investigates which dynamics are important to be accounted for when calculating DP capabilities by analyzing the sensitivity of the vessel station-keeping capability to different dynamic effects.

## 1 INTRODUCTION

In the last decade, the number of DP (Dynamic Positioning) vessels has increased dramatically driven by an increased offshore activity. Operations such as deep-water drilling, diving, subsea construction and maintenance, pipe-laying, shuttle offloading, platform supply and flotels rely heavily on the vessel capability to maintain its position and heading, typically also after a single failure.

For these operations, where the stakes are high both regarding cost and safety, it is essential that the vessel operators can rely on realistic estimates of the vessel DP capability to determine the weather operational window. This is particularly important when not all the equipment is available and after the Worst-Case Single Failure (WCSF). Decisions made on wrong assumptions may compromise the safety of the operation.

Dynamic Positioning (DP) performance is today demonstrated by Capability plots, which estimate the vessel DP capability for different weather conditions and vessel configurations.

The importance of DP capability is steadily increasing as the industry is moving into harsher environments, and focus on risk management and HSE is increasing.

The current industrial standard for DP capability analysis is described in ISO 19901-7, IMCA M140 (2000) and DNV ERN (2013), aiming to enable a direct comparison of individual vessel's performance and provide an indication of station keeping capability in a common and understandable format. However, there are significant limitations in these standards, and the trustworthiness of the current capability analyses are often questioned; Are they conservative or non-conservative? Can they be compared? Do they convey a realistic picture of a vessel's station-keeping capability in dynamic operating conditions?

The traditional DP capability (hereafter called DPCap) analysis is typically based on the IMCA M140 specification; lately also the DNV ERN standard has been adopted more frequently. Both are based on static balance of the maximum obtainable thruster force against a resultant mean environmental force due to wind, wave drift, current, and possible other loads. This is done for the full angle-of-attack envelope (0–360 deg). The results of such analyses are presented in form of polar plots termed wind envelopes, where the maximum wind speed at which the vessel can maintain position and heading is plotted for each angle of attack, typically given with 10–15 degree spacing. In addition, results may also be presented as thrust envelopes showing the thruster utilization for a given design condition at different wind angles of attack. Figure 5 show examples of typical wind envelopes.

The IMCA M140 specification is quite basic allowing the analysis to be computed with environmental forces from non-vessel-specific coefficients, thruster forces from generic rules-of-thumb and without including specifications on DP control system and thrust allocation. It is possible to extend the analysis with more realistic assumptions and models. This can be done for example by using actual vessel model data such as wind, current, and wave-drift coefficients, realistic thruster models, and realistic static thrust allocation including e.g.

forbidden zones and thrust loss effects based on actual allocated thrust. However, such extensions are not standardized. Due to the lack of precise requirements, capability plots computed from different suppliers may differ significantly when computed for the same vessel.

The DNV ERN standard on the other hands presents a quite precise method, which does not give much space for using of different thruster and environmental force models. The drawback of this standard is the simplicity of the mathematical models to be used; for example the thrust loss factor is fixed to 10% of the nominal thrust regardless of thruster position and type. In real life this can vary significantly due to ventilation, thruster-thruster and thruster-hull interactions. See for example Smogeli (2006), Phillips & Muddesetti (2006), and Bulten & Stoltenkamp (2013a), (2013b).

One of the strongest assumptions in these methods is that the vessel is considered at rest. It is not possible to include the dynamic loads from waves, wind and current, and the corresponding dynamic response of the vessel with its DP system. Hence, the DPCap analysis can only balance the static (mean) environmental forces with the mean thruster forces, meaning that a certain (assumed) amount of thrust must be reserved to counteract the unknown dynamic forces and vessel motion. Typically 15%–20% of the thrust is reserved for dynamic loads. This is often referred to as dynamic allowance. Furthermore, the 6-Degree-Of-Freedom (6DOF) vessel motion and the related thrust losses, as well as all other dynamic effects in the propulsion system like rate limits are usually neglected.

Another significant shortcoming of the quasi-static DPCap analysis it is that it cannot account for the transient conditions during a failure and recovery after a failure. Even if the quasi-static capability plots show that the vessel can maintain position and heading both in intact condition and after a single failure, nothing can be said about the motion of the vessel from the time the failure occurs until the desired position and heading has been regained. Especially after a worst case single failure for a DP2 or DP3 vessel, where as much as half of the thrust capacity may be lost, reallocation of thrust can take significant time due to limitations in rise time for propellers as well as rudder and azimuth angle rates. For a safety-critical DP operation such as diving or vessel-to-vessel replenishment or personnel transfer, the allowance for such transient motion can be very limited.

Smogeli et al. (2013) showed that dynamic station-keeping capability (DynCap), the next level DP capability analysis tool based on systematic time-domain simulations, produces DP capability estimates much closer to reality compared to the quasi-static methods. This is obtained by employing a complete 6 DOF vessel model, including dynamic wind and current loads, 1st and 2nd order wave loads including slowly-varying wave drift, a complete propulsion system including thrust losses, a power system, sensors, and a DP control system with observer, DP controller, and thrust allocation. This result is also proven in the model test validation of DynCap carried out in Børhaug (2012).

The aim of this paper is to highlight the importance of the inclusion of the complete vessel dynamics, environmental forces, and control system dynamics for obtaining realistic results. This is done by investigating the sensitivity of the vessel station-keeping capability to relevant dynamic effects and comparing results between the traditional DPCap and the new DynCap analysis.

## 2 CONCEPT

The main purpose of the DynCap analysis is to calculate the station-keeping capability of a vessel based on systematic time-domain simulations. The station-keeping capacity is calculated by searching for an environment limit at which the vessel is still able to satisfy a set of user defined acceptance criteria. In the IMCA specification for DP capability, the acceptance criterion is being able to keep the position and heading. The analysis can be performed for collinear environmental loads (wind, current and waves attacking from the same direction) or non-collinear loads.

The analysis is typically performed for intact condition where all the equipment is available and for the Worst Case Single Failure (WCSF) condition. In addition, the analysis can be run with any thruster and power setup to evaluate the station-keeping capability when not all the equipment is available (for example due to maintenance).

One of the advantages of the DynCap analysis, compared to a traditional DPCap, is that the limiting environment can be computed by applying a set of user defined acceptance criteria. The position and heading excursion limits can be set to allow a wide or narrow footprint, or the acceptance criteria can be based on other vessel performance characteristics such as sea keeping, motion of a crane tip or other critical point, dynamic power load, or tension and/or angle of a hawser or riser. In this way the acceptance criteria can be tailored to the requirements for each vessel and operation. An example of position and heading acceptance criteria is shown in Figure 1. In this case, the station-keeping capacity is found by searching for the maximum wind speed in which the vessel footprint stays within the predefined position and heading limits.

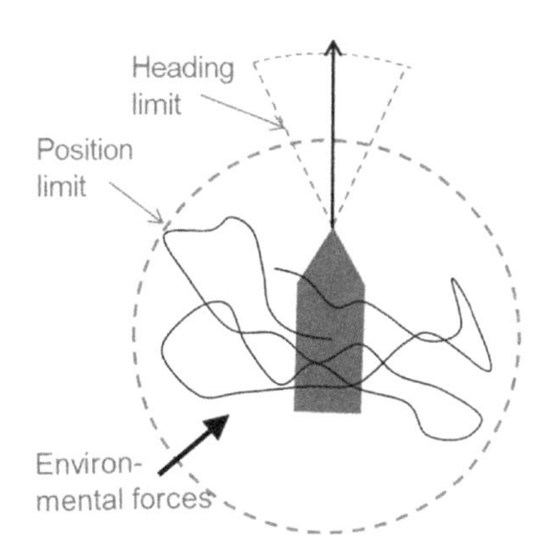


Figure 1. Example of heading and position acceptance limits.

By considering the complete vessel dynamics it is also possible to identify temporary position and/or heading excursions due to dynamic and transient effects. As an example, a vessel may stay in position without one thruster according to the traditional DPCap, but the loss of that thruster during station-keeping may cause a temporary excursion outside the positioning acceptance limits.

During DP operation, the vessel position and heading motion is characterized by two components:

- The motion displayed on a DP screen is checked towards positioning limits in the DP (watch circles). This is a filtered, low-frequency motion, which is due to the mean wave drift, thruster, wind and current forces. In literature, this is also referred to as the Low-Frequency (LF) motion.
- The harmonic (wave) motion due to first-order wave loads, which is oscillating about the LF motion. In literature, this is also referred to as the Wave Frequency (WF) motion.

The actual motion of the vessel is the sum of these two components; see Figure 2 for an example. Depending on the requirements to the operation, either the LF motion or the total vessel motion (LF + WF motion) can be used to check if the position acceptance criteria are satisfied in the DynCap analysis.

The DynCap results can be provided in various formats depending on purpose and simulation setup:

- Wind envelopes, directly comparable to the results obtained with a traditional DPCap study.
- Thrust envelopes, directly comparable to results obtained with a traditional DPCap study.
- Yearly operability at a given location, based on metocean data
- Yearly fuel consumption.

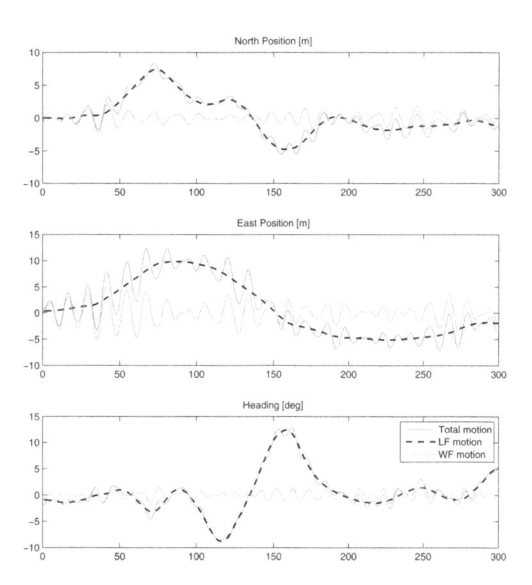


Figure 2. Dynamic motion of a vessel.

## 3 VESSEL SIMULATOR

DynCap is based on systematic time-domain simulations with a complete 6DOF closed loop vessel model. A block diagram describing the vessel simulator is shown in Figure 3. By allowing the vessel to move, the strongest assumption for the traditional DPCap analysis is removed. This facilitates inclusion of dynamic wind and current loads, 1st and 2nd order wave loads including slowly-varying wave drift, as well as the dynamics of the propulsion system and power system. A model of the Power Management System (PMS) is also included to simulate relevant functionality for DP operations such as black-out prevention, load limitation and sharing, and auto-start and auto-stop of generators. To close the loop a model of the full DP control system is included with an observer to estimate position and velocity with performance comparable to a standard Kalman filter, nonlinear PID-controller with wind feed-forward action, thrust allocation, sensors, and position reference systems. The complete propulsion system model includes actuator rate limits and computation of dynamic thrust loss effects such as the interaction between thrusters, interaction between thrusters and hull, ventilation, out-of-water effects, and transversal losses based on empirical models. More details on the vessel model can be found in Nguyen et al. 2013.

By considering the vessel, environmental loads and DP system dynamics, it is not necessary to reserve a certain amount of thrust for dynamic loads as for the traditional DPCap analysis.

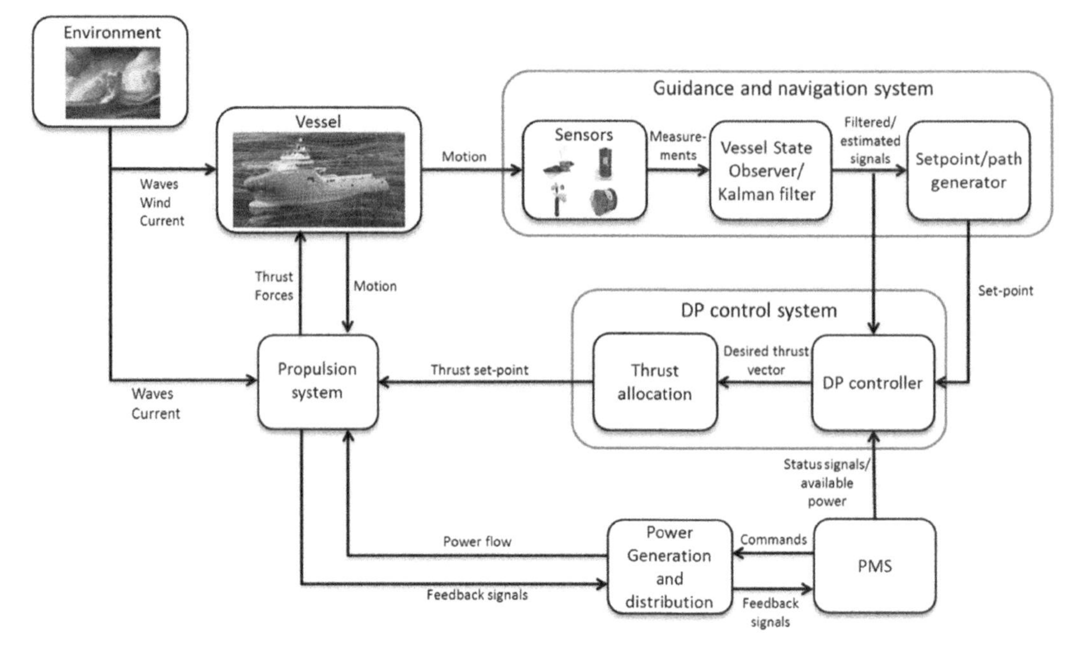


Figure 3. Closed-loop time-domain vessel simulator.

DynCap utilizes all the available thrust capacity like the vessel would do in real life.

In addition, the DP system model includes functions that can be found in the majority of DP control systems available today, such as black-out prevention and load limitation. If the required power for maintaining position and heading exceeds a preset limit, the thruster loads are limited such that those limits are not exceeded.

When running time-domain simulations, it is necessary to perform multiple runs. This is because for a given directional wave spectrum, there are many realization in time-domain that can results in the given spectrum. Each realization can result in different capability results, therefore the wind envelopes from DynCap analyses are presented as the average of five runs with different wave frequency and direction components random seeds.

## 4 CASE STUDY: PLATFORM SUPPLY VESSEL

In next sections, comparisons between DPCap and DynCap wind envelopes are presented for a typical platform supply vessel. The main particulars of the vessel are described in Table 1.

The hydrodynamic coefficients such as added mass, potential damping, hydrostatic coefficients, and 1st and 2nd-order wave load coefficients are computed using WAMIT. WAMIT is a 3D potential theory computer program capable of analyzing wave interactions with offshore platforms and other structures or vessels. The input to the program is a 3D geometry file represented by panels, as shown in Figure 4.

Table 1. Vessel main particulars.

| | |
|---|---|
| Length between perpendiculars | 80.0 m |
| Breadth | 20.0 m |
| Draught | 7.0 m |
| Displacement | 7500 tons |
| *Propulsion system—4 thrusters* | |
| Thruster 1: type bollard pull | Bow tunnel 1<br>172 kN |
| Thruster 2: type bollard pull | Bow azimuth<br>143 kN |
| Thruster 3: type bollard pull<br>forbidden sector* | Main azimuth port<br>280 kN<br>From −105° to −75° |
| Thruster 4: type bollard pull<br>forbidden sector* | Main azimuth stbd<br>280 kN<br>From 75° to 105° |
| *Power system—4 generators of 2000 kW each in 2-split configuration* | |

*Forbidden sector angle is defined as 0° alongship towards the bow and increasing clockwise.

The waves in this analysis are simulated using a JONSWAP wave spectrum with peak parameter $\gamma$ = 3.3 and spreading factor $s$ = 1 as recommended by DNV 2010. The wind-wave relationship is adopted

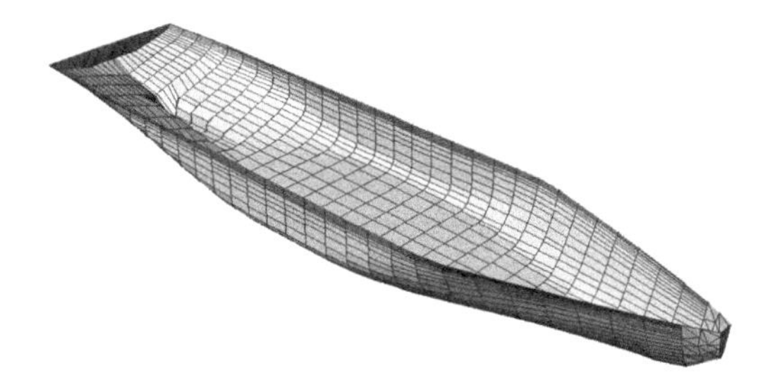

Figure 4. 3D hull geometry.

Table 2. DP gain settings.

| | Undamped period | Relative damping ratio |
|---|---|---|
| Surge | 80 s | 0.7 |
| Sway | 100 s | 0.7 |
| Yaw | 70 s | 0.7 |

from the North Sea data in IMCA M140. The environmental loads are set as collinear (wind, current and waves have the same direction). The power system operational philosophy is two-split switchboard. The Worst Case Single Failure (WCSF) is defined as loss of one switchboard.

### 4.1 *Nominal run setup*

In order to run a sensitivity analysis, a nominal analysis setup has been defined. This includes both vessel model parameters and analysis parameters. In the nominal analyses of both the traditional DPCap and DynCap, the equipment such as thruster, generators, etc., is modelled based on the nominal values received from the manufacturer.

The nominal DPCap analyses included also:

- All static thrust losses: Coanda, inline, interaction and transverse losses.
- 15% dynamic allowance.

For DynCap nominal analysis, the DP system model used in the simulations is configured and tuned according to industrial standards. The DP gain has been tuned such that the DP vessel in closed loop acts as a mass-spring-damper system with undamped natural periods and relative damping ratios as specified in Table 2. The thrust allocation for both DPCap and DynCap analysis is implemented such that the azimuth thrusters are free to rotate and the two main azimuth thrusters have forbidden zones (Table 1) to avoid flushing to each other.

The nominal DynCap analyses includes also:

- All dynamic thrust losses: Coanda, inline, interaction, transverse and ventilation losses.
- All dynamic effects.
- The wind and current magnitudes are modeled by considering an average speed and a random effect (wind gusts and current fluctuations). The current speed is set to 0.75 m/s.
- The low-frequency motion (see Fig. 2) is used to check whether the vessel is able to stay within position and heading limits.
- The acceptance criteria for position and heading are 5 m and 5 deg.

### 4.2 *Results*

Figure 5 shows five wind envelope cases for intact vessel conditions. The DPCap envelopes were computed with the nominal setup (15% dynamic allowance and static thrust losses) and with 0% dynamic allowance and no thrust losses. The latter case yields the maximum wind envelope and the one with 15% dynamic allowance and static thrust losses as expected results in smaller wind envelope. The environment limits from the DPCap analysis appear unrealistic with 55 to 65 m/s wind speed for head seas. In the same plot, three DynCap results obtained with three different acceptance criteria are depicted.

The DynCap results appear more realistic with 30 to 40 m/s wind speed for head sea. As expected the wind envelope shrinks with increasingly strict acceptance criteria, however even the widest acceptance criteria (20 m/20 deg) yield a smaller envelope than the DPCap with 15% dynamic allowance and thrust losses in head seas. Similar results are obtained after the WCSF, see Figure 6.

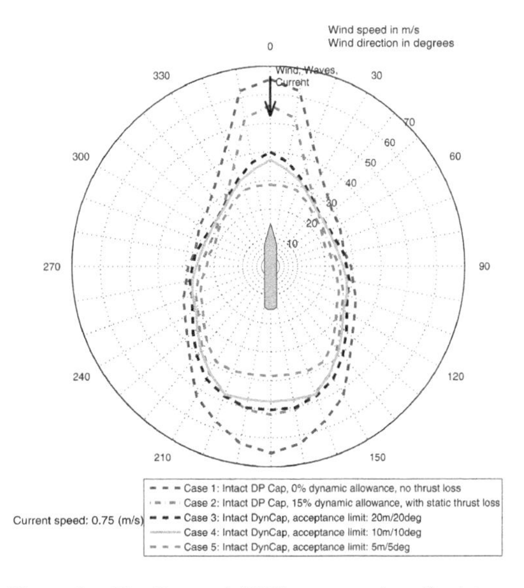


Figure 5. DynCap and DPCap comparison for intact vessel conditions.

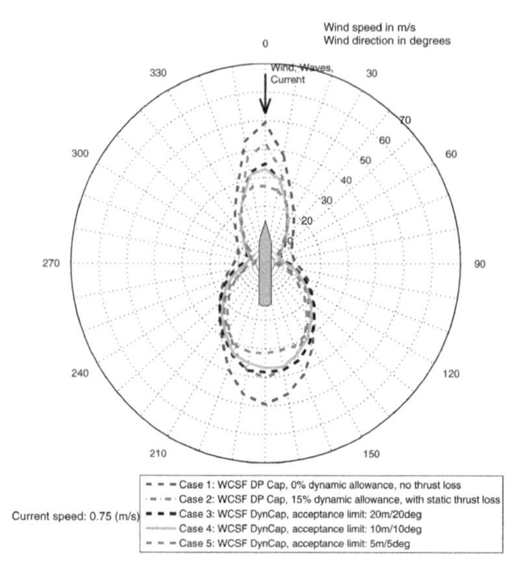


Figure 6. DPCap and DynCap comparison after WCSF.

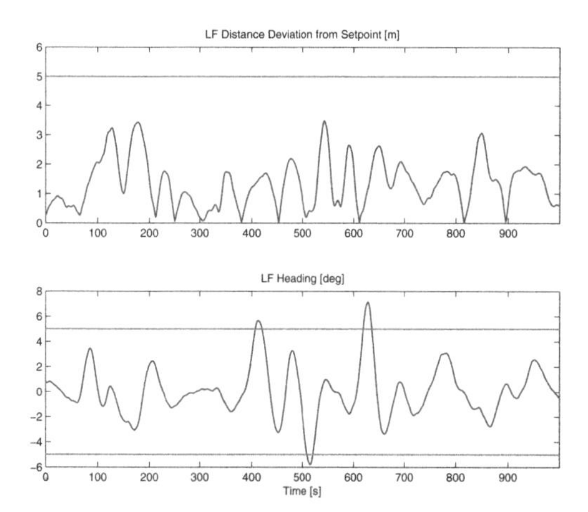


Figure 7. LF position and heading time-series for limiting wind speeds at 0 deg heading.

The difference for head and following seas is mainly due to two effects. When closing in on the limiting condition, the environmental forces give significant heading motion due to the difference in frontal and side vessel wind and current areas. The bow thrusters struggle to control the heading, where the tunnel switches between positive and negative force. Due to the thruster dynamics, the thrusters cannot produce force immediately, thus limiting the vessel heading controllability. The other limitation is the tunnel thrust loss due to ventilation caused by large vessel pitch motion. A time series for head sea (0 deg) is given in Figure 7. The average thrust utilization calculated from DynCap is about 20% of the total available while for the traditional analysis it is the 62%. The utilization in the DPCap is less than the configured 85% as the tunnel thruster is not used to its maximum. The thruster utilization from the DynCap analysis is highly affected by the average thrust from the tunnel thruster, which continuously changes direction. In addition, as the heading controllability is the limiting factor, the other thrusters are not used to their full capacity to counteract surge and sway forces.

At the limiting wind speeds for beam seas, see Figure 9, the tunnel thruster does not switch direction as for head seas and the total average thruster utilization is larger, about 52%. The motion in the north-east coordinate is also reduced compared to head seas resulting in a condition closer to the quasi-static method. This explain the smaller difference in the wind envelopes between DPCap and DynCap.

To better understand what are the factors that lead to the difference in the capability results, a DPCap run was performed by considering 25% of dynamic allowance and 20% less bollard-pull thrust on each thruster. The increase of dynamic allowance is to reserve more thrust for dynamic effects as we have seen that they are one of the main factors. The decrease in the bollard-pull thrust is to account for dynamic losses. The results are shown in Figure 8. Considering that the static loss model employed for the DPCap analysis is quite sophisticated, even with 20% less bollard-pull thrust and

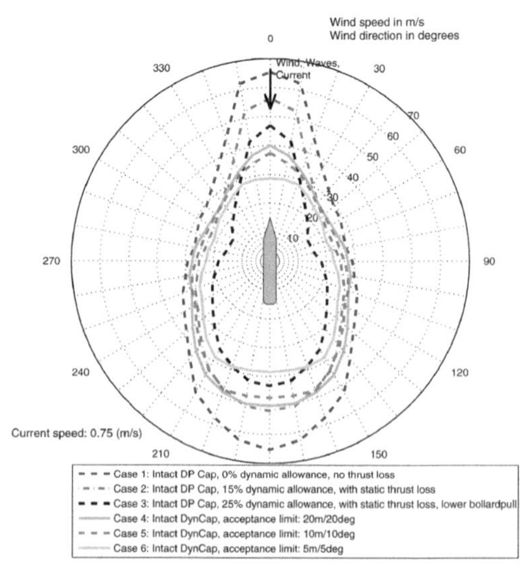


Figure 8. Comparison between DPCap with reduced bollard-pull and more dynamic allowance and DynCap.

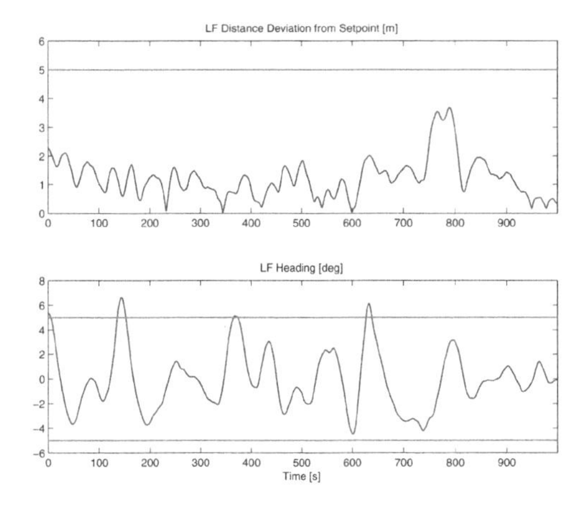


Figure 9. LF position and heading time-series for limiting wind speeds at 90 deg heading.

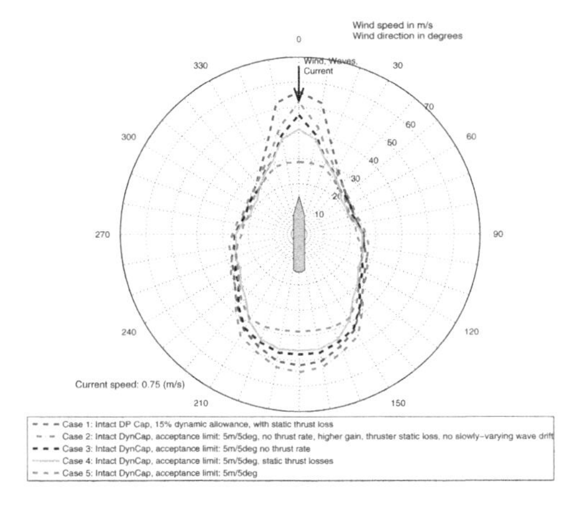


Figure 10. The effects of thruster dynamics, dynamics losses and slowly-varying wade drift on DynCap analysis.

more dynamic allowance, the traditional method still produces larger wind speed limits for head sea. For following sea the results from DPCap are in between the 5 m/5 deg and 20 m/20 deg limits. On the contrary, for other headings, the DPCap gives a smaller wind envelope. This result confirms the conclusion drew from the time series given in Figure 7 and Figure 9, where it was shown that the vessel dynamics is more significant for head and following seas than for other headings. This means that the dynamic allowance should not be constant but heading dependent.

Figure 10 shows the results from different DynCap analyses considering the effect of different dynamic phenomena.

These are compared to the nominal DynCap computed for 5 m/5 deg acceptance criteria (red dashed line) and DPCap with 15% of dynamic allowance (blue dashed line).

When the thrust dynamic losses such as ventilation and wave-induced velocity are not included the DynCap wind envelope increases as expected (green line in Fig. 10). The increase is significant but not enough to explain the difference for head seas with respect to the DPCap result.

Similar results were obtained when excluding the slowly-varying wave drift forces (mean wave drift forces are included).

To investigate the effect of thruster motor and azimuth angle dynamics, DynCap was run considering very fast dynamics (and correspondent very fast power generation). Without increasing the DP control gains, the vessel performance do not improve that much. This result is not reported Figure 10. With increased DP control gains, the capability is improved but for head seas the results are still far from the DPCap wind limits (black line in Fig. 10). It is important to note that the DP control gains cannot be increased too much as the position and velocity estimates from the Kalman filter contain noise. This shows that also the DP control system dynamics must be modelled with accuracy and the DP controller tuned with precision.

Last run was carried out with very fast thruster dynamics, no dynamic thrust losses, increased DP controller gains and no slowly-varying wave drift (only mean wave drift forces), see the magenta line Figure 10. The wind envelope for head sea is closer to the nominal DPcap envelope. In this condition the DP control system is able to use the thrusters more than for the other cases (60% of max thrust) resulting in a condition very similar to the quasi-static one. This proves that it is important to include the thruster and power generation dynamics as well as the dynamic thruster losses and slowly-varying wave drift forces otherwise the DP capability estimate may be too non conservative.

## 5 CONCLUSION

Estimating the vessel station-keeping performance has been always a challenge for vessel design and operation. The traditional DP capability analyses have been shown to have significant shortcomings. Dynamic Capability (DynCap) analysis has been developed as a new method to give more accurate estimates of the station-keeping capability, employing systematic time-domain simulations with a sophisticated closed-loop vessel simulator. Most of the limiting assumptions needed for the traditional DPCap analysis were removed, yielding results that are expected to be much closer to reality.

The paper has showed the importance of the inclusion of the complete vessel, environmental forces, and control system dynamics for obtaining realistic results. This was done by investigating the sensitivity of the vessel station-keeping capability to relevant dynamic effects and comparing results from the traditional DPCap and the new DynCap analysis. Future work is to improve the wave model for harsh seas, e.g. inclusion of breaking waves, and full scale validation.

## REFERENCES


Børhaug B. 2012. Experimental Validation of Dynamic Stationkeeping Capability Analysis. Master Thesis. NTNU.

Bulten N. & Stoltenkamp P. 2013a. DP-capability of tilted thrusters. Marine Technology Society (MTS) Dynamic Positioning Conference, Houston, US.

Bulten N. & Stoltenkamp P. 2013b. Full Scale Thruster-Hull Interaction Improvement Revealed With CFD Analysis. OMAE Conference, Nantes, France.

Det Norske Veritas (DNV). 2010. Recommended practice DNV-RP-C205: Environmental Conditions and Environmental Loads. October 2010.

Det Norske Veritas (DNV). 2013. Rules For Classification Of Ship, Part 6, Chapter 7, July 2013.

International Marine Contractors Association (IMCA). 2000. Specification for DP Capability Plots. IMCA M 140 Rev. 1 June 2000.

Nguyen D., Pivano L., Børhaug B., & Smogeli Ø. 2013. Dynamic station-keeping capability analysis using advanced vessel simulator. 54th SIMS conference on Simulation and Modelling, Bergen, Norway.

Phillips D. & Muddesetti S. 2006. A Practical Approach to Managing DP Operations. Marine Technology Society (MTS) DP Conference, Houston, US.

Pivano L., Børhaug B., Smogeli Ø. 2013. Challenges in estimating the vessel station-keeping performance. European Dynamic Positioning Conference, London, UK.

Smogeli Ø, Nguyen D., Børhaug B., Pivano L. 2013. Marine Technology Society (MTS) Dynamic Positioning Conference, Houston, US.

Smogeli Ø. 2006. Control of Marine Propellers. From Normal to Extreme Conditions. PhD-thesis. NTNU.

WAMIT. 2008. WAMIT User Manual. Versions 6.4, 6.4PC, 6.3S, 6.3S-PC. WAMIT Inc.

*Maritime-Port Technology and Development – Ehlers et al. (Eds)*
*© 2015 Taylor & Francis Group, London, ISBN 978-1-138-02726-8*

# Possibilities to determine design loads for thrusters and drivetrain components using the flexible multibody-system method

B. Schlecht & T. Rosenlöcher
*Institute of Machine Elements and Machine Design, Chair of Machine Elements, Technische Universität Dresden, Germany*

ABSTRACT: The usage of modern thrusters allows the combining of the drive and the ship rudder in one unit. The propeller can be driven directly or indirectly. The present paper concentrates on indirect drives where the driving torque is transferred by bevel gear stages and shafts from the motor to the propeller. Due to their closed and inaccessible construction, high reliability has to be achieved. Especially for the design of the highly-loaded bevel gear stages accurate information of the occurring loads is required. The available experience of the operation of thrusters shows, that primarily rarely occurring special load cases must be considered in the design process. Such operational conditions can only be determined by expensive long-term measurements. By means of a detailed multibody-system simulation model of the thruster, it is already possible to develop a basic knowledge of the dynamic properties of the drivetrain and to determine design loads for drivetrain components.

## 1 INTRODUCTION

The different drive train and ship concepts, the complicated operational conditions and the high demand on the reliability lead to many different tasks and conditions which have to be considered in the design process of thrusters. Therefore the occurring operational conditions are analyzed using simple torsional oscillation models of the drivetrain up to now (DNV 2014). In addition to the typical concept where the fixed propeller is driven by a long shaft, water jet engines, thrusters and also special solutions like the Voith-Schneider drive are used. The thrusters are mounted underneath the ship hull and the thruster housing can rotate around the vertical axes, so that they can be used as pushing or pulling drive and also as ship rudder. The driving power can directly be supplied by an electrical motor, installed in the nacelle (ABB, Rolls-Royce). Alternatively the driving torque is transferred by gearboxes and long shafts from the driving unit in the ship hull to the propeller of the thruster (Schottel, Rolls-Royce, Wärtsilä). Thruster which are driven by an electrical motor or a combustion engine using a gearbox are able to operate with a constant driving speed if the provided thrust is adjustable by pitchable propeller blades. Due to the good maneuverability thrusters are used in ferries and tug boats. Thrusters are also often used if high demands on the positioning accuracy are required by ships for gas- and oil production as well as for scientific marine research.

In comparison to the typical driving concepts using a long shaft, changed design loads have to be considered due to the combined function of driving and steering as well as the paired arrangement on both sides of the ship (Fig. 1).

Further the discontinuously operation and the area of application can have influence on the design process. Next to the occurring torques also bending moments around the vertical axes resulting from the steering movements of the thruster have to been taken into account for the different operational conditions. A relevant design load case results from the positioning of the thrusters on both sides of the rear. In high waves an emersion of

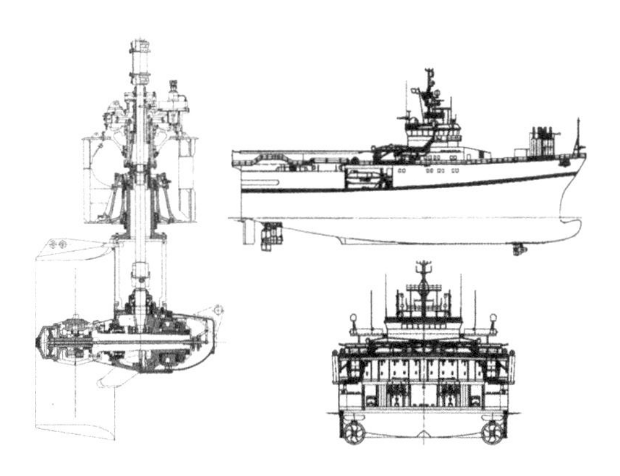

Figure 1. Positioning and design of the thruster.

propeller can occur so that during the immersion the blade tips are slamming on the water surface. This leads to short time overloads which have to be transferred and supported by the drivetrain without damages.

The determination of the occurring drivetrain component loads and analysis of the dynamic behavior can be achieved either by a complex measurement setup in the thruster nacelle or with the aid of detailed simulation models (Schlecht et al. 2013). The challenges of a measurement campaign are the difficult environmental conditions and missing accessibility to install sensors after the assembly of the thruster. So, a measurement setup is time consuming and expensive. An availability of detailed measurement results for different drivetrain components will be an individual case and not be applicable to design thrusters. The determination of the component loads using the simulation results of complex drivetrain models can already be performed during the product development process.

## 2 BASICS OF THE DRIVETRAIN SIMULATION

The analysis of drivetrains operating under high dynamic loads presupposes the assembly of a detailed simulation model which is able to represent the dynamic behavior of the drivetrain in the frequency and time domain. Even if high performance computers are available the level of detail of the simulation model has to correspond to the formulated question to ensure a feasible calculation effort. Despite the currently given possibilities of the simulation software the modelling process is very time-consuming. Basing on the present data of the drivetrain a discrete simulation model has to be assembled. A successive and modular assembly of fully parameterized simulation models allows a clear and reproducible modelling process compared to the combination of all drivetrain components in one unstructured model.

The modular concept requires in a first step the decomposition of the drivetrain into its substructures. According to this approach a simulation model of a thruster consists of the following substructures: propeller, propeller shaft, coupling, motor and an additional subdivided gearbox. Further the gearbox can be subdivided in different spur and helical gear stages, bevel gear stages and planetary gear stages whereby in the analyzed thruster only bevel gear stages are used (Fig. 2). Each substructure consists of model components which can be subdivided into shafts, gear stages, bearings and supporting structures. The combination of single substructures finally leads to

Figure 2. Different kinds of substructures.

the complete simulation model of the thruster. Dependent on the present requirements an adjustment of the level of detail and the needed degrees of freedom for each submodel can be performed. Compared to the work with one single simulation model for the complete drivetrain, the usage of different submodels enables an easy verification of the function and accuracy (Rosenlöcher 2012), (Schlecht & Rosenlöcher 2013).

## 3 ASSEMBLY OF THE SIMULATION MODEL OF THE THRUSTER

The assembly and functions of a thruster should be shown exemplary using the sectional drawing in Figure 1, left. The nacelle with the function of the gearbox and cover of the drivetrain is mounted rotatable around the vertical axes in the ship hull and can be turned by an additional drive. The main driving machine in the ship hull transfers the required torque for a horizontal positioned driving aggregate by an elastic coupling and a bevel gear stage. For a vertical mounted driving machine the torque is transferred directly by a coupling to the vertical driveline in the nacelle. The segmented driveline is supported by several bearings in the housing. The shaft segments as well as the pinion of the bevel gear stage are connected by geared couplings. The wheel of the bevel gear stage is mounted on a carrier. The carrier is directly connected to the propeller shaft, which is supported by a roller and a sliding bearing. The axial mounted hub is used to support the four pitchable propeller blades, which can be positioned by a hydraulically acting linkage.

The dynamic behavior of the drivetrain is mainly characterized by the large motor- and propeller side mass moment of inertias as well as the high flexibility. Dependent on the required thrust propeller diameter up to 5 meters are installed. The occurring torque and bending moment during operation presuppose a stiff design of the propeller shaft. By contrast small diameter shafts are used in the vertical driveline because the gear stage ratio lowers the torque and the resulting stress. The lower torsional and bending stiffness of these shafts has to be taken into account. Further the motor is connected by an elastic coupling with the drivetrain. In a simplified manner the complete system can be described by two large mass moments of inertia, connected by a soft torsional stiffness. Additionally all drivetrain components are supported in the nacelle, which is also an elastic system and only connected at the top to the ship hull. Under consideration of the acting forces, the torque and the bending moments a simulation model representing the torsional degrees of freedom can only be used for a simplified and rough analysis of the dynamic behavior. A comprehensive investigation of the entire system requires a detailed modelling of all relevant degrees of freedom in a simulation model.

All shafts of the drivetrain have to be modelled by taking into account the torsional and bending stiffness, the mass and mass moment of inertia as well as the rotatory and translatory degrees of freedom. The determination of the mass parameters can be performed using common three-dimensional CAD-software or by means of simple analytical approaches. A higher effort is demanded to calculate the stiffness of the components. The torsional stiffness of the drivetrain is mainly characterized by the flexibility of the shafts. Especially thin shafts with their elastic properties have to be considered. Additionally the bending stiffness of such shafts can have non negligible influences on the dynamic behavior and the occurring displacements. A simulation model which is required to represent shafts can be assembled by means of the method of discretization, by the implementation of beam models or by using modally reduced elastic structures (Fig. 3), (Claeyssen & Soder 2003), (Dresig 2006), (Ruge & Birk 2007), (Wünsch & Carcia del Castillo 1986).

The consideration of axial and radial degrees of freedom supposes the modelling of the bearings. Essentially the modelling of the bearings is realized by a force element which introduces the reaction forces in the axial and radial directions as well as

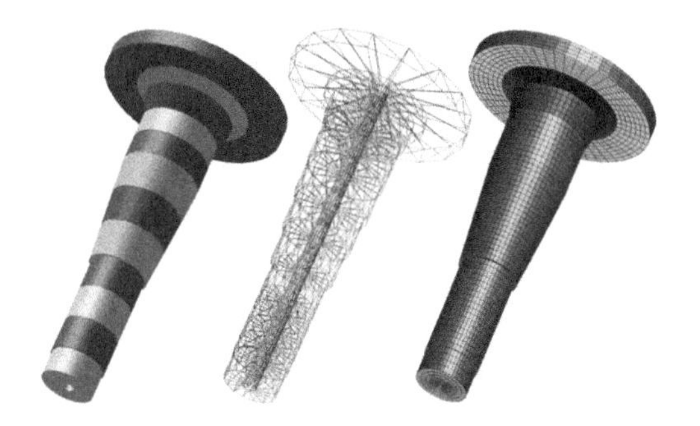

Figure 3. Possibilities to model the elasticity of a shaft.

the reaction moments if necessary. The bearing properties can be described by average bearing stiffness, characteristic curves or complex models imported as DLL's (Wiche 1967).

To support the shafts in the thruster housing the bearings are modelled as translatory spring-damper elements, whereby the load dependent bearing stiffness is implemented, so that for all occurring load cases the approach can be used. Also the properties of the elastically and geared couplings are described by spring-damper elements. For the motor side coupling information of the stiffness characteristics is required which must be provided by the manufacturer. According to the current knowledge the stiffness of geared couplings can only be determined by using analytical approaches (Benkler 1970), (Fleiss 1977), (Heinz 1976). Information on the radial stiffness and stiffness against inclination due to the comprehensive influence factors and uncertain calculation methods is not available. As first approach all possible degrees of freedom are locked by constraints or high stiffness and the occurring influences on the dynamic behavior have to be determined by sensitivity analysis.

The model of the bevel gear stage between the vertical driveline and the propeller shaft must describe the transfer behavior for the torque as well as for all force components so that the dynamic properties of the complete drivetrain can be represented correctly. Next to the description of the nonlinear characteristic of the stiffness resulting from the changing contact conditions, the backlash has to be considered during the calculation of the acting forces. For each step of the integration the determination of the equilibrium between the acting forces, displacements, the inclination of the shafts and the inner gearing forces must be ensured. The simulation software SIMPACK offers the tool boxes GEARWHEEL and GEARPAIR to model gearings and to describe the transfer behavior in detail.

An alternative modelling approach offers the mathematical description of the resulting forces in the gearing by means of user routines. Based on the calculation of the tooth normal force in the ideal pitch point, the complete tooth contact is simplified and described in one point. The tooth normal force consists of stiffness and damping dependent parts. Information about the displacements and velocities in tangential, radial and axial directions resulting from the relative position of the gears can be determined by the joint states and the corresponding trigonometric relationships. The gearing stiffness can be considered as average contact stiffness according to DIN 3990 and variable gearing stiffness over the path of contact using Fourier coefficients (Fig. 4).

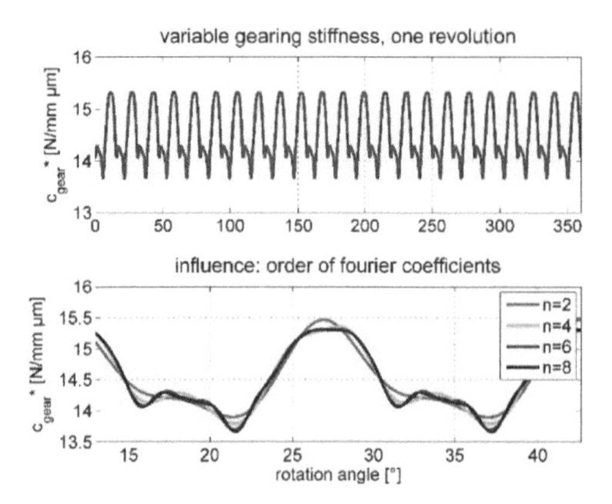


Figure 4. Variable gearing stiffness over the contact path using Fourier coefficients.

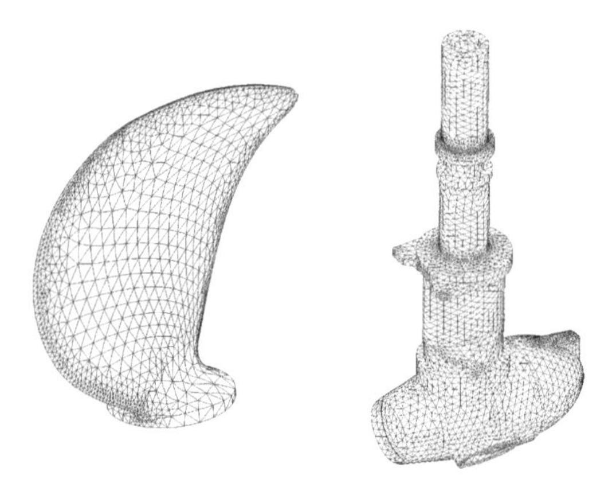

Figure 5. Finite-element model of the propeller blade and the thruster housing.

The rigid modelled hub is mounted axially on the propeller shaft and supports the four pitchable propeller blades. To consider the flexibility and resulting deformations of the blades under the high loads, the material and shape dependent elasticity has to be considered. Due to the complex geometry the method of discretization or beam approaches cannot be used so that on basis of a detailed finite-element model and the dynamic reduction the propeller blades are represented by modal reduced elastic structures (Fig. 5, left).

The implementation of a flexible structure in SIMPACK is based on a meshed finite-element model of the component geometry and the definition of the material properties. Additionally the modelling of the connection points between the elastic structure and the rigid bodies of the multibody-system model is required. The connection points to the supporting spring-damper elements can be modelled by means of multipoint constraints (MPC). However, the resulting FE-model is assembled by many shell or solid elements and has therefore much more degrees of freedom as necessary to describe the rigid body motions in the MBS model. Because such complex models cannot be handled by a classic MBS solver, the level of detail of the finite-element model has to be reduced to the transfer behavior between the connection points. All additional information to the displacement of nodes, which are not used as connection points in the MBS model, are not available in the reduced model of the structure. The application of the reduction approach according to Craig-Bampton requires the definition of the connection points between the flexible structure and the rigid bodies. The mode shapes of the reduced model are used to determine the deformation under load (Craig & Bampton 1968), (Guyan 1965). The number of natural frequencies chosen for the modal reduction defines the valid frequency range and the accuracy of the model, which is also influenced by the choice of frequency response modes in SIMPACK (Heckmann et al. 2006).

A comparable proceeding has to be performed to represent the elastic properties of the thruster housing. The large mass of the propeller and the propeller shaft with the bevel gear wheel as well the propeller side loads have to be supported by the structure of the thruster housing and have to be transferred to the large bearing in the ship hull. The expected deformations under the load will have an influence on the dynamic behavior of the drivetrain. Based on the geometry of the thruster housing a finite-element model can be assembled (Fig. 5, right). The connection points for the support in the ship hull and the positions of the bearings for the drivetrain components are linked by constraints to a number of surface nodes in the area of the bearing seats. After the implementation of the reduced finite-element model the spring-damper elements which are representing the bearings will be defined between these connection nodes and the body marker of the MBS model. So all introduced loads are directly transferred as torque or supported by the bearings in the thruster housing and the ship hull.

## 4 ANALYSIS OF THE THRUSTER DRIVETRAIN IN THE FREQUENCY DOMAIN

To realize the described modularization consequently, the simulation model of the thruster consists of the submodels motor, coupling, the vertical driveline, the bevel gear stage, the propeller shaft and the propeller. All components are assembled in a complete model of the thruster and supported in the modally reduced finite-element model of the thruster housing. The release of all degrees of freedom and consideration of all supporting and connecting spring-damper elements allows by comparison of natural frequencies and excitation frequencies the determination of critical operational speeds and the analysis of the dynamic behavior of drivetrain components and the supporting structure (Fig. 6). Possible excitations are the rotation frequency of all drivetrain components in the first and second order, the gear meshing frequency of the bevel gear stage with the first order and higher harmonics, the rotation frequency of the propeller with the first order and higher harmonics corresponding to the number of installed blades and disturbance of the flow due to the nacelle design. The named sources can excite torsional, bending, radial and axial mode shapes which has to be analyzed for each determined critical operational speed. Especially the propeller side excitations have an important influence on the dynamic behavior of the complete system because the torque as well as the acting forces can lead to resonances with different harmonics of the rotation frequency of the propeller shaft.

Figure 7 shows exemplary the Campbell diagram for the thruster with all natural frequencies and the first (1p), second (2p), third (3p), fourth (4p) and eighth (8p) order excitation of the propeller rotation speed as well as the first (1p) and second (2p) order excitation of the gear meshing frequency up to 140 Hz. The first natural frequency of the complete system at 10 Hz is characterized by a bending mode shape of the thruster housing against the support in the ship hull. The fourth order of

Figure 6. Mode shapes of the thruster (10 Hz, 11 Hz).

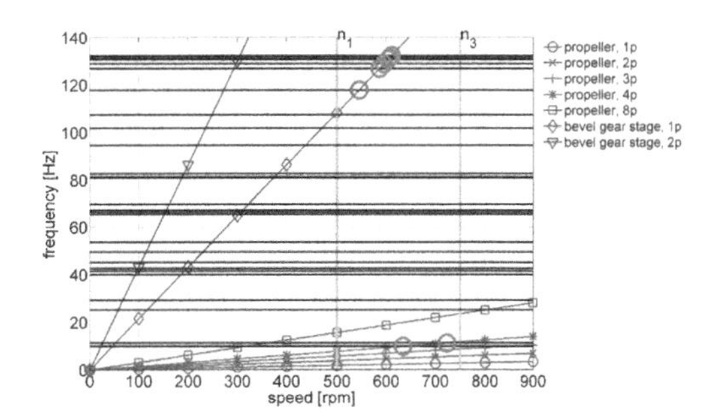


Figure 7. Campbell diagram for the flexible multibody-system model of the thruster.

the propeller rotation frequency could cause a resonance with this mode shape at an operational speed of 635 rpm (Fig. 6, left). In addition to the stiffness of the housing the stiffness of the bearing which supports the thruster in the ship hull has an influence on the mode shape, too. The first torsional mode shape of the drivetrain at 11 Hz is superposed by a second bending mode shape of the housing. This natural frequency can also be excited by the fourth order of the propeller rotation frequency at an operational speed of 720 rpm (Fig. 6, right). The mentioned excitation frequency is caused by flow disturbance, which occurs when a blade passes the thruster housing. The changing torque, bending moments and forces can lead to an excitation of both mode shapes.

Also the higher natural frequencies of the drivetrain are characterized by the superpositioning of housing and drivetrain mode shapes and can be excited by higher harmonics of the propeller rotation frequency. For the analyzed drivetrain only the fourth and eighth order of the propeller rotation frequency are relevant. For the further investigations the interesting frequency/speed range is limited by the lower border of the operational speed and the gear meshing frequency of the bevel gear stage at approximately 110 Hz.

In the analysis of the higher frequency range all possible intersections between the gear meshing frequencies and the determined mode shapes have to be taken into account. Due to the exciting gearing force components in tangential, radial and axial direction an excitation of torsional, axial and bending mode shapes as well as mode shapes against the shaft support are possible. In addition to the theoretically investigations using the Campbell diagram, a comprehensive evaluation of the excitability can only be performed by the simulation of a slow run up and a detailed analysis of the simulated velocities, accelerations and torques.

## 5 ANALYSIS OF THE THRUSTER DRIVETRAIN IN THE TIME DOMAIN

Besides the analysis of possible excitations of natural frequencies in the range of the operational speed by means of the detailed simulation model the occurring loads for all drivetrain components and different operational conditions can be analyzed. This requires an enlargement of the mechanical model to characterize the acting motor and propeller side loads in detail.

The description of the electric motor can be realized by modelling the different control loops in MATLAB/SIMULINK. Important for a realistic motor model is the knowledge of all motor parameters. These parameters must be provided by the manufacturer of the electric motor. In the case of the presented thruster only some rough information about the motor is available, so that by means of speed-torque characteristics a simplified model is used to describe the motor behavior. The modelling of the propeller side loads requires a comprehensive discussion of the proper modelling approach. To analyse a simple torsional vibration model the information regarding the occurring torque is sufficient and allows already a first evaluation of the dynamic behavior and testing of the model. The simplified consideration of the torque neglects the important influences of the thrust forces and bending moments which are also applied on the propeller. These load components cause the bending of the thruster housing, the displacements of the propeller shaft and have an impact on the contact conditions in the bevel gear stage, too. Next to the analysis of operational states under full load for different flow angles, especially extreme load cases like the immersion of the propeller at maximum input power can be seen as critical for the reliable operation of the thruster and should be investigated with the model (Mork 2006). Until today no comprehensive measurement results for such thrusters are available so that the occurring loads during the immersion and emersion of the propeller were analyzed only with scaled models in water tanks. When the propeller approaches the water surface, the surrounding water is already mixed with air, so that the thrust and the acting torque decrease. The speed of the motor increases due to the lower resisting torque. A further emersion of the propeller causes at first a water free movement of the upper blade. At the moment of the blade immersion the resisting force increases suddenly which leads to a short-time increase of the torque in the drivetrain (Koushan 2007), (Koushan et al. 2009), (Kozlowska et al. 2009).

To determine the occurring component loads during such load cases, a very detailed propeller force model is necessary. The introduction of the water

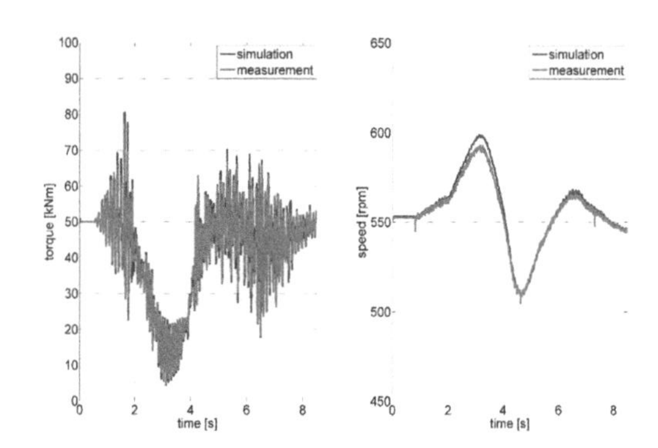


Figure 8. Time series for the immersion and emersion of the thruster.

resisting forces is carried out by modelling a discrete load distribution over the blade length for the tangential and axial force component, separable for each blade. The calculation of the acting forces for each load introduction point occurs dependent on the rotation angle, the distance from the water surface and measurement based assumptions for the force progression by means of a MATLAB/SIMULINK model. The introduction of the resulting forces on the propeller blades in the MBS model allows a first analysis of the occurring loads for shafts, bearings and gearings. Figure 8 shows the torque of the propeller shaft and the motor speed over the simulation time as a comparison between measurement and simulation results.

Further possibilities for the description of the propeller side loads are given by the computational fluid dynamics. The method is used to analyses the ship hull and the interactions between ship hull and thruster already during the design process. Because of the detailed, computational intensive models a CFD simulation can only be done for single revolutions of the propeller and defined environmental conditions. Due to the long simulation time by using the currently available computer performance a combination of CFD- and MBS-simulation is not applicable for the dynamic simulation. But the propeller forces and torques can be precalculated for defined quasi static load cases and introduced in the MBS model using force elements (Polter 2010). To improve the propeller force models and to validate the mechanical simulation model the measurement of the different real occurring operational loads by means of an extensive measurement setup over a long period is mandatory.

## 6 CONCLUSION

The described methods can be used to model the mechanical components as well as the acting propeller loads, so that basic statements on the dynamic behavior of the thruster drivetrain can be given. The comparatively simply constructive design of the drivetrain is characterized by the high flexibility of the driveline and the thruster housing. This leads in combination with the large propeller and motor mass moment of inertias as well as the acting forces to high dynamic states in the drivetrain. If a sudden increase or decrease of the propeller side torque occurs, first the twist of the drivetrain must be resolved, before the backlash in the gearings or couplings can affect the dynamic drivetrain behavior. The occurrence of back flank contact in the bevel gear stage gets possible, if the propeller load changes with an amplitude and for the duration, so that a twist free drivetrain exists. The motor side coupling can reduce overloads at the motor. Damages in the drivetrain can be avoided only by an overload protection at the propeller shaft. The challenge is the design of an overload protection which can limit the torque reliable. It has to be small enough for an installation in the available space in the thruster housing.

## REFERENCES


Benkler, H. 1970. *Berechnung von Bogenzahnkupplungen.* Fortschritt-Berichte VDI, Reihe 1, Nr. 27. Düsseldorf: VDI Verlag.

Claeyssen, J.R. & Soder, R.A. 2003. A dynamical basis for computing the modes of Euler-Bernoulli and Timoshenko beams. In *Journal of Sound and Vibration* 259(4), p. 986–990.

Craig, R.R. & Bampton, M.C. 1968. Coupling of Substructures for Dynamic Analyses. In *AIAA Journal* (1968), Vol. 6, No. 7, p. 1313.

DET NORSKE VERITAS AS 2014. *Nauticus Machinery Torsional Vibration.* Høvik: DNV.

Dresig, H. 2006. *Schwingungen mechanischer Antriebssysteme.* Berlin: Springer.

Fleiss, R. 1977. *Das Radial- und Axialverhalten von Zahnkupplungen.* PhD dissertation, Technische Hochschule Darmstadt.

Guyan, R.J. 1965. Reduction of Stiffness and Mass Matrices. In *AIAA Journal*, Vol. 3, No. 2, p. 380.

Heckmann, A. et al. 2006. *The DLR FlexibleBodies library to model large motions of beams and of flexible bodies exported from finite element programs.* Wien: Modelica 2006.

Heinz, R. 1976. *Untersuchung der Kraft- und Reibungsverhältnisse in Zahnkupplungen für große Leistungen.* PhD dissertation, Technische Hochschule Darmstadt.

Koushan, K. 2007. Dynamics of Propeller Blade and Duct Loadings on Ventilated Ducted Thrusters Due to Forced Periodic Heave Motion. In *Proceedings of International Conference on Violent Flows.* Kyushu University, Fukuoka.

Koushan, K. et al. 2009. Experimental Investigation of the Effect of Waves and Ventilation on Thruster Loadings. In *First International Symposium on Marine Propulsors.* SMP'09, Trondheim.

Kozlowska, A.M. et al. 2009. Classification of Different Type of Propeller Ventilation and Ventilation Inception Mechanism. In *First International Symposium on Marine Propulsors.* SMP'09, Trondheim.

Mork, L.R. 2006. *Slamming on propeller blade.* Pre project thesis, NTNU.

Polter, M. 2010. *Dynamische Analyse eines Strahlruderantriebes unter Berücksichtigung detaillierter Lastannahmen auf Grundlage von CFD-Berechnungen.* Diploma thesis, Technische Universität Dresden.

Rosenlöcher, T. 2012. *Systematisierung des Modellierungsprozesses zur Erstellung elastischer Mehrkörpersystem-Modelle und dynamischen Untersuchung von Großantrieben.* PhD dissertation, Technische Universität Dresden.

Ruge, P. & Birk, C. 2007. A comparison of infinite Timoshenko and Euler-Bernoulli beam models on Winkler foundation in the frequency- and time-domain. In: *Journal of Sound and Vibration* (304), p. 932.

Schlecht, B. et al. 2013. *Comprehensive Analysis of Thruster Drivetrains to Determine Reliable Load Assumptions.* Hamburg: International Conference on Computational Methods in Marine Engineering—MARINE 2013.

Schlecht, B. & Rosenlöcher, T. 2013. *Möglichkeiten zur Ermittlung maximaler Beanspruchungen für Antriebsstrangkomponenten in Strahlruderantrieben durch Einsatz elastischer Mehrkörpersystem-Modelle.* Dresden: Dresdner Maschinenelemente Kolloquium.

Wiche, E. 1967. Radiale Federung von Wälzlagern bei beliebiger Lagerluft. In: *Konstruktion* (19), Heft 5, p. 184–192.

Wünsch, D. & Carcia del Castillo, L. 1986. *Experimentelle Modellfindung und modellhafte Ermittlung dynamischer Belastungen torsionsschwingungsfähiger Systeme.* Frankfurt am Main: FVA Nr. 95, Forschungshefte 213 und 214 der Forschungsvereinigung Antriebstechnik e.V.

*Maritime-Port Technology and Development – Ehlers et al. (Eds)*
*ISBN 978-1-138-02726-8*

# Computational Fluid Dynamics simulations of propeller wake effects on seabed

V.T. Nguyen, H.H. Nguyen & J. Lou
*Institute of High Performance Computing, ASTAR, Singapore*

L. Yde
*DHI Water and Environment, Singapore*

ABSTRACT: During operations of propellers in restricted areas such as near port, coastal areas bounded by the seabed, propeller jets may cause seabed scouring as well as damage to port structures. This phenomena is also known as propeller wash, which strongly affects sediment dynamics in most navigable estuaries and rivers; causing other environmental impacts, such as sea bed erosions, sediment re-suspension, that damages corals and other water lives. In this work Computational Fluid Dynamics (CFD) simulations is employed as a tool to investigate the dynamics of propeller jets under restricted operation conditions; thus shedding light on investigations of mechanism of sediment pick-up, re-suspension, scour as well as providing quantitative assessment of erosion and scour caused by propeller jets.

## 1 INTRODUCTION

Propellers are designed to provide main thrust to accelerate ships by drawing waters and then discharging this water body downstream, known as propeller jets. Energy from these propeller jets dissipates into the surrounding waters as the ship moves along. When a ship maneuvers in shallow water under restricted areas such as near port, coastal areas, the seabed thus causing seabed scouring as well as damaging to port structures restricts the movement of the jet. This phenomenon is also known as propeller wash. It is believed that the propeller wash strongly affects sediment dynamics in most navigable estuaries and rivers; thus causing other environmental impacts, such as seabed erosions, sediment re-suspension that damages corals and other water lives.

Understanding of propeller wash induced scour and its consequences are crucial to engineers and authorities in management and planning of ports and harbors. Propeller wash problem is complex, large scale in nature and research related to this area has been actively carried out to gain better understanding of the phenomena. There have been a number of works reported in literature concerning propeller wash and its effects on seabed scouring such as (Hamill 1987), (Hashmi 1993), (Hamill & McGarvey 1996), (Stewart et al. 1991) and (Lam et al. 2011). However, understanding of its impact is far from complete and it is still limited to mostly model scale experiment and empirical relations. In this current work we propose to use Computational Fluid Dynamics (CFD) simulations as a tool to investigate the dynamics of propeller jets under restricted operation conditions. Advances in numerical modeling techniques as well as computational resources enable us to build a full-scaled propeller model for studying water jets generated from propellers while interacting with rudder as well as seabed. Therefore, it is able to shed lights on the understanding of the propeller wash impact on related sediment transport phenomena. The main objective of the work will be on simulations of propeller hydrodynamics; thus enabling investigations of mechanism of sediment pick-up, re-suspension, scour as well as providing quantitative assessment of erosion and scour caused by propeller jets.

In this work, various models were firstly explored for simulations of propeller jets with sliding mesh and multiple reference frame approach. These models were employed for simulations of different scenarios of propeller rotating speed, rudder position, ship hull wakes as well as keel clearance in order to understand the impact of jet flows on sediment transports. Validation and comparison of different models for simulations of propeller-rudder interactions were conducted with various rotational speed measured in Revolutions Per Minute (RPM), rudder angle and keel clearance.

## 2 METHODOLOGY

Complexity of flow fields over propellers is attribute to the presence of rotating parts and its interaction with stationary ones such as ship hull and rudder.

The rotation of rotor blades dominated by mutual hydrodynamics interferences of blades ensures unsteadiness in the flow field. The rotors, at the same time, interact with rudder and hull in a complex, non-linear way making it challenging for numerical simulations. Many approaches have been developed for simulations of propellers including the earlier work on potential flows dated back in 1980s. The more advanced and detailed simulations are obtained by solving full Navier-Stokes equations taking account the rotation of the blades and its interaction with rudder and hull. The flow is modeled as viscous incompressible fluids governed by the incompressible Navier-Stokes equations that express the conservation of mass and momentum. The water is considered as Newtonian fluid of density $\rho = 1000\ kgm^{-3}$ and kinematic viscosity of $\nu = 1.0 \times 10^{-6}\ m^2\ s^{-1}$. The flow is characterized by Reynolds number, ($Re = U_{ref}L_{ref}/\nu$) where $U_{ref}$ and $L_{ref}$ are the reference velocity and length scale. The equations are closed with appropriate boundary conditions imposed on the boundary of the domain. Details on boundary conditions for the current applications are provided in the subsequent section.

In this work, the Reynolds Averaged Navier Stokes (RANS) approach was used and appended with suitable turbulence models for resolving of turbulence quantities. In RANS turbulence modeling, turbulence fluctuations in the flow is averaged out and regarded as part of the turbulence. Although known as a less detailed turbulence simulation method, RANS solutions show reasonable resolutions to turbulence flows, especially for relatively large Reynolds number. At the given length and time scale of the applications in this project, we believe that a RANS approach provides sufficient resolutions of turbulence quantities for the current simulations. In the context of RANS simulations, there are many different turbulence models developed in literature (Wilcox 1998) from algebraic models to one-equation and two-equation models. Among those, two-equation models have been the industry stand of turbulence modeling for many years. Most of these two-equation models solve a transport equation for turbulent kinetic energy (k) and a second transport equation that allows a turbulent length scale to be defined. The most common forms of the second transport equation solve for turbulent dissipation $k - \varepsilon$ models or turbulent specific dissipation ($\omega \sim \varepsilon/k$), known as $k - \omega$ models.

### 2.1 *Propeller modeling*

Flows in the presence of propeller as well as its interactions with other stationary part like rudder, seabed is complex in nature. The analysis of flow fields in the presence of propeller requires a three-dimensional flow solver capable of modeling the actual motion of rotors and its interaction with stationary components. Generally, the Navier-Stokes equations are discretized and solved on either structured or unstructured grids where the relative motion between propeller, rudder and seabed needs to be modeled. There are effectively two separate domains in this type of simulations, one moving domain attached to the propeller and the other stationary domain containing the computational domain with the rudder, seabed. To model the relative motion of propeller version rudder, there are various techniques; notably Multiple Reference Frame (MRF) and sliding mesh approach.

In MRF approach, the Navier-Stokes equations are solved on different reference domains for rotating and stationary parts. It is apparent that the Navier-Stokes equations for a fluid in the inertial (laboratory) reference frame cannot be directly applied to the reference frame spinning with the rotor. Thus, a co-relation between the two reference frames needs to be established and simultaneously changes must be made to the Navier-Stokes equations. This approach is simple in treating rotating domain since it only requires an addition of Corriolis term to take into account the effect of rotating frame. Due to this simplification, it can only predict the steady state flows and ignore unsteady effects arising from rotating parts.

In this approach, the interface between the two parts is called the sliding interface. At each time step, in order to account for the rotation, the neighboring cells of the sliding interface are moved. The General Grid Interaface (GGI) (Beaudoin & Jasak 2008) is a coupling interface used for joining two non-conformal meshes in which patches and the mesh points on the either sides do not match with each other. For example, consider the case of flow of a liquid in a mixer vessel. The interface in between the rotor and the stator parts can be a GGI interface. Instead of re-meshing, GGI uses weighted interpolation methods to evaluate and transmit the flow variables across the coupled patches. This moving mesh approach is well able to provide full details of unsteady flow features of propellers in operations, as it is able to explicitly capture the boundary layer around the blades as well as propeller-rudder interactions. On the other hand, it is very computationally intensive to handle the overlapping and sliding meshes, especially for unstructured grids. It is increasingly more expensive when Reynolds Average Navier-Stokes (RANS) is considered for turbulence modelling. Results from sliding mesh model simulations will be applied to verify and modify the simple source term model. Results of these simulations will also be applied to develop a model for values of the turbulence model variables, in the source term model.

In this work, we employed the two main solvers in OpenFOAM for simulations of propeller and rudder interactions. The **turbDyMFoam** solver based on sliding mesh approach was used for unsteady simulations of open water test to provide benchmark results as well as thrust and torque coefficients for the momentum source model. The **MRFSimpleFoam** solver based on multiple reference frame approach was employed for steady state simulations of propeller-rudder interactions as well as propeller-rudder-seabed interactions. Incompressible NS flows were first initialized by potential flows using **potentialFoam** solver before the steady and unsteady simulations.

### 2.2 *Computational domain and mesh*

Hybrid meshes were used for steady and unsteady simulations in which the computational domain is partitioned into several regions including the rotor region, the rudder box, the seabed region, the upstream and downstream area as shown in Figure 1. In the rotor and rudder regions, unstructured meshes are used to cater for complex geometries of the propeller and rudder, while structure grids are used for the rest of the domain to save computational time. For unsteady simulations using sliding mesh approach, rotor and stator parts were generated separately with non-matching mesh topology. The General Grid Interface (GGI) is used to couple the two parts of the mesh. In order to appropriately resolve turbulent quantities at the seabed, a boundary layer was generated; see Figure 2 such that the condition of y+ is satisfied for the use of the selected turbulent model and wall function.

### 2.3 *Boundary conditions*

The computational domain includes boundary patches of inlet, outlet, and bottom defined as seabed, blades and rudder as well as side and top planes as shown in Figure 2. In present computation, inlet velocity condition is applied at inlet patch, outlet pressure for outlet patch, symmetric boundary conditions applied for side and top patches, noslip wall boundary conditions for bottom seabed as well as propeller and rudder surface.

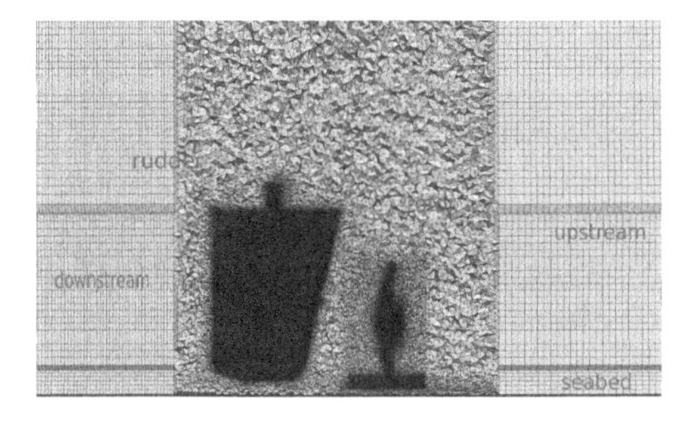


Figure 1. Computational domain is composed of several regions for mesh generation.

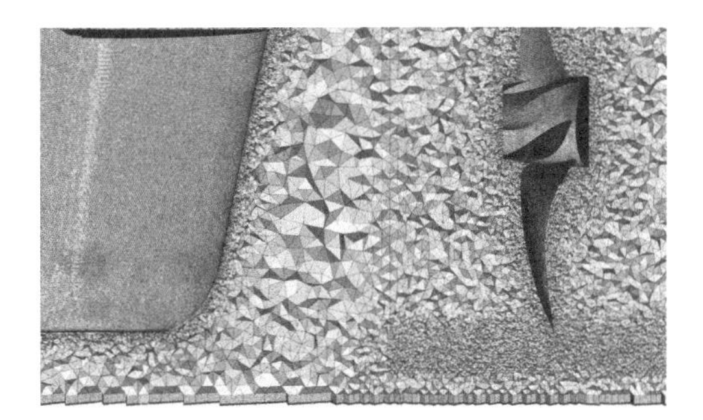

Figure 2. Boundary layer generated at the seabed to appropriately capture turbulent quantities.

**Inlet:** The velocity fields at the inlet are supplied and, for consistency, the boundary condition on the pressure is zero gradient. In addition, the turbulence kinetic energy as well as its dissipation rate is specified. The turbulence kinetic energy and dissipation rate are computed from the inlet velocity.

**Outlet:** The pressure field at the outlet is supplied and a zero gradient boundary condition on velocity is specified. Furthermore, zero gradient boundary conditions are applied to turbulence kinetic energy as well as its dissipation rate.

**Symmetry plane:** The boundary condition for any field is obtained by considering its mirror image, which simplifies to a zero normal gradient condition for scalar fields.

**No-slip wall:** The velocities of the fluids are equal to that of the wall, i.e. a fixed value condition is specified. The pressure is specified to be zero gradient since the flux through the wall must be zero. Zero gradient boundary conditions are also used for the additional scalars if appropriate. For turbulence kinetic energy and dissipation rate, the wall function boundary condition is applied.

Zero velocity is applied on velocity field on the rudder as well as propeller blades in the case of using MRF approach. In the sliding mesh approach, the rotating speed of the propeller is prescribed as velocity at the blades. As for the seabed, it is assummed in this project that the seabed is moving with the ship velocity while propeller and rudder is stationary. Therefore, ship velocity is prescribed at the seabed for velocity field. The propeller and rudder are considered as smooth wall, while seabed is prescribed as rough wall boundary for all of the simulations in this work.

## 3 NUMERICAL RESULTS

### 3.1 *Propeller in open water study*

First we conducted the study of propeller wake in open water set-up. This is to validate the models and compare with experimental data mostly done in a similar condition in labs. The propeller of Wageningen B series of 5 blades was considered in this case. This B series propeller were designed and tested at the Netherlands Ship Model basin in Wagenigen (Bernitsas et al. 1981). In this model, the propeller is placed in a channel of width and height of 3D where D is the diameter of the propeller. The length of the domain $L = D_i + D_w$ varies from 6D to 20D where $D_i$ and $D_w$ are the distance of inlet and wake area.

The performance of Wagenigen propeller was first investigated using CFD simulations. The sliding mesh approach (GGI) was employed for simulations of flows in the presence of rotating propeller. The propeller is placed in an open water set up of with $D_i$ = 2D, $D_w$ = 4D with different operating conditions of rotating speeds and incoming water velocity. To characterize performance of the propeller the advanced coefficient J is defined as

$$J = \frac{V_a}{nD}$$

where n is the rotating speed in revolutions per second, D is the propeller diameter and Va is the inlet velocity. Simulations were conducted for a range of advanced coefficient J = 0.1, 0.2, 0.4, 0.6, 0.88, 1.0 where the rotation speed of the propeller is fixed at RPM = 90.

Table 1 shows the grid sensitivity for flows over the B5 Wagenigen propeller at J = 0.88. The mesh is refined by changing the mesh size around the blade ($h_b$) and mesh size of the wake area ($h_w$). The results showed convergence of thrust and torque coefficients when the mesh is refined from a corse mesh of 1 million elements to the finest mesh of 5 millions elements. It can be seen that the torque and thrust coefficient are converged in refining the mesh, at $h_b$ = 0.1 m and $h_w$ = 0.2 m, the difference in toque and thrust is within 1% compared to the finest mesh. Therefore, in this study $h_w$ = 0.2 m and $h_b$ = 0.1 m were used for all the simulations.

Results for thrust, torque coefficients and efficiency of the propeller are shown in Figure 3 for the range of advanced coefficient. The simulated results are compared with experiment (Bernitsas et al. 1981) with good agreement on a reasonable mesh resolution.

Table 1. Torque and thrust coefficients in grid independence study for simulations of 5-blade Wagening propeller.

| Mesh | $h_w = 0.5$ $h_b = 0.25$ | $h_w = 0.25$ $h_b = 0.15$ | $h_w = 0.15$ $h_b = 0.10$ | $h_w = 0.2$ $h_b = 0.10$ | Exp. |
|---|---|---|---|---|---|
| N. Elem | 0.9580 | 2.4900 | 4.95 | 3.2 | – |
| $K_t$ | 0.1088 | 0.1171 | 0.1172 | 0.1184 | 0.1022 |
| $K_q$ | 0.0253 | 0.0265 | 0.02615 | 0.02615 | 0.0205 |

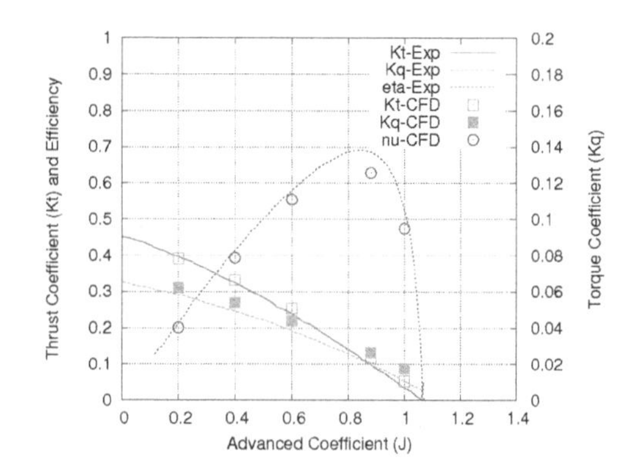


Figure 3. Comparison of thrust, torque coefficient and efficiency between simulations and experiment for the 5-blade Wagen propeller.

### 3.2 *Propeller wake structures*

Propeller wake plays an important role in ship design and ocean engineering. Complex structures of wake behind a propeller have been a subject of intense study over the years. As it is described in (Lam et al. 2011), "the regions within a jet can be divided into two zones, the Zone of Flow Establishment (ZFE) which lies close to the propeller face, followed by the Zone of Established Flow (ZEF)". For details of the wake structure, one can refer to (Lam et al. 2011) and references cited therein.

The beginning of zone of flow establishment is identified by area of maximum velocity, called efflux velocity $V_0$. $V_0$ is the maximum velocity right at propeller face. In general form, the efflux velocity is obtained as

$$V_a = \alpha n D \sqrt{C_t}$$

where n is the rotating speed of propeller in revolution per second, D is the diameter of the propeller and $C_t$ is the thrust coefficient and α is the scaling coefficients. There are various models of prediction for efflux velocity resulting in different scaling coefficient as shown below:

a. Axial momentum theory α = 1.59
b. Semi-empirical models:

- Stewart model (Stewart et al. 1991): propeller characteristics dependence:

$$\alpha = \zeta = D^{-0.0686}(P/D)^{1.519}BAT^{-0.323}$$

- Hashmi model (Hashmi 1993): hub diameter effect:

$$\alpha = E_0 = \left(\frac{D}{D_h}\right)^{-0.403} C_t^{-1.79} BAR^{0.744}$$

In those expression BAR is the blade area ratio, P/D is the pitching ratio and $D_h$ is the hub diameter. Position of the maximum efflux velocity is predicted as

$$R_{mo} = 0.67(R - R_h)$$

where $R_{mo}$ is the radial distance of maximum velocity dependent on propeller (R) and hub ($R_h$) radius. Length of flow establishment zone is characterized by the ratio of axial distance and propeller radius x/D = 2 − 6.2 and the limit of the establishment zone is defined as

$$\frac{x_0}{D_{or}} = \frac{1}{2C}$$

where $D_{or}$ is the diameter of the orifice. It is understood that the maximum velocity is decaying in the flow establishment zone. In the zone of developed flow, there is only one peak in velocity profiles and the maximum velocity is decaying radially as described by different empirical formulae proposed by various authors in (Hamill 1987; Stewart et al. 1991; Hashmi 1993).

Figure 4 shows velocity profiles in the wake of the propeller at different advance ratio. A typical wake structure behind the propeller is observed from simulations where two separate zones of flow development with 2 peaks in velocity profile and zone of developed flow with single peak in velocity profile. This is consistent with observation from previous studies of propeller wakes including Lam et al. (Lam et al. 2011) and references cited there in.

As flows travel from the Zone of Flow Establishment (ZFE) to the Zone Of Developed Flow (ZDF), the maximum velocity at the peaks decreasing. Figure 4 shows the decaying of maximum velocity in the wake of propeller for different advance ratio and comparison of maximum velocity against empirical formula. For comparison, the prediction of maximum velocity by Hamill (1987) for the zone of flow establishment and zone of developed flow is plotted against the current CFD results. It is observed from the empirical formulations that the decaying of maximum velocity is depending on the prediction of efflux velocity ($V_0$). In equation (2) the efflux velocity is defined as maximum flow velocity at the face of the propeller in the case of maneuvered ships at very low velocity. Therefore we made a simple correction to the efflux velocity by adding the ship velocity, $\bar{V}_0 = V_0 + V_\alpha$ and this efflux velocity is used for the prediction of maximum velocity as shown in Figure 4. It can be seen that the CFD results show a good agreement in trend and limits with empirical data.

Next the effect of outlet boundary condition was studied by extending the computational domain from 8D to 20D in length while the height of the domain is increased from 6D to 8D. Figure 6 showed the velocity profile in the wake of Wagenigen B5 propeller at advanced coefficient of J = 0.88 using GGI model for different domain sizes of W = 8D and W = 20D. The effect of domain size is minimal in the simulations as shown in Figure 6. Due to the use of outflow boundary conditions, it is observed that the outlet boundary has little effect upstream.

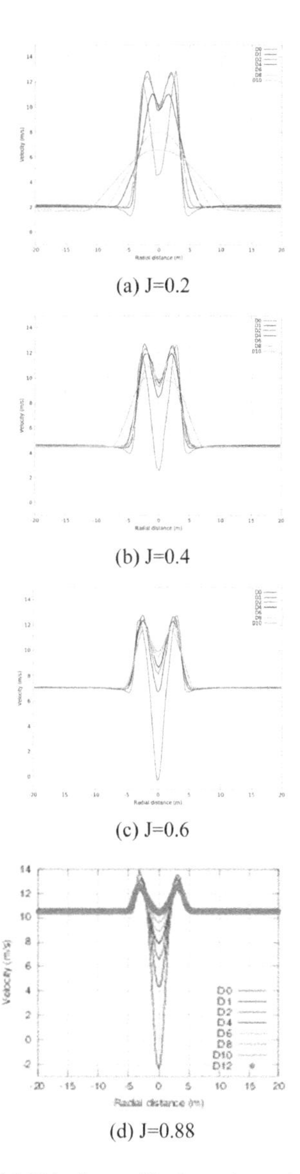


(a) J=0.2

(b) J=0.4

(c) J=0.6

(d) J=0.88

Figure 4. (a) Velocity profile downstream in the wake of the B5 Wagenigen propeller at different advance ratio J = 0.2, 0.4, 0.6, 0.88. The propeller wake structure changes from two peaks to single peak as reported in various previous studies.

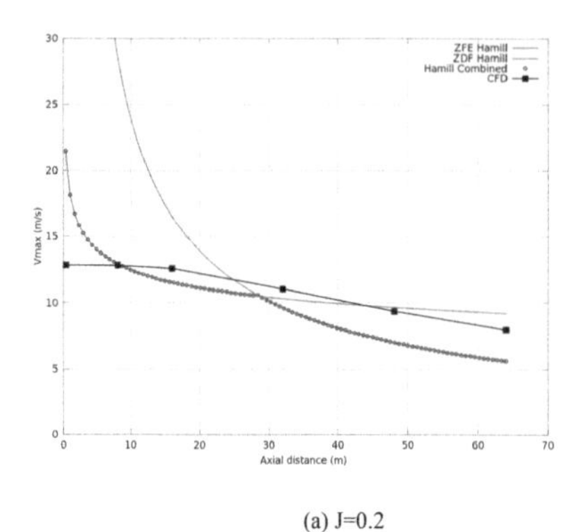


(a) J=0.2

(b) J=0.88

Figure 5. Decaying of maximum velocity in the establishment and developed zones. Comparison of the present numerical results with empirical formula shows a good agreement in trend and limits of the decaying.

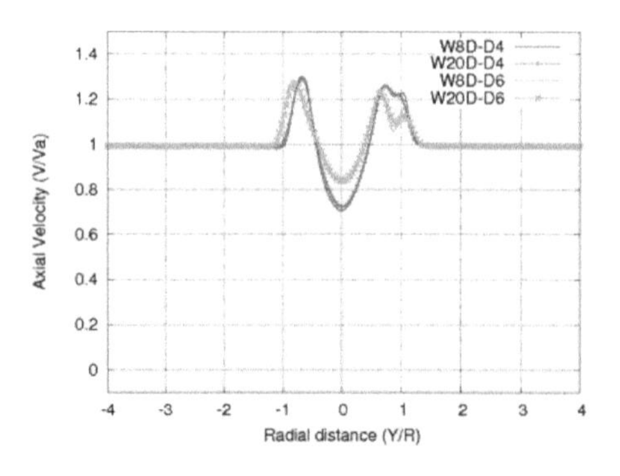


Figure 6. Velocity profile in the wake of B5 Wagenigen propeller at advanced coefficient of J = 0.88 using GGI model for different domain sizes of W = 8D and W = 20D. The comparison shows small difference in velocity profile between two domain sizes.

Figure 7 shows comparison of velocity profiles in the wake of the Wagenigen B5 propeller for different propeller models, namely MRF and GGI. If one takes the sliding mesh approach as the most accurate one, multiple reference frames provides a good approximation in the range of 10% for errors in peak velocity. In addition, the MRF model is about 10 times faster than sliding mesh computation, it is more efficient to run MRF models for subsequent simulations where the computation is getting much more expensive with the presence of rudder and seabed.

### 3.3 *Propeller-rudder interaction*

Next, propeller jets were studied in the presence of a rudder placed at 1D behind the propeller in the computational domain of 20D in length and 5D in width and height. The multiple reference frame solver was employed for simulations of the flows with the wall boundary conditions at propeller and rudder as defined in the previous section. It has been shown in (Hamill & McGarvey 1996), (Hamill et al. 1998) that in the presence of

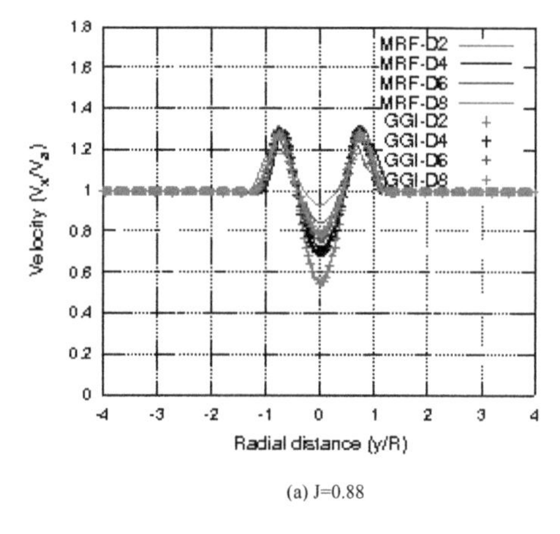


(a) J=0.88

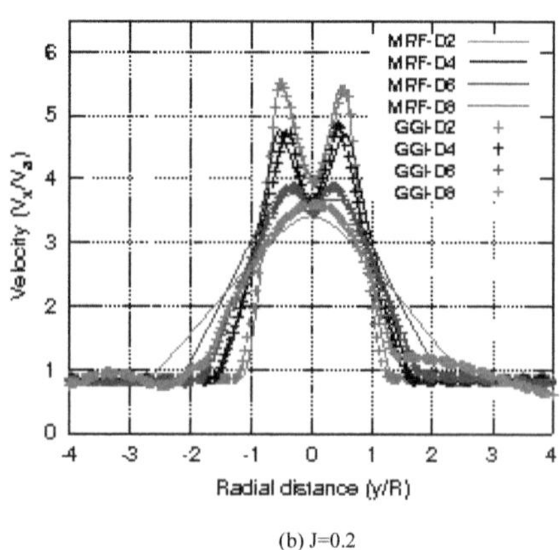


(b) J=0.2

Figure 7. Velocity profile downstream in the wake of the B5 Wagenigen propeller at (a) J = 0.88 and (b) J = 0.2. Comparison on velocity profile from GGI and MRF model shows that the difference between the two models gets larger in moving away from the propeller and MRF model under-predicts velocity peak of about 10–20%.

the rudder, it splits the flow into two high velocity streams, one directed towards the surface and the other directed towards the bottom. While the presence of the rudder insignificantly affects the efflux velocity, it was found to change the wake structure dramatically when compared to the case of propeller jets without a rudder. In the subsequent figures, the results of propeller wake structures for different rudder angle of $\theta = -20, 0, 20$ were presented.

As the rudder approaches at zero angles, the formation of two separate jets is clearly shown in Figure 8. Though the presence of the rudder breaks the symmetry of the flow, maximum velocity of the top and bottom jet in this case are in the same magnitude. Figure 9 shows details of wake profiles at various distance from the propeller for three different rudder angles of $\theta = 0, -20, 20$. As the rudder turns towards left-side of the ship ($\theta = -20°$), the maximum velocity in the bottom jet tends to be larger than the top jet; thus resulting in a stronger bottom jet. This is consistent with earlier study reported in (Hamill & McGarvey, 1996). As the rudder turns towards the right side of the ship, the top jet becomes more dominant as opposed to the case of turning left.

### 3.4 *Propeller-rudder-seabed interaction*

With understanding from propeller and rudder interactions, we moved on to study the effect of seabed and its interaction with propeller and rudder on the wakes. Similar to the previous study, the propeller and rudder were placed in the computation domain of length 20D while the bottom boundary is restricted by seabed with a defined kneel clearance measured from the tip of the propeller to the seabed. In this study the seabed is assumed to be flat and non-deformable. The roughness of the seabed is modeled as rough surface boundary condition as described earlier in section 3. In this case, the seabed is modeled as a moving, rough wall with moving velocity taken as the ship velocity.

To fully investigate the interaction of propeller, rudder and seabed on the propeller wakes, a number of configurations with different conditions of rudder angle, kneel clearance as well as propeller speed and ship velocity were chosen. Figures 10 to 12 show the results of wake profiles behind the rudder at various rudder angles. From the simulation results, a few remarks on the interactions of propeller, rudder and seabed are as follows:

In the presence of the seabed, it obstructs the development of the flow jets thus changing the wake structure and inducing more stress on the seabed.

At the same kneel clearance, ship velocity and rudder angle, varying of rotating speed affects wall shear stress level on the seabed. Increase in advance ratio (J) induces more wall shear stress on the seabed.

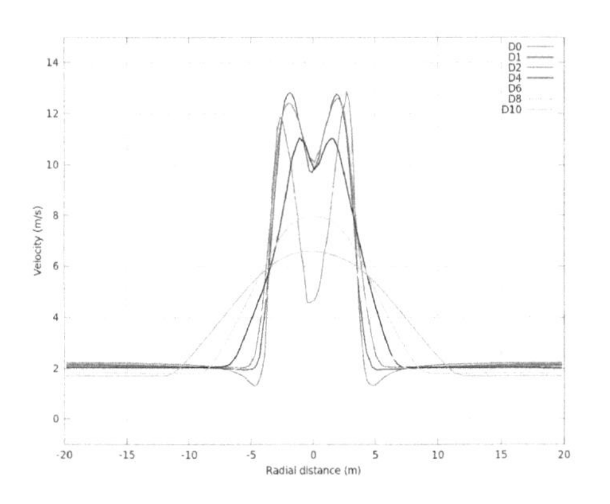


Figure 8. Results of velocity profiles for RPM = 90, ship speed $V_a = 12$ knots, rudder angle $\theta = 0$.

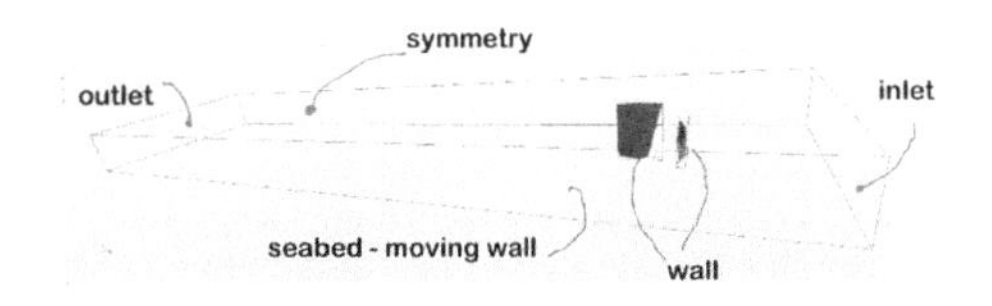


Figure 9. Computational domain for simulations of propeller-rudder-seabed interaction: Maersk propeller D = 9 m, domain size of H = 6D, $D_e = 3D$, $D_w = 20D$, KC = 1, 3, 5 m.

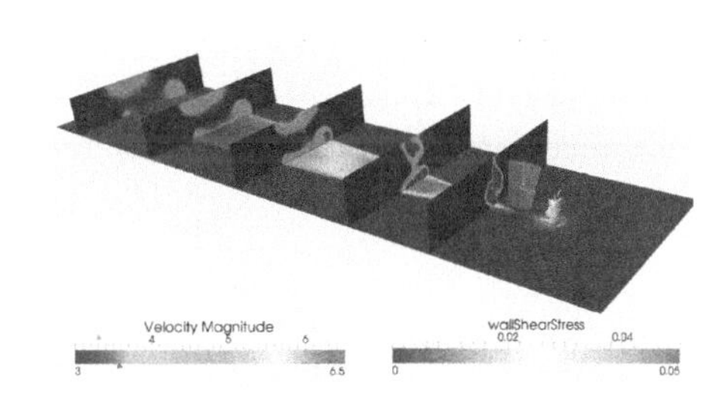


Figure 10. Results of flow stream and velocity profile in the wake of propeller and rudder with seabed. Kneel clearance KC = 1 m, RPM = 60, ship speed $V_a = 6$ knots, rudder angle $\theta = 0°$.

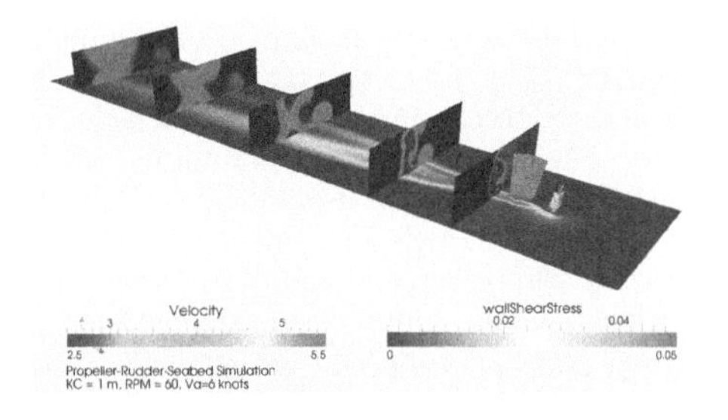


Figure 11. Results of flow stream and velocity profile in the wake of propeller and rudder with seabed. Kneel clearance KC = 1 m, RPM = 60, ship speed $V_a = 6$ knots, rudder angle $\theta = -20°$.

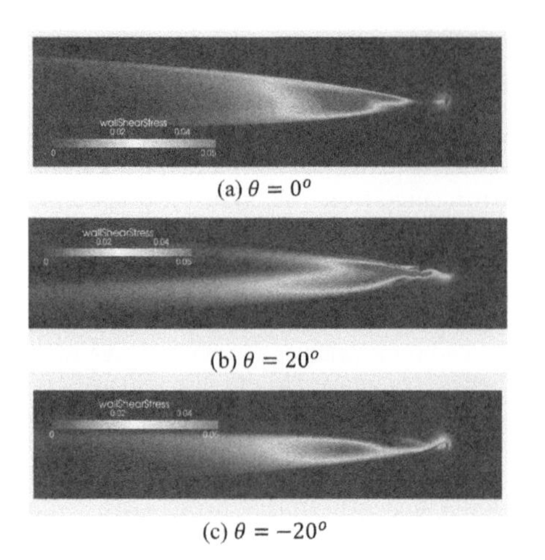

(a) $\theta = 0^o$

(b) $\theta = 20^o$

(c) $\theta = -20^o$

Figure 12. Effect of rudder angle on wall shear stress level on the seabed for KC = 1 m, RPM = 60, ship speed $V_a$ = 6 knots.

At the same kneel clearance, ship velocity and advance ratio, ship turning left ($\theta = -20°$) results in minimal shear stress to the seabed. This is counter-intuitive as it's shown that as turning left the bottom jet is more dominant?

The effect of kneel clearance is most straightforward as it can be seen from the simulation results that the higher the clearance, the less damage on the seabed.

In this study, the seabed is modeled as a rough surface, which uniform roughness. As the roughness decreases, the wall shear stress level is reduced on the seabed.

## 4 CONCLUSIONS

In this work, we have investigated the propeller wake under interactions with rudder and seabed using Computational Fluid Dynamics (CFD) tools. Advantages of CFD has been fully explored to study wake structures behind propeller and rudder in an unprecedented way where it is close to impossible to carry out by on-field measurement or even in-vitro experiment. The simulations were carried out on high performance computing facilities at the Institute of High Performance Computing, ASTAR. High fidelity CFD model using steady and unsteady flow simulation models were employed first to characterize the propeller performance and validate with experimental data. Good agreement with available data has proved accuracy of the simulation models and provided confidence for full-scale simulations of propeller, rudder and seabed interactions. For full scale simulations, a real size propeller and rudder were modeled with and without the presence of seabed to investigate the wake structures as well as the effect on the seabed. Simulations were carried out with various configuration setups of propeller speeds, ship velocities, rudder angles, kneel clearance and roughness of the seabed. The wake structures obtained from simulations were found to be consistence with observation from earlier studies including model-testing results. The effect of wake structures on the seabed are characterized by the level of wall shear stress on the bed and effects of different parameters were evaluated; the trend of influence of those parameter generally agree with reported results as well as field observations. However, it is necessary to qualify those mentioned agreement in details and possibly conduct a full parametric study to derive relationship between those operating conditions and the resulted damage on the seabed. This remains as future work in this project.

## REFERENCES


Beaudoin, M. and Jasak, M. 2008. Development of a Generalized Grid Interface for Turbomachinery simulations with OpenFOAM, Open Source CFD International Conference.

Bernitsas, M.M., Ray, D. and Kinley, R. 1981. Kt, Kq and efficient curves for the Wagenigen B-series propellers, Report No. 237, University of Michigan, 1981.

Lam, W., Hamil, G.A., Song, Y.C. Robinson D.J. and Raghunathan S. 2011. A review of the equations used to predict the velocity distribition within ship's propeller jet, Ocean Engineering 38 (2011) 1–10.

Hamill, G.A. 1987. Characteristics of the screw wash of a manoeuvring ship and the resulting bed scour. Thesis submitted to the Queens University of Belfast for the degree of Doctor of Philosophy, 1987.

Hamill, G.A., McGarvey, J.A. 1996. Designing for propeller action in harbours. 25th International Conference on Coastal Engineering. Orlando, Sept. 1996.

Hamill, G.A., McGarvey, J.A. and Mackinnon, P.A. 1998. A Method for Estimating the Bed Velocities Produced by a Ships Propeller Wash Influenced by a Rudder, Coastal Engineering, Part V: Coastal, Estuarine and Environmental Problems, pp. 3624–3633.

Hamill, G.A., McGarvey, J.A., Hughes, D.A.B. 2004. Determination of the efflux velocity from a ships propeller. Proceedings of the Institution of Civil Engineers: Maritime Engineering 157(2), 8391.

Hashmi, H.N. 1993. Erosion of a granular bed at a quay wall by a ships screw wash. Thesis submitted to the Queens University of Belfast for the degree of Doctor of Philosophy, 1993.

Menter, F.R. 1994. Two-Equation Eddy-Viscosity Turbulence Models for Engineering Applications, AIAA Journal, vol. 32, no 8. pp. 1598–1605.

Stewart, D.P.J., Hamill, G.A., Johnston, H.T. 1991. Velocities in a ships propeller wash, Proceedings of International Symposium on Environmental Hydraulics, Rotterdam.

Wilcox D.C. 1998. Turbulence modeling for CFD, DWC Industries, Second Edition.

*Maritime-Port Technology and Development – Ehlers et al. (Eds)*
*© 2015 Taylor & Francis Group, London, ISBN 978-1-138-02726-8*

# Enhanced electrochemical performance in merged olivine structured cathode materials

S. Saadat
*Energy Research Institute at NTU (ERI@N), Singapore*

R. Yazami
*Energy Research Institute at NTU (ERI@N), Singapore*
*TUM CREATE, Singapore*

ABSTRACT: First report of elevated electrochemical performance in mixture of two olivine structured cathode materials, with the general formulae $LiMPO_4$; where M = Fe, Mn, of identical or different compositions (Fe and Mn contents), as compared to their individual components. The two constituents are prepared under different synthesis conditions and have different characteristics. The first category materials (A) are commercially available olivines with the composition of C-$LiFePO_4$ and C-$LiFe_{0.3}Mn_{0.7}PO_4$. The second material (B) is carbon coated $LiFe_{0.3}Mn_{0.7}PO_4$, which is fabricated in house by a solid state method. The weight fraction of B in the A+B mixture can range between 5% and 95%. The electrochemical properties of the cathodes are examined by charge-discharge tests. It is observed that compared to single component cathodes (A/B), merged cathodes (A+B) demonstrate superior performance. It is suggested that such elevation in battery properties of A+B cathodes is a result of synergetic kinetics effects between the two components.

## 1 INTRODUCTION

Currently environmental considerations have high significance in marine and shipping industry. The aim is to reduce the greenhouse gas emissions from ships and minimize the growth of the industry's portion of international emissions. An attractive choice is application of full or hybrid electric systems. Battery technologies can provide interesting properties for fabrication of cost-effective and efficient marine systems (Crowell, 2006).

Lithium Ion Batteries (LIB) are known as rechargeable storage devices which have high energy and power density and operate in high operation voltages. Compared to other battery types, lower self-discharge, higher cycle life and energy density and design flexibility are a few of many advantages of LIB. These advantages make LIB a promising candidate for application in mobile electronic devices and electric and hybrid vehicles.

Experimental Among the different types of cathode materials, layered lithium transition metal oxides; $LiMO_2$ (M: Mn, Co, and Ni) has achieved great success. (Kraytsberg et al, 2012). The superiority of this group of cathode materials is the highly accessible ion diffusion pathways. However, with the introduction of $LiFePO_4$, olivine cathode materials have received great attention as an encouraging cathode material (Zaghib et al, 2013); (Ellis et al, 2010); (Aravindan et al, 2013). This category of iron-based compounds are naturally abundant and inexpensive and compared to vanadium, cobalt, and nickel compounds are more environmentally friendly (Laffont et al, 2006).

In this work we investigate a combination cathode of two olivine structured materials, each of them having a general formulae $LiMPO_4$; where M = Fe, Mn, are of identical or different compositions (Fe and Mn contents), prepared under different conditions and having different characteristics. The aim is to study possible improved energy storage performances of combination cathode as compared to its individual materials. Here the first material A is a commercially available material and the second material B is produced by high energy ball milling in our laboratories. The weight fraction of B in the A+B mixture can range between 5% and 95%. In this paper the results of in house/commercial powder ratio of 40/60 is demonstrated and the battery electrochemical performance of mixed and individual olivine materials is compared.

## 2 EXPERIMENTAL

### 2.1 *Preparation of C-$LiMn_{0.7}Fe_{0.3}PO_4$ using ball milling*

The C-$LiMn_{0.7}Fe_{0.3}PO_4$ compound was synthesized by a solid-state reaction between $MnCO_3$,

$Fe(C_2O_4)_2.2H_2O$, and $LiH_2PO_4$, obtained from sigma Aldrich. The precursors were mixed thoroughly in stoichiometric ratio with 10 wt.% carbon black (acetylene black) in a glovebox filled with argon and transferred to ball mill jars with stainless steel balls. A combination of large and small balls were used and the ball to powder ratio was kept constant at 30. The mixture was ground by high energy ball milling with 300 rpm speed for 7 hours. The samples were pressed into pellets and sintered at 700°C for 10 h. The sintering was performed under Ar-$H_2$ atmosphere. After sintering, the powders were grinded manually using mortar and pestle.

### 2.2 *Preparation of mixed olivine cathode*

To fabricate the mixed powder, ball milled C-LFMP products were mixed with commercial carbon coated $LiFePO_4$ and $LiMn_{0.7}Fe_{0.3}PO_4$. Commercial C-LFP and C-LFMP powders; products of Clariant were used as received. The weight fraction of in house/commercial in the A+B mixture can range between 5% and 95%. In this paper results of in house/commercial ratio of 40/60 is demonstrated. For optimum mixing, the powders were dispersed in ethanol and stirred overnight. The powders were dried at 80°C under vacuum condition and the mixed powders were grinded before slurry preparation.

### 2.3 *Sample characterization*

The sample morphology was examined using a field-emission scanning electron microscopy (FESEM; JEOL, JSM-7600F). The elemental compositions of the samples were characterized with Energy-Dispersive X-ray spectroscopy (EDX) which is attached to the SEM instrument. Crystallographic data of the specimen was collected using powder X-ray diffractometer (Bruker, Cu KR radiation with $\lambda$ = 1.5406 Å). For Transmission Electron Microscopy (TEM) and High Resolution Transmission Electron Microscopy (HRTEM) characterization, the powders were dispersed in ethanol and ultrasonicated for 10 minutes. After ultrasonication, the solution was drop cast onto carbon coated 200 mesh Cu grids. TEM/HRTEM was obtained by using a JEOL 2010 system operating at 200 kV.

### 2.4 *Electrochemical measurements*

To investigate the electrochemical properties of olivine products, 80 wt% of each sample were mixed with 10 wt% carbon black (acetylene black) and 10 wt% Polyvinylidene fluoride (PVDF) in a mortar. Then N-Methyl-2-Pyrrolidone (NMP) was added to prepare slurry. The slurry was coated on a piece of Al foil using doctor blade equipment. The thickness of the coated thin films was controlled at 50 µm. The coated foils were punched to 1.4 cm circles and dried at 110°C for 6 hours under nitrogen atmosphere. The prepared cathodes were pressed and the mass of active material was accurately measured. The coin cells were assembled inside an Ar-filled glove box with oxygen and moisture content less than 1.00 ppm. The prepared electrodes were used as the working electrode. The lithium foils were used as counter/reference electrodes and the electrolyte was a solution of 1 M $LiPF_6$ in Ethylene Carbonate (EC)/Dimethyl Carbonate (DMC) (1/1, w/w). For the electrochemical measurement coin battery cells were installed and galvanostically tested using a NEWARE battery tester between 2.7 and 4.4 V (vs. Li/$Li^+$).

## 3 RESULTS AND DISCUSSION

### 3.1 *Microstructural examination*

XRD examination of in house product, fabricated by ball milling is shown in Figure 1a, which demonstrates formation of olivine $LiMn_{0.7}Fe_{0.3}PO_4$ compounds with an orthorhombic structure (space group Pnmb, JCPDF 4-014-3741). To

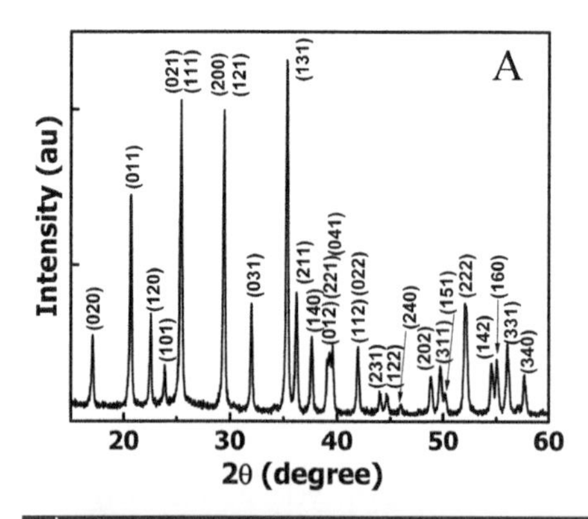


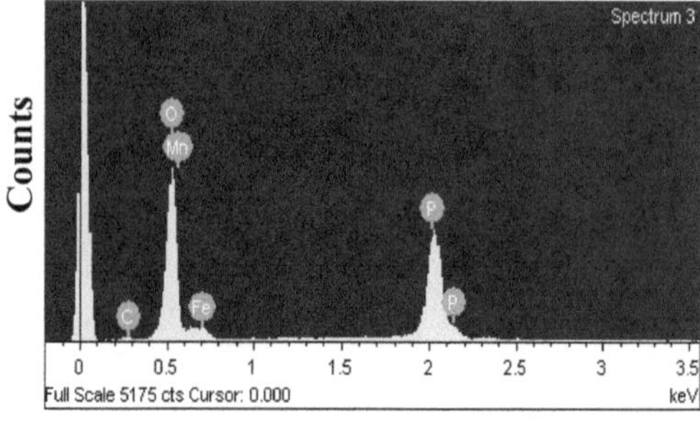


Figure 1. (a) XRD pattern, (b) EDS spectrum of ball milled C-$LiMn_{0.7}Fe_{0.3}PO_4$ particles.

accurately calculate the amount of Fe/Mn ratio in the fabricated $LiMn_XFe_{1-X}PO_4$ (LFMP) powder, energy-dispersive X-ray spectroscopy (EDX) were used. The EDS of the LFMP samples is shown in Figure 1b. Peaks of Fe, Mn, O, C and P can be observed in the EDS spectrum and exposes average Fe:Mn ratio of 70:30.

The morphology of the samples can be observed in Figure 2. The ball mill powder is composed of agglomerated particles with a broad size range of ~40–170 nm diameters (see Fig. 2a-b). During the battery testing the nanosized particles reduce the solid-state diffusion path, thus expediting the lithium-ion transport (Liu et al, 2010). The agglomeration is not severe and the particles can be distinguished apart. Based on the TEM examination, each powder agglomerate is composed of several nanoparticles with porosity in between (see Fig. 2c). To achieve high specific capacity, especially at high current densities, existence of porosity is necessary to enable penetration of electrolyte in to the structure and reduce the diffusion distance (Yu et al, 2010). Based on HRTEM investigation, each particle is highly crystalline with a uniform layer of 3–6 nm carbon coating, uniformly covering each nanoparticle (see Fig. 2d). A uniform carbon coating of particle surface is necessary to enhance electronic conductivity (Oh et al, 2010); (Shin et al, 2006). It was also observed that lattices have an equal interfringe spacing of 0.25 nm, corresponding to the (211) plane of orthorhombic olivine (JCPDF 4-014-3741).

XRD examination of commercial powders are shown in Figure 3(a-b). Figure 3a demonstrates formation of olivine $LiFePO_4$ phase with an orthorhombic structure (space group Pnma, JCPDF 04-016-2401). Based on the supplier's

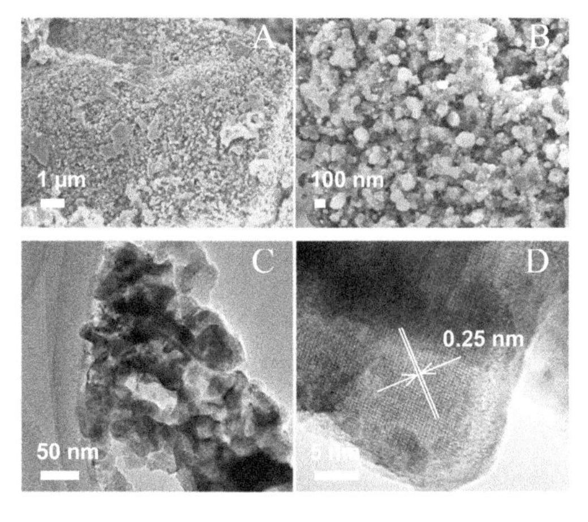


Figure 2. (a-b) SEM (c) TEM, (d) HRTEM of ball milled C-$LiMn_{0.7}Fe_{0.3}PO_4$ coated using carbon black.

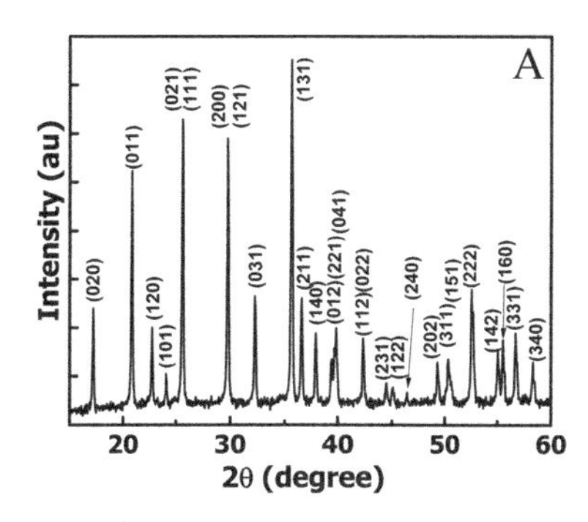


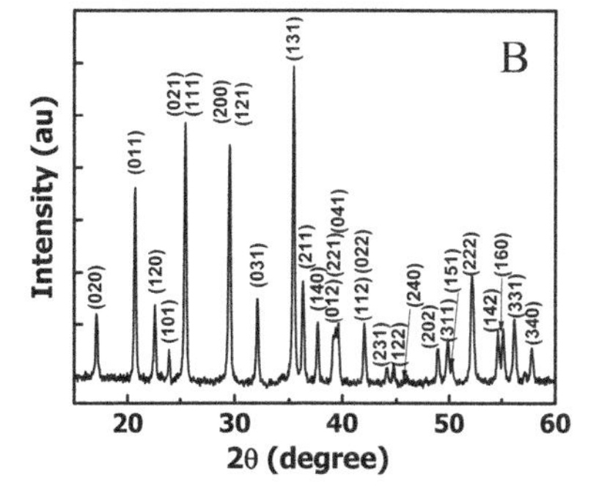


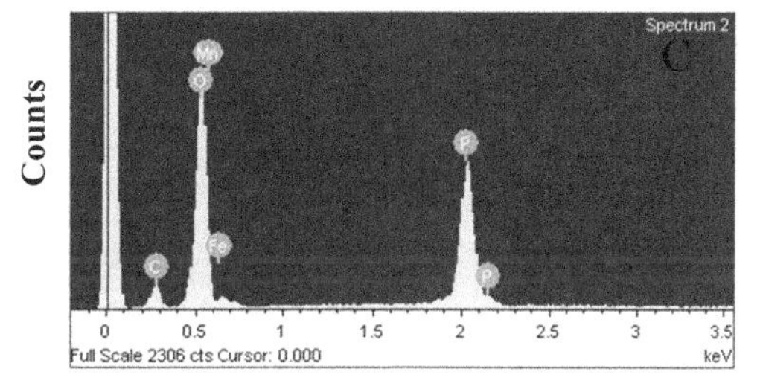


Figure 3. XRD patterns of the (a) Commercial C-$LiFePO_4$, (b) Commercial C- $LiMn_{0.7}Fe_{0.3}PO_4$, (c) EDS spectrum of commercial C- $LiMn_{0.7}Fe_{0.3}PO_4$ particles.

product description report, the LFMP olivine powder has $LiMn_{0.67}Fe_{0.33}PO_4$ composition. The XRD spectrum is in good agreement with this claim and demonstrates formation of olivine $LiMn_{0.7}Fe_{0.3}PO_4$ compounds with an orthorhombic structure (space group Pnmb, JCPDF 4-014-3741). To confirm the composition, energy-dispersive X-ray spectroscopy (EDX) were used (see Fig. 3c). Peaks of Fe, Mn, O, C and P can be observed in the EDS spectrum and expose average Fe:Mn ratio of 70:30, which is very close to the company's claim.

The morphology of commercial C-LFP particles is shown in Figure 4a-b. It can be seen that the powder is made of bars and sphere particles. The particles have a diameter of 100–300 nm, while the bars have a length of approximately 200–600 nm. The TEM presents a better understanding

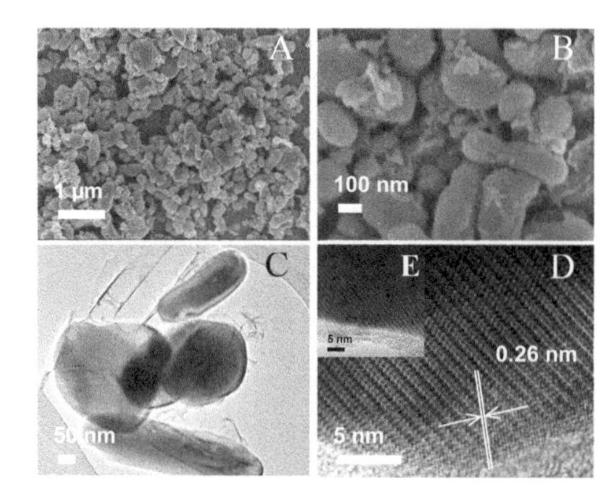


Figure 4. (a-b) SEM, (c-e) TEM and HRTEM of commercial C-$LiFePO_4$.

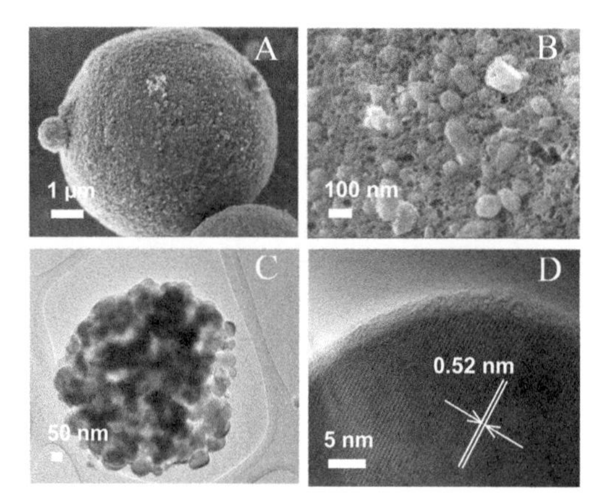


Figure 5. (a-b) SEM, (c-d) TEM of commercial C-$LiMn_{0.7}Fe_{0.3}PO_4$.

of the shape and size of the fabricated particles (see Fig. 4c). In this TEM image, both sphere and bar particles can be observed. The HRTEM image (see Fig. 4d) shows that the particles are single crystals with high crystallinity. The LFP lattices have an interfringe spacing of 0.25 nm, corresponding to the (131) plane of orthorhombic olivine (space group Pnma, JCPDF 04-016-2401). In Figure 4e a uniform carbon coating layer of 3–6 nm thickness covers the LFP particle.

The microstructure of commercial C-LFMP powders is shown in Figure 5a-b. It can be observed that the powder is composed of 5–10 μm granules. Each granule is composed of numerous fine round particles in the range of 20–150 nm. In the TEM image (see Fig. 5c) it can be seen clearly that the round C-LFMP granule is made up of numerous fine particles. Its noteworthy that according to HRTEM (Fig. 5d) each fine LFMP particle is a single crystal with interfringe spacing of 0.52 nm, corresponding to the (020) plane of orthorhombic olivine (space group Pnmb, JCPDF 4-014-3741) and it is coated with a uniform thin layer of carbon coating with thickness of approximately 3–6 nm.

As discussed the hypothesis of this work is to evaluate whether mixture of A and B olivine powder can electrochemically perform better than individual A and B. It is hoped that mixing generates a synergetic effect between the components. The procedure is to mix C-$LiFe_{0.3}Mn_{0.7}PO_4$ powder fabricated in lab using solid state method with commercial grade C-LFP and C-LFMP powder. This was performed with the aim to elevate the battery performance.

Two sets of samples were prepared. The first set was fabricated by mixing in house fabricated C-$LiFe_{0.3}Mn_{0.7}PO_4$ with commercial C-$LiFePO_4$ and the second set by mixing it with commercial C-$LiMn_{0.7}Fe_{0.3}PO_4$. As can be seen in SEM images of mixed powders shown in Figure 6, both constituents can be clearly seen and recognized in both set of samples. Figure 6a-b show set 1, while Figure 6c-d displays set 2 mixture. It is important to emphasize that complete mixing of the two components is essential. To ensure complete blending, wet mixing by dispersing powders in ethanol or acetone and stirring until the mixture is uniform is of utmost importance.

### 3.2 *Electrochemical investigation*

Figure 7a illustrates the charge/discharge voltage profiles of in house fabricated C-$LiFe_{0.3}Mn_{0.7}PO_4$, tested at 17 mA/g (0.1C) between 2.7–4.4 V (vs $Li/Li^+$). The charge/discharge plateaus at 4.1 V is

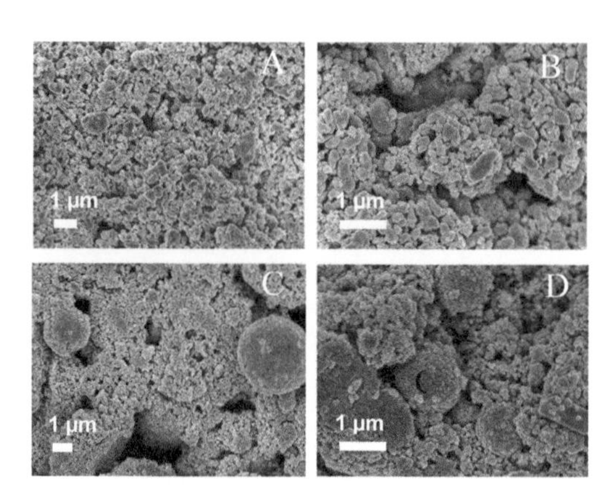


Figure 6. (a-b) SEM of mixed commercial C-$LiFePO_4$ and in-house C-$LiMn_{0.7}Fe_{0.3}PO_4$ (c-d) SEM of mixed commercial C-$LiMn_{0.7}Fe_{0.3}PO_4$ and ball milled C-$LiMn_{0.7}Fe_{0.3}PO_4$.

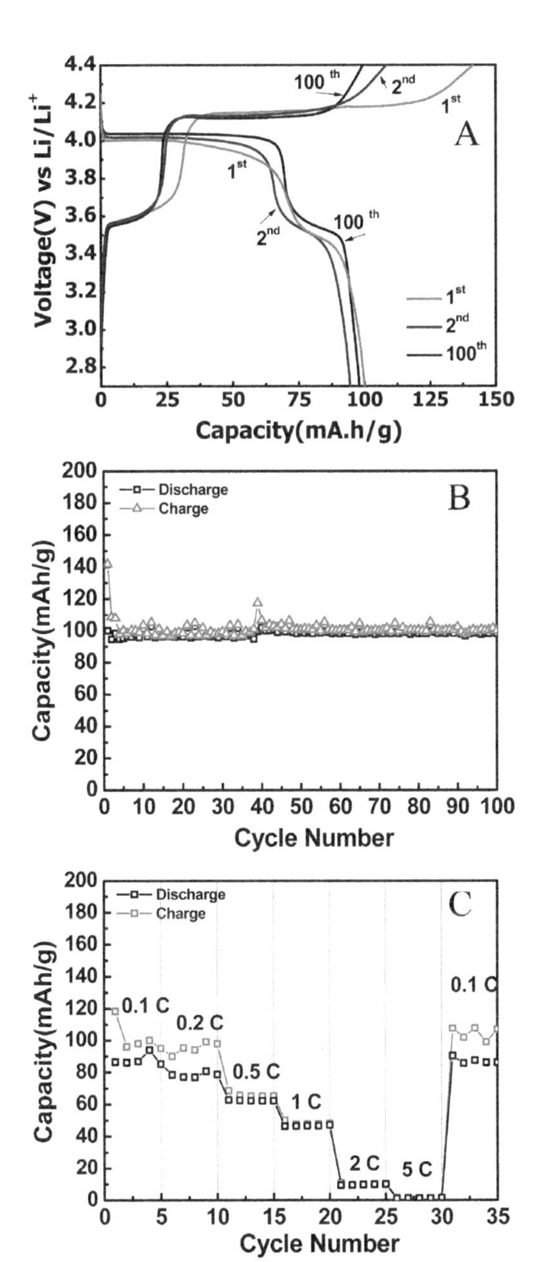


Figure 7. (a) Charge/discharge voltage profiles at 0.1C, (b) Charge/discharge cycling performance at 0.1C, (c) Plot of the discharge and charge capacity vs. cycle number at various C rates of in-house ball milled C-$LiMn_{0.7}Fe_{0.3}PO_4$ between 2.7–4.4 V (vs $Li/Li^+$).

related to the $Mn^{2+}/Mn^{3+}$ redox couple, and plateaus at 3.6 V is related to the $Fe^{2+}/Fe^{3+}$ redox couple. The first cycle gives a charge capacity of 141 mA h/g and a subsequent discharge capacity of 100 mA h/g, this result gives an initial coulombic efficiency of 71%. In the second cycle, the charge capacity reaches 108 mA h/g with a corresponding discharge capacity of 95 mA h/g, leading to a high columbic efficiency of 88%. Based on the charge/discharge cycling results shown in Figure 7b, the sample depicts good cycling stability, delivering a charge and discharge capacity of 99 and 97 mAh/g, respectively, with 98% coulombic efficiency during the 100 cycle.

The rate capability of in house ball milled C-LFMP electrode was further examined at current densities from 17 mA/g (0.1 C) to 85.5 mA/g (5C) (see Fig. 7c). The 1st-cycle charge capacities are 118 mAh/g, 90, 69, 50, 12 and 2 mAh/g at 0.1 C, 0.2 C, 0.5 C, 1 C, 2 C and 5 C rates, respectively. The performance of C-LFMP electrode at each rate is stable, but capacity drop with increase in rate, especially at higher current densities is high. This demonstrates the poor Li storage properties of C-LFMP electrode at high cycling rates. When decreasing the charge/discharge rate to 0.1C, we find that the discharge capacity can recover back to 107 mAh/g, but the coulombic efficiency is low.

Figure 8a illustrates the charge/discharge cycling of pure commercial C-$LiFePO_4$ at 0.1 C. The first cycle gives a charge capacity of 195 mA h/g and a subsequent discharge capacity of 155 mAh/g. This results in a good initial coulombic efficiency of 79%. The extra capacity in the initial cycle may be assigned to the solid electrolyte formation (SEI) and electrolyte decomposition (Morales et al, 2004); (Laruelle et al, 2002). In the second cycle, the charge capacity reaches to 160 mA h/g with a corresponding discharge capacity of 156 mA h/g, leading to a high coulombic efficiency of 97.5%. It can be observed that the sample depicts very good cycling stability until about 80 cycles, thereafter some gradual capacity drop can be observed, resulting in a discharge and charge capacity of 139 and 137 mAh/g, respectively, with coulombic efficiency of 98.5% during the 100th cycle.

The rate capability of pure commercial C-$LiFePO_4$ was also examined at current densities from 0.1–0.5 C (see Fig. 8b). The 1st-cycle charge capacities of C-LFP are 159, 140, 117, 99 and 39 mAh/g, at 0.1 C, 0.2 C, 0.5 C, 1 C and 2 C. It can be seen that the performance of C-LFP electrode at each rate is stable, but capacity drop with increase in rate, especially at higher current densities is high. At 5 C rates, the capacity drops a lot, delivering almost no capacity. This demonstrates that at high current densities the commercial C-LFP displays poor electrochemical performance. However, when decreasing the charge/discharge rate to 0.1 C, it was observed that the charge capacity can recover back to 96 mAh/g and subsequently reach 140 mAh/g.

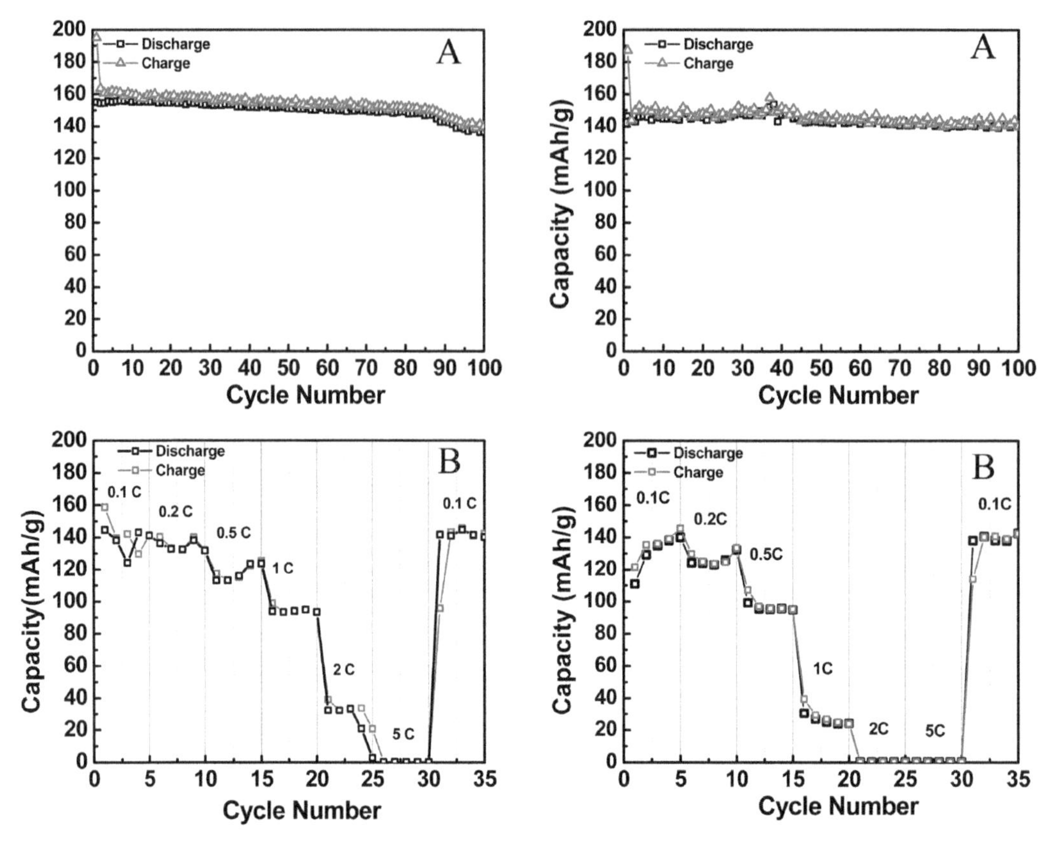


Figure 8. (a) Charge/discharge cycling performance at 0.1 C (b) Plot of the discharge and charge capacity vs. cycle number at various C rates of commercial C-$LiFePO_4$ between 2.7–4.4 V (vs $Li/Li^+$).

Figure 9. (a) Charge/discharge cycling performance at 0.1 C (b) Plot of the discharge and charge capacity vs. cycle number at various C rates of commercial C-$LiMn_{0.67}Fe_{0.33}PO_4$ between 2.7–4.4 V (vs $Li/Li^+$).

In comparison to C-LFP, commercial C-LFMP demonstrates a lower delivered capacity (see Fig. 9a). As can be observed, the olivine cathode demonstrates an initial charge capacity of 187 mAh/g and discharge capacity of 146 mAh/g, which leads to 78% coulombic efficiency. In the second cycle charge capacity reduces to 144 mAh/g, with 143 mAh/g discharge capacity. The electrode exposes good cycling stability, delivering charge capacity of 140 mAh/g with 99% coulombic efficiency at the 100 cycle. By comparing the electrochemical properties of C-LFP and C-LFMP cathodes it can be concluded that although C-LFP electrode initially delivers a higher charge capacity than C-LFMP electrode, but the C-LFMP exposes better cycling stability, delivering almost the same charge capacity as the C-LFP electrode after 100 cycles.

In terms of rate capabilities of commercial C-LFMP, an initial charge capacity of 121 mAh/g with 90% coulombic efficiency was achieved, which improved to 135 mAh/g in the second charge cycle. Thereafter, a stable capacity of 130, 107 and 40 mAh/g was delivered at 0.2, 0.5 and 1 C respectively. However, at 2 and 5 C the delivered capacity is very low. However, after applying such high rates to the electrode, when the rate was again decreased to 0.1 C, the C-LFMP cathode recovers and delivers a high charge capacity if 114 mAh/g and stays almost unchanged thereafter. This observation demonstrates that although the C-LFMP does not function well at high rates, the structure is stable and it can recover well after testing at high current densities.

Figure 10a illustrates the cycling graph of mixed commercial C-$LiFePO_4$ and in house fabricated C-$LiMn_{0.7}Fe_{0.3}PO_4$, optimized and produced by ball milling process. The initial cycle results in a charge capacity of 144 mAh/g and a subsequent discharge capacity of 126 mAh/g, this gives an initial coulombic efficiency of 87%. In the second cycle, the charge capacity decreases to 120 mAh/g, with a

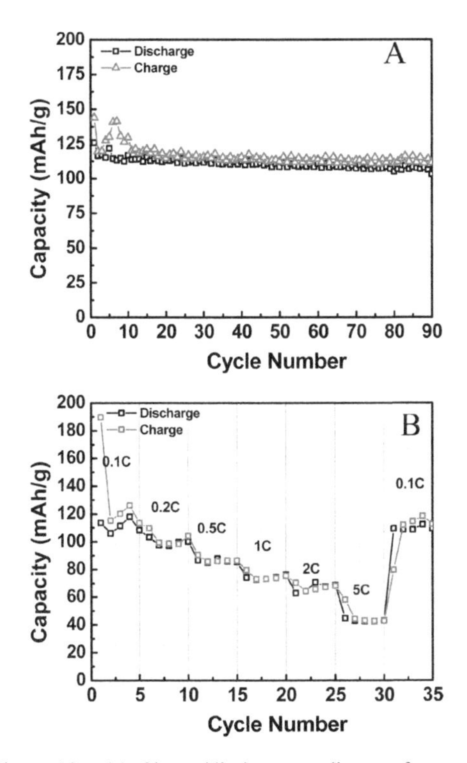


Figure 10. (a) Charge/discharge cycling performance at 0.1 C (b) Plot of the discharge and charge capacity vs. cycle number at various C rates of mixed commercial C-$LiFePO_4$ and in-house ball milled C-$LiMn_{0.7}Fe_{0.3}PO_4$ between 2.7–4.4 V (vs $Li/Li^+$).

corresponding discharge capacity of 117 mAh/g, leading to a higher coulombic efficiency of 97%. After the third cycle, the charge capacity increases and coulombic efficiency is unusually low. After ten cycles the delivered capacity stabilizes and coulombic efficiency increases to over 92%. It can be suggested that such behavior is a result of combining two different olivine products, with different composition and morphology characteristics. It can be observed that the electrode exhibits good cycling stability and delivers charge capacity of 112 mAh/g after 90 cycles with 93% coulombic efficiency.

The rate capability of the mixed electrode is also encouraging (see Fig. 10b), showing superior electrochemical performance compared to pure commercial and in house fabricated olivine products. The electrode was examined at current densities from 0.1–5 C. The 1st-cycle charge capacities are 190, 110, 91, 79, 70 and 58 mAh/g at 0.1 C, 0.2 C, 0.5, 1, 2 and 5 C. Additionally, when the charge/discharge rate is decreased to 0.1 C, we find that the discharge capacity can recover back to 80 mAh/g and subsequently reach 112 mAh/g charge capacity with high coulombic efficiency. This result can be considered promising both in terms of cyclability and rate capability. Especially since the cycling stability is maintained for 90 cycles and the delivered capacity is satisfactory at high current densities. A good rate performance which was not observed in pure commercial grade C-LFP or in house fabricated C-LFMP electrodes.

The second set of mixed electrode was prepared by mixing commercial C-$LiMn_{0.7}Fe_{0.3}PO_4$ and in house fabricated C-$LiMn_{0.7}Fe_{0.3}PO_4$; produced by ball milling process. The first cycle gives a charge capacity of 184 mA h/g and a subsequent discharge capacity of 107 mA h/g, this results in an initial coulombic efficiency of 58% (see Fig. 11a). In the second cycle, the charge capacity drops to 111 mA h/g with a corresponding discharge capacity of 92 mA h/g, leading to a relatively high coulombic efficiency of 83%. It can be observed that the capacity continues to drop gradually in the initial 20 cycles. Thereafter the charge capacity suddenly rises similar to set 1 mixed sample, resulting in low coulombic efficiency. The exact reason for such behavior is still under investigation. After about 30 cycles, the capacity stabilizes and the

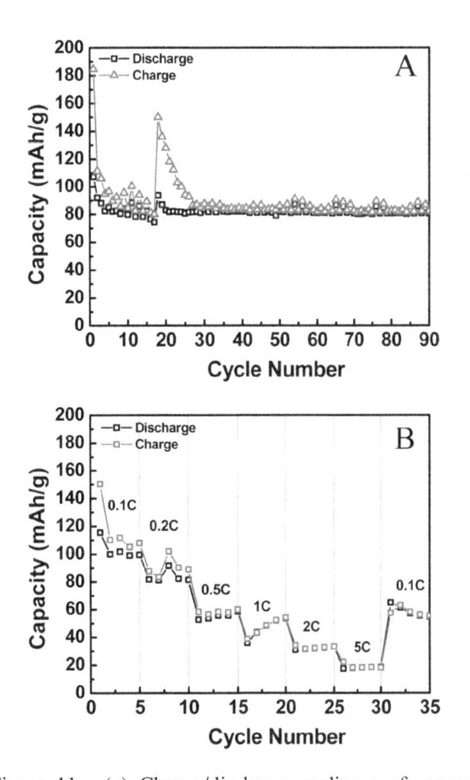


Figure 11. (a) Charge/discharge cycling performance at 0.1 C (b) Plot of the discharge and charge capacity vs. cycle number at various C rates of mixed commercial C-$LiMn_{0.7}Fe_{0.3}PO_4$ and in-house ball milled C-$LiMn_{0.7}Fe_{0.3}PO_4$ between 2.7–4.4 V (vs $Li/Li^+$).

Table 1. A review of electrochemical performance of in house, commercial and mixed olivine cathodes (1st C stands for 1st delivered charge capacity, 1st C.E% stands for initial coulombic efficiency, Final C and Final C.E% stands for final cycle charge capacity and coulombic efficiency, 1st C At 1C and 1st C At 2C is the 1st delivered charge capacity at 1C and 2C rate during rate capability testing).

| | 1st C [mAh/g] | 1st C.E % | 2nd C [mAh/g] | 2nd C.E % | Final C [mAh/g] | Final C.E % | 1st C at 1C [mAh/g] | 1st C at 2C [mAh/g] |
|---|---|---|---|---|---|---|---|---|
| In-House LFMP | 141 | 71 | 108 | 88 | 99 | 98 | 50 | 12 |
| Industrial LFP | 195 | 79 | 160 | 97.5 | 139 | 98.5 | 39 | 99 |
| Industrial LFMP | 187 | 78 | 144 | 99 | 140 | 99 | 40 | <10 |
| Mix LFP + LFMP | 144 | 87 | 120 | 97 | 112 | 93 | 79 | 70 |
| Mix LFMP | 184 | 58 | 111 | 83 | 82 | 98 | 39 | 34 |

cycling stability is good thereafter, delivering a discharge and charge capacity of 82 and 81 mAh/g, respectively, with coulombic efficiency of 98% at the 90th cycle.

The rate capability of the mixed C-LFMP electrode was also further examined at 0.1–5 C (see Fig. 27d). The 1st-cycle charge capacities are 150, 87, 58, 39, 34 and 22 mAh/g at 0.1 C, 0.2 C, 0.5 C, 1 C, 2 C and 5 C rates, respectively. The performance of mixed C-LFMP electrode at each rate is stable and capacity drop with increase in rate is small. A behavior which is highly desirable for many applications. However, when decreasing the charge/discharge rate to 0.1 C, we find that the charge capacity cannot be recovered to the initial quantity and charge capacity of 57 mAh/g is delivered. By comparing the cycling performance and rate capability of two mixed electrodes it can be concluded that mixed commercial and in house C-LFMP demonstrate interesting and promising properties. It is highly probable that by further optimizing mixed ball milled and commercial olivine product the electrochemical performance can further be improved.

## 4 CONCLUSION

In an attempt to elevate electrochemical performance of olivine cathodes, C-$LiMn_{0.7}Fe_{0.3}PO_4$ powder is fabricated by solid state method. When physically mixed with industrial grade C-$LiFePO_4$ and C-$LiMn_{0.7}Fe_{0.3}PO_4$ battery performance of both set of mixed cathodes is studied in details and based on cycling performance and rate capability of two mixed electrodes it can be concluded that mixed industrial grade olivine and in house C-LFMP deliver superior performance compared to their constituents. A review of discussed electrochemical properties of single component and mixed olivine cathodes are displayed in Table 1. In the next step different ratios of in-house C-LFMP and commercial C-LFMP and C-LFP cathodes will be fabricated and electrochemically analyzed to achieve the combination electrode with the best delivered battery properties.

## REFERENCES


Crowell, J., 2006, Lithium ion battery power for marine devices. *Sea Technology*. 47(7): p. 29–31.

Ellis, B.L., K.T. Lee, and L.F. Nazar. 2010. Positive electrode materials for Li-Ion and Li-batteries. *Chemistry of Materials*. 22(3): p. 691–714.

Kraytsberg, A. and Y. Ein-Eli. 2012. Higher, Stronger, Better. A Review of 5 Volt Cathode Materials for Advanced Lithium-Ion Batteries. *Advanced Energy Materials*. 2(8): p. 922–939.

Laffont, L., et al. 2006. Study of the $LiFePO_4/FePO_4$ two-phase system by high-resolution electron energy loss spectroscopy. *Chemistry of Materials*. 18(23): p. 5520–5529.

Liu, Y. and C. Cao. 2010. Enhanced electrochemical performance of nano-sized $LiFePO_4$/C synthesized by an ultrasonic-assisted co-precipitation method. *Electrochimica Acta*. 55(16): p. 4694–4699.

Laruelle, S., et al. 2002. On the origin of the extra electrochemical capacity displayed by MO/Li cells at low potential. *Journal of the Electrochemical Society*. 149(5): p. A627-A634.

Morales, J., et al. 2004. Nanostructured CuO thin film electrodes prepared by spray pyrolysis: a simple method for enhancing the electrochemical performance of CuO in lithium cells. *Electrochimica Acta*. 49(26): p. 4589–4597.

Oh, S.-M., et al. 2010. High-Performance Carbon-LiMnPO4 Nanocomposite Cathode for Lithium Batteries. *Advanced Functional Materials*. 20(19): p. 3260–3265.

Shin, H.C., W.I. Cho, and H. Jang. 2006. Electrochemical properties of carbon-coated $LiFePO_4$ cathode using graphite, carbon black, and acetylene black. *Electrochimica Acta*. 52(4): p. 1472–1476.

Yu, F., et al. 2010. Porous micro-spherical aggregates of $LiFePO_4$/C nanocomposites: A novel and simple template-free concept and synthesis via sol-gel-spray drying method. *Journal of Power Sources*. 195(19): p. 6873–6878.

Zaghib, K., et al. 2013. Review and analysis of nanostructured olivine-based lithium recheargeable batteries: Status and trends. *Journal of Power Sources*. 232(0): p. 357–369.

*Maritime-Port Technology and Development – Ehlers et al. (Eds)*
 *ISBN 978-1-138-02726-8*

# Sea trials for validation of shiphandling simulation models—a case study

Ø. Selvik & T.E. Berg
*Department of Ship Technology, Norwegian Marine Technology Research Institute (MARINTEK), Trondheim, Norway*

S. Gavrilin
*Department of Marine Technology, Norwegian University of Science and Technology (NTNU), Trondheim, Norway*

ABSTRACT: Shiphandling simulation models are used for a number of purposes such as prediction of a ship's manouevring characteristics, verification of port and fairway design and training for pilots, masters and deck officers in shiphandling. The quality of the simulation model may vary, depending on the particular application of the model. Most training simulators utilise generic models for different types and sizes of vessel. Such models will not have the quality needed for studies of the manoeuvring performance of a specific vessel or for the investigation of collisions or groundings. In general, the validation process for simulation models used by the maritime community has tended to be neglected by simulator operators. This paper offers an overview of validation processes for simulation models and describes how the validation process has been specified and performed for the research vessel "R/V Gunnerus", which is owned by the Norwegian University of Science and Technology. Sea trials were performed in order to document IMO standard manoeuvres as well as low-speed manoeuvring tests specified by the master of the case vessel in collaboration with MARINTEK research staff. The quality and repeatability of sea trial measurements is discussed and illustrated by the outcomes of the IMO zig-zag tests. The final part of the paper describes how experience from the "RV Gunnerus" tests can be employed in further validation studies of sea-going vessels.

## 1 INTRODUCTION

Shiphandling simulation models are used in ship design, engineering studies related to the design of ports and fairways, development and verification of complex manoeuvres and training of ship-bridge staff. Model validation must be prioritised in efforts to obtain high-quality results from simulation studies and training activities.

This paper describes some of the R&D involved in the project "Sea Trials and Model Tests to Validate Shiphandling Simulation Models" (for short SimVal), which is briefly described in section 2. The project has selected a number of case vessels for which model tests and sea trials will be performed. Sea trials will consist of standard IMO manoeuvring tests (IMO, 2002) and case vessel-specific manoeuvres proposed by senior deck officers with experience of operating vessels similar to the case vessels. Section 3 discusses the concept of validation. Sections 4–6 describe one case vessel, the "R/V Gunnerus", its preparation for sea trials, how the trials were performed and some initial analysis of test results and sources of uncertainty. Section 7 looks at application of sea trial results in validation studies. Finally, section 8 presents lessons learned and specifies improvements that should be done prior to and during sea trials with the other case vessels in the project.

## 2 PROJECT DESCRIPTION

The Research Council of Norway (RCN) is through the MAROFF programme funding the Knowledge-building Project for Industry "Sea Trials and Model Tests to Validate Shiphandling Simulation Models". MARINTEK is the project owner, and scientific project partners come from Norway (NTNU), Belgium (Flanders Hydraulics Research and Ghent University), Brazil (University of Sao Paulo, Institudo SINTEF do Brazil), Japan (Tokyo University of Science and Technology) and Singapore (Singapore Maritime Academy). The web site http://www.sintef.no/Projectweb/SimVal/ lists the project partners and describes the project structure. Our Figure 1 shows the structure of the project. The shipping company partners have nominated

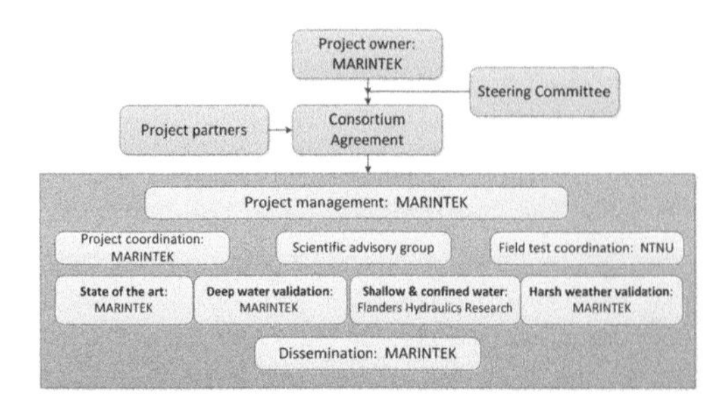


Figure 1. R&D project organisation.

case vessels to be investigated, while some of the scientific partners' own research vessels will also be used as case vessels. The Norwegian University of Science and Technology (NTNU) research vessel "R/V Gunnerus" has been used as the pilot vessel for planning, performing and analysing seatrial data. A model of the vessel has been used in captive testing and free-sailing tests in MARINTEK's towing tank and ocean basin respectively.

## 3 VALIDATION—A CHALLENGE FOR SIMULATION MODELS

Since the 1990s, government agencies and professional societies have been focusing increasingly on Verification, Validation and Accreditation (VV&A). A paper by Page (2004) stated that there is a variety of formal definitions of VV&A terms. For shiphandling simulation models there is a lack of commonly accepted definitions of the VV&A concepts. The 25th ITTC Manoeuvring Committee (ITTC, 2008) presented an overview of the accuracy and costs of various methods of documenting manoeuvring performance, and discussed their pros and cons. It was concluded that more work was needed to validate mathematical models, especially for non-standard manoeuvres at low speed or in shallow water. Two of the general conclusions from the workshop were:

- There is a general need for more quantitative verification and validation.
- There is a general need for a definition of how to validate a manoeuvring prediction method, i.e. what degree of accuracy is acceptable? If possible a "prediction quality index" should be defined.

For the project it was decided that definitions of the terms verification, validation (and accreditation) should as far as possible follow the new definitions that will be included in the 2014 version of the ITTC Terms of Reference. The ITTC procedure 7.5–02 06–03, rev. 2 (ITTC, 2011) should be used as a baseline for documentation of validation studies of numerical simulation models.

## 4 CASE VESSEL "R/V GUNNERUS"

NTNU's research vessel "R/V Gunnerus" is used as one of the case vessels in the SimVal project. The ship was built in 2006. NTNU's web-site states the following (http://www.ntnu.edu/marine/gunnerus): *"Gunnerus is equipped with the latest technology for a variety of research activities in biology, technology, geology, archaeology, oceanography and fisheries research. The ship is fitted with a dynamic positioning system. This provides optimal conditions for ROV operations and the positioning of any deployed equipment".*

Gunnerus is equipped with a diesel-electric system with 2 × 500 kW electric propulsion, powering twin 5-bladed ducted propellers (diameter 2 meters) with flap rudders, plus one 200 kW bow tunnel thruster. The ship has twin skegs with ice fins on the outer side of each skeg. The main dimensions of "R/V Gunnerus" are given in Table 1.

## 5 SEA TRIAL PLANNING

### 5.1 *Test planning*

The sea trials had four main objectives:

1. Document speed and manoeuvring capabilities to capture differences in the planned change of propulsion system on "R/V Gunnerus".
2. Generate data for comparison between full-scale, model tests and simulations of ship manouevres at cruising speed and low speed.
3. Contribute to the development of standard test procedures to characterise ship manoeuvring capabilities at low speeds.

Figure 2. "R/V Gunnerus" during the sea trials.

Table 1. "R/V Gunnerus" main dimensions.

| | Symbol | Value |
|---|---|---|
| Length overall | $L_{OA}$ | 31.25 m |
| Length betw. perp. | $L_{PP}$ | 28.9 m |
| Breadth moulded | B | 9.6 m |

4. Develop techniques for measuring environmental forces on ships at low speed.

Point 1 above relates to the innovation project "R/V Gunnerus"—a full-scale laboratory for testing future marine technology in a close collaboration between industry and academia", funded by the RCN. Points 2 and 3 were studied in the SimVal project funded by the RCN.

Point 4 relates to research on dynamic positioning.

### 5.2 *Measurement system*

During the full-scale sea trials both the ship's own instrumentation and external instrumentation were used. Data from the ship's system were recorded using the vessel's Kongsberg DP system and permanently installed Seapath 330+ from Kongsberg Seatex. This utilises a dual GPS antenna and a motion reference unit (MRU5+) to measure position, speed and motions. The output includes surge and sway speed, position, heading and ship motions such as heave, roll, pitch and yaw, in addition to angular speeds. They refer to a point close to the assumed centre of gravity (12.64 m forward of AP, on the centreline and 3.456 m above BL).

Three-axis accelerometers measured acceleration in a body-reference system in longitudinal (x-axis), transverse (y-axis) and vertical directions (z-axis). The measured acceleration includes the effect of change in the g-component due to a tilt angle (roll and/or pitch) of the z-axis.

Rudder angles were measured by wire position measurements. Before the sea trials, the rudder angle measurements were calibrated by reading the rudder angle on the rudder stock.

Propeller RPM was measured by optical readings of pulses from the shaft.

During the calm-water sea trials, a Wave Scan buoy from Fugro Oceanor was used to measure wind, waves and surface current. The buoy measures the following oceanographic parameters: wave height, spectrum and direction, surface current (depth approx. 1.5 m) and direction, in addition to CTD (Conductivity, Temperature and Depth) data from the surface. The following meteorological parameters were measured: Wind speed and direction, air temperature and barometric pressure.

## 6 SEA TRIALS AND RESULTS

### 6.1 *Sea trial test programme*

In order to meet the objectives listed in section 5.1, a comprehensive test matrix was prepared. Several other tests were performed in addition to standard IMO manoeuvres such as zig-zags, turning circles and spiral test. The manoeuvres were repeated several times in order to assess the effects of uncertainty in sea trials. Comprehensive speed trials were performed. Low-speed tests, such as zig-zag using the bow thruster, acceleration, crabbing and reversing were performed, as was a bollard pull test. A DP test programme was included to aid the development of techniques for measuring the effects of environmental forces on low-speed ship motions.

### 6.2 *Environmental conditions during trials*

Important environmental influences on the results of sea trials are wind, waves and currents. Wind speed during trials was relatively stable, normally in the range 4–7 m/s, which correspond to Beaufort Scale 3–4 and heading 250°–300°. Wave height was typically 0.1–0.3 m, corresponding to sea state 2. The most changeable effect was current (both direction and speed). Hourly measurements of current direction during the second day of trials (from 8 to 20 hours) are shown in Figure 3. Current speed was in the range 0–0.5 m/s. Both the tides and two rivers flowing into the fjord near the area of trials influenced currents.

### 6.3 *Test results and analysis*

In order to investigate repeatability, twelve 10°/10° zig-zag tests under autopilot control (seven to starboard and five to port side) were carried out. The results are presented in Figures 4 and 5.

Mean values and standard deviations are shown in Table 2. A relatively wide scatter in results can be observed. It is worth noting that the first overshoot angle is generally smaller during the first turn to portside (PS) and larger during the first turn to starboard (SB). A narrower scatter during series 200× than in 10xx can also be seen.

The influence of engine power on the result can be illustrated by the outcomes of 20°/20° zig-zag

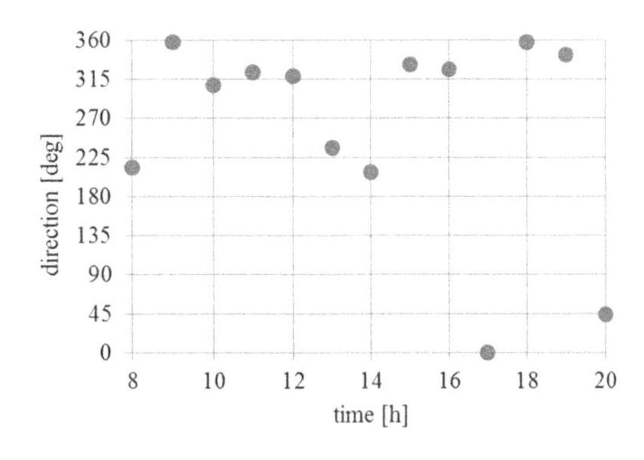


Figure 3. Current direction measured by wave buoy during second day of trials.

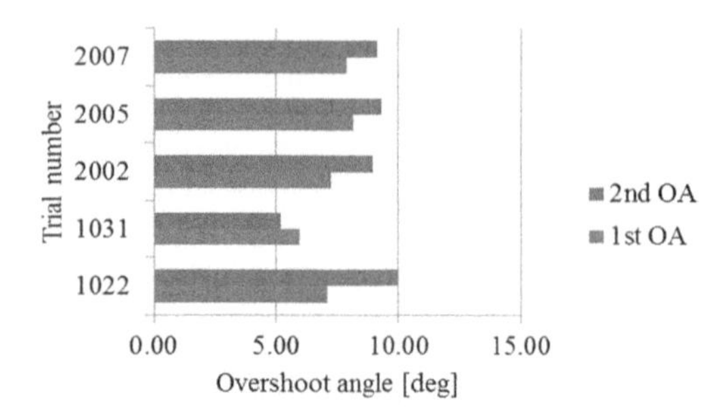


Figure 4. Overshoot angles during 10°/10° port zig-zag tests.

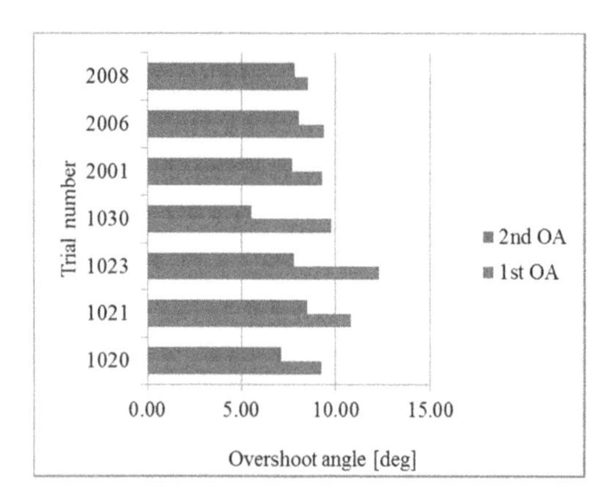


Figure 5. Overshoot angles during 10°/10° starboard zig-zag tests.

Table 2. Results of 10°/10° zig-zag tests.

| | 1st OA | | 2nd OA | |
|---|---|---|---|---|
| Side | PS | SB | PS | SB |
| Mean [deg] | 7.24 | 9.89 | 8.51 | 7.45 |
| σ [%] | 11.5 | 12.9 | 22.3 | 12.6 |

tests performed at four different power settings. Lists of tests with corresponding power settings and results are shown in Table 3 and Figure 6.

## 6.4 *Sources of uncertainty*

Analysis of full-scale zig-zag trials revealed a relatively wide scatter of the results. This scatter could be due to several different factors.

Among others sources of uncertainty in the test results, a significant effect may be due to the method of calculation of parameters of interest.

Table 3. 20°/20° zig-zag trials at different power settings.

| Run N | Side | Power [kW] |
|---|---|---|
| 2025 | SB | 423 |
| 2030 | SB | 368 |
| 2029 | SB | 367 |
| 2014 | PS | 244 |
| 2012 | SB | 242 |
| 2011 | SB | 239 |
| 2019 | PS | 99 |
| 2018 | SB | 99 |

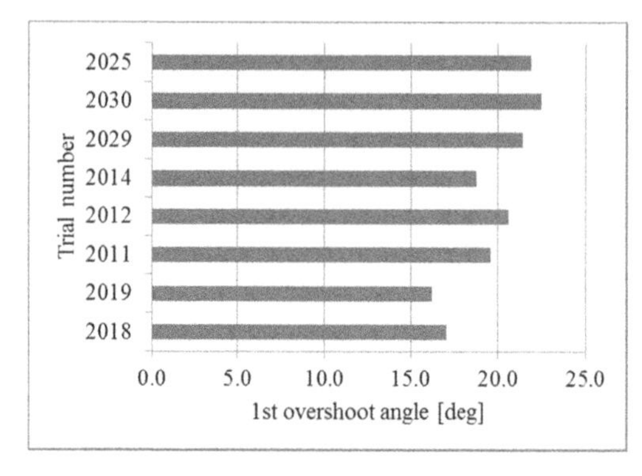


Figure 6. Comparison of first overshoot angles from 20°/20° zig-zag trials at different power settings.

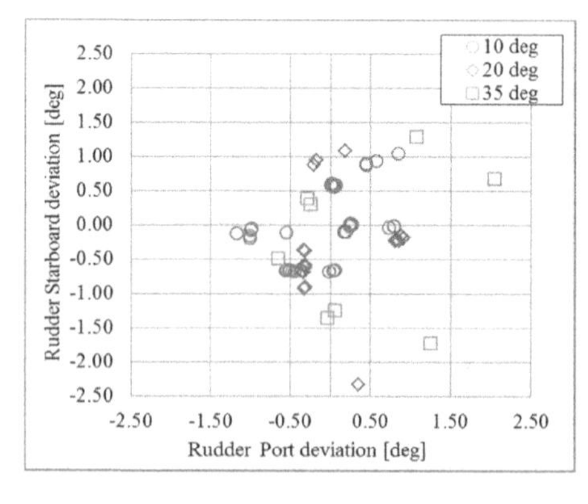


Figure 7. Rudder deviations from commanded values during all trials.

When tests are maintained by autopilot during approach, the autopilot tries to maintain a constant heading which diverts the rudder from its neutral position. The amplitudes of these deviations observed during trials could be as much as

Table 4. Mean approach engine power and velocity and wind data during 10°/10° zig-zag tests.

| Side | Run N | Power [kW] | Approach speed [kn] | Mean wind speed [m/s] | Relative wind direction [deg] |
|---|---|---|---|---|---|
| Starboard | 1020 | 248 | 9.9 | 4.6 | –2 |
| | 1021 | 251 | 9.7 | 6.7 | 7 |
| | 1023 | 241 | 10.4 | 5.7 | 169 |
| | 1030 | 242 | 10.4 | 4.1 | 157 |
| | 2001 | 244 | 10.9 | 5.6 | 180 |
| | 2006 | 243 | 9.6 | 5.9 | –12 |
| | 2008 | 241 | 10.8 | 4.7 | 165 |
| Portside | 1022 | 250 | 9.7 | 5.7 | 285 |
| | 1031 | 240 | 10.4 | 4.4 | 85 |
| | 2002 | 243 | 11.3 | 4.9 | 35 |
| | 2005 | 243 | 9.5 | 6.1 | 181 |
| | 2007 | 240 | 10.8 | 4.4 | 24 |

5 degrees. It may therefore be quite difficult to define the initial heading at the time of the first execute, as this time is not clear. Another way to define the approach heading is to derive it from headings during second and third execute, but then the reaction time of autopilot needs to be known. From the experience of 10°/10° zig-zag trials it follows that various approaches to calculation of overshoot angles can lead to a difference of up to 2°.

It was also observed that the measured rudder angle could significantly deviate from the value commanded. The combined results for all manoeuvres are shown in Figure 7 for both starboard and port turns. Preliminary modeling of rudder effect on overshoot angles during 10°/10° zig-zag tests shows that a deviation from the commanded value of both rudder angles of only one degree leads to change of overshoot angle value on 12%. This indicates that uncertainty in rudder angles is one of the important contributions to the total uncertainty of the result.

Propulsion characteristics such as velocity, RPM and engine power also have important effects on results. Measurements show that power during the approach phase prior to starting the test manoeuvres was in the range of 240 – 250 kW for each propeller, even though the desired power was 245 kW. Sometimes the power was slightly different for the port and starboard propellers. Vessel speed through the water during this phase is another parameter of interest, but this is difficult to define because the current velocity is unknown. Vessel speed in this phase can also be influenced by the actual wind speed and direction.

Environmental forces also contribute to the uncertainty of results. Modeling shows that at some relative wind directions, winds of Beaufort Scale 4 can change the values of overshoot angles by up to 50% compared to values under calm conditions. This means that even during relatively calm conditions, wind effects should be taken into account.

Information on mean approach engine power and velocity as well as wind data during 10°/10° zig-zag tests are shown in Table 4.

## 7 USING SEA TRIAL DATA FOR VALIDATION STUDIES

The results of the sea trials may be used to validate a simulation model. The sea trials showed the test results have significant scatter. This scatter needs to be taken into account when the output of simulations is compared with measured values. However, it may not be economically feasible to repeat tests with the same parameters in order to determine the uncertainty of sea trial results. Instead, the effects of different factors that affect the uncertainty can be estimated using mathematical models. Instead of estimating the uncertainty of sea trials due to presence of environmental effects and others factors, it is possible to use results of trials "as are" (although these trials should be performed under conditions as close to "ideal" as possible) and then reproduce all the trial conditions in the simulator. This would limit the uncertainty of any sea trial itself to the precision of measurements of parameters of interest (for example overshoot angles for zig-zag manoeuvres). However, the uncertainty of the simulator output increases because of the limited ability of models to reproduce all physical effects. In such cases, a validation procedure can be based on two conditions: that the results of full-scale trials lie within a chosen confidence interval of the simulation output, and that this confidence interval is narrow enough to be useful.

Captive and free-sailing model tests with a model of "R/V Gunnerus" were performed and are currently being analysed and compared with the sea-trials data. The free-sailing tests in MARINTEK's Ocean basin performed without environmental effects such as wind, waves and currents will be compared with sea-trial data. This comparison will be completed in October 2014. The captive PMM model tests will be used to generate manoeuvring coefficients for MARINTEK's Vessel Simulator, VeSim, which will simulate the sea trial manoeuvres in late 2014. The validity of the simulations will be studied by means of a two-step procedure. The first step will be a comparison with the free-sailing model test results while the second will be a comparison with the sea trial measurements.

## 8 LESSONS LEARNED AND IMPLEMENTATION FOR LATER SEA TRIALS

Sea trials are expensive, and to gain the most from sea trials careful planning is essential. Good observation data of the environmental parameters are also important. Lessons learned from previous projects are that there can be large differences between predicted, measured and observed (by ship crew) environmental data.

We need to be aware of the sampling frequency of the data-storage system in order to ensure that transient ship motions are captured. This is more important for small manoeuvrable ships (such as "R/V Gunnerus") than for large slowly moving vessels.

The rudder angle's zero position and some extra angles should be checked against a manual reading on the rudder stock.

More attention should be paid to the effects of wind during zig-zag tests. As far as possible the vessel should be heading into the wind at the start of the manoeuvre. Preliminary simulations for "R/V Gunnerus" show that the vessel's initial heading with respect to wind direction had a fairly strong effect on the values of overshoot angles, as the vertical area of the superstructure is large compared to the underwater area of this vessel.

The analyses of the sea trial data drive home the need for writing a comprehensive test log. This can best be done by a dedicated researcher with previous experience of documenting sea trials. A template for the log should be prepared on the basis discussions with the test vessel's deck officers.

## ACKNOWLEDGEMENTS

The authors thank the Norwegian University of Science and Technology for allowing us to use the University's research vessel "R/V Gunnerus" as one of the case vessels for this project. Special thanks are due to the crew on the vessel for their assistance and patience during the sea trials. The project is funded through the MAROFF research programme of the Research Council of Norway (project no. 225141).

## REFERENCES

Berg, T.E and Ringen, E. 2011: Validation of shiphandling simulation models. Proceedings of the 30th International Conference on Ocean, Offshore and Arctic Engineering, paper no. OMAE 2011-50107, Rotterdam, The Netherlands, June 2011.

IMO 2002: Standards for ship manoeuvrability, Resolution MSC137(76), IMO, London 2002.

ITTC. 2008: Report of the Manoeuvring Committee. 25th International Towing Tank Conference. Fukuoka, Japan, 2008.

ITTC, 2011: Validation of manoeuvring simulation models. ITTC—Recommended procedures and guidelines, 7.5-02 06-03, Revision 02, ITTC 2011.

Pace, D.K: Modeling and simulation verification and validation challenges. John Hopkins APL Technical Digest, Vol. 25, no. 2 (2004), p. 163–172, Maryland, USA.

*Maritime-Port Technology and Development – Ehlers et al. (Eds)*
 *ISBN 978-1-138-02726-8*

# Anti-corrosion method for marine steel structure by the calcareous deposits

T. Iwamoto, Y. Suzuki, K. Ohta & K. Akamine
*IHI Corporation, Yokohama, Kanagawa, Japan*

Y. Yamanouchi
*IHI Asia Pacific Pte Ltd., Singapore*

ABSTRACT: We have developed our original anti-corrosion system for repair on the marine steel structures using the calcareous deposits. Experiments were performed under the actual environment in Singapore. As a result, by cathodic electrolysis test in the actual environment, we verified that the calcareous deposits could be formed as the laboratory tests. Moreover, we found that the calcareous deposits have effective for anti-corrosion, this process is prospective as economic anti-corrosion method for repair on the marine steel structures.

## 1 INTRODUCTION

The Marine steel structures are important infrastructures and effective anti-corrosion method are needed for the maintenance and repair. So we need select the proper anti-corrosion method from them in consideration of many different kinds of factors such as structure's importance, service life, difficulty of repair, life-cycle cost.

In this method, we set the electrodes counter to the ferrous structures in seawater so as to form calcareous deposits on the steel surface by external power supply. Here if the cathodic current is impressed to the steel structure, pH around the steel surface is increased, then calcareous deposits composed of $CaCO_3$ and $Mg(OH)_2$ is formed onto the surface as the following equations (Cox 1940, Humble 1948):

$$H_2O + 1/2O_2 + 2e \rightarrow 2OH^- \quad (1)$$

or

$$2H_2O + 2e \rightarrow 2OH^- + H_2 \quad (2)$$

$$OH^- + HCO_3^- \rightarrow H_2O + CO_3^{2-} \quad (3)$$

$$CO_3^{2-} + Ca^{2+} \rightarrow CaCO_3 \quad (4)$$

and

$$2OH^- + Mg^{2+} \rightarrow Mg(OH)_2 \quad (5)$$

These calcareous deposits consist of two layers, $Mg(OH)_2$ layer play the role to prevent corrosion of steel by alkali, $CaCO_3$ serves a function to decrease the diffusion of dissolved oxygen in sea water onto the steel surface (Akamine & Kashiki 2003).

After the formation of calcareous deposits, it is maintained stably for a long term by conventional cathodic protection. But the protective current density becomes extremely small by formed calcareous deposits. Therefore, an economical anti-corrosion design is enabled, because of galvanic anode life extension, reduction of the anode setting amount compared to conventional cathodic protection. It means that this process has merit in terms of life cycle cost.

Moreover by the forming of calcareous deposits initially, the combination with cathodic protection makes it possible to prevent the area of structures at tidal zone from corrosion (Hamada et al. 2005). This advantage is characteristic of this process and this combination of calcareous deposits and cathodic protection may become alternative to painting in the anticorrosion process to the area at tidal zone.

This system is known as one of anti-corrosion method for port steel structures, we had some experiments to establish the processing condition such as the electric condition, thickness of deposits in domestic harbor (Hamada et al. 2005, Coastal Development Institute of Technology 2009).

Also this system is so much efficient and economical in the case of that underwater work by diver is difficult. The construction of conventional cathodic protection in the repair on old structures needs installation of sacrificial anodes by divers. But in severe marine conditions such as high tide speed and deep water area, the operation is so difficult and unsafe. Especially in this case, the effect of reduction of installed anode is much more economical.

But the chemical composition of deposits related to anti-corrosive performance and growth rate are influenced by pH and chemical composition of water, tide condition and so on (Akamine & Kashiki 2004).

So we inspected the growth rate and anti-corrosive performance of the calcareous deposits under actual environment in Singapore, then verified the applicability of this system to the offshore structure in tropical area. Here we report the results of these experiments.

## 2 EXPERIMENTAL

### 2.1 *Experimental site*

Experiments were performed on the one quay in Jurong Shipyard Pte Ltd, which was located in the south west side in Singapore (Fig. 2). The salinities and pH of the water in the area were 2.8–3.0% and 7.7–7.9 respectively. But the chemical composition of deposits related to anti-corrosive performance and growth rate are influenced by pH and chemical composition of water, tide condition and so on (Akamine & Kashiki 2004).

There is little tide in the area. The protective current density to bare steel was about 0.100–0.300 A/m$^2$ in the potential of –900 mV vs. Ag/AgCl. In recommended practice by DNV, the initial protective current density under tropical and shallow condition is 0.150 A/m$^2$ (DET NORSKE VERITAS 2010), the values measured in the area are bigger, corrosive environment in this site is estimated to be more severe than usual tropical sea.

Figure 1. Calcareous deposits to port steel sheet piles in Japanese port (Research Group for Corrosion Protection and Repair Method 2013).

Figure 2. Experimental site before installation of test bed facilities.

### 2.2 *Experimental setup*

The designs of facilities for these experiments are shown in Figure 3. For these experiments, we have manufactured davit structures to set the steel test pieces in sea water and installed them on the land side of the quay (Fig. 4).

One of them is used for the investigation of the corrosive environment, other is used for the

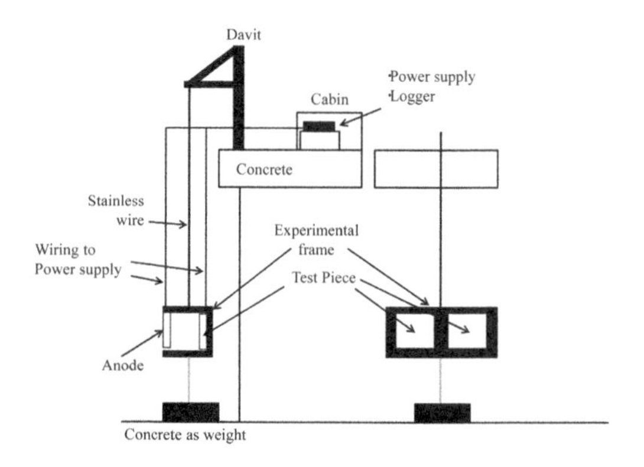


Figure 3. The images of experimental facilities.

Figure 4. Appearances of test site after installation of test bed facilities.

Figure 5. Appearances of the experimental frame (The brown plastic pole in the right picture is Ag/AgCl/seawater reference electrodes).

Figure 6. Appearances of the surface of the test pieces after electro-deposition (The current density: 2.0 A/m$^2$, deposition term: 142 hours).

conventional cathodic protection test. Others are used to evaluate the anti-corrosive performances of calcareous deposits.

Each experimental frame has two pairs of anode and test pieces (cathode), the wirings were set between each anodes and cathodes for the application of electric current.

Test pieces are SS400 carbon steel plates (SS400 means Japanese Industrial Standard (JIS) material designation. Rolled steel for general structure, tensile strength: about 400 MPa), whose sizes are 350 × 400 × 6 mm and the just one side of these plates were used as the test area (that is, test area is 0.14 m$^2$). The surfaces were prepared by grid blast in ISO Sa2 1/2 grade.

All sides of these plates except the test areas were painted by the epoxy-paint for the insulation and anti-corrosion in these experiments. The sizes and materials of anodes are same as the test pieces (Fig. 5).

Then, in order to determine the relative potential against Ag/AgCl/seawater reference electrodes on the surface of each test piece in the experiments, the reference electrodes are set around each test piece (Fig. 5).

First, in order to examine the growth rate in the forming of calcareous deposits, we formed some test pieces with calcareous deposits. We set the cathodic current density condition in calcareous deposition to 1.0 A/m$^2$ and 2.0 A/m$^2$ from the results of laboratory tests (Akamine & Kashiki 2002).

Next, we determined the protective current density in the case of that the potential is retained to −900 mV vs. Ag/AgCl for the evaluations of the anti-corrosive performance of the calcareous deposits.

## 3 RESULTS AND DISCUSSION

### 3.1 *Calcareous deposit formation*

Figure 6 shows the appearances of the surfaces on the test pieces after the electro-deposition. Here, the current density and deposition term were set 2.0 A/m$^2$, 142 hours respectively. The white calcareous deposits are formed on the test area of steel plates, the peeling and blister of the deposits on the test surface were few. If the current density is too high, the evolution of hydrogen from the surface of the steel occurs more intensely and it leads to the peeling and blister of the deposits (Akamine & Kashiki 2002).

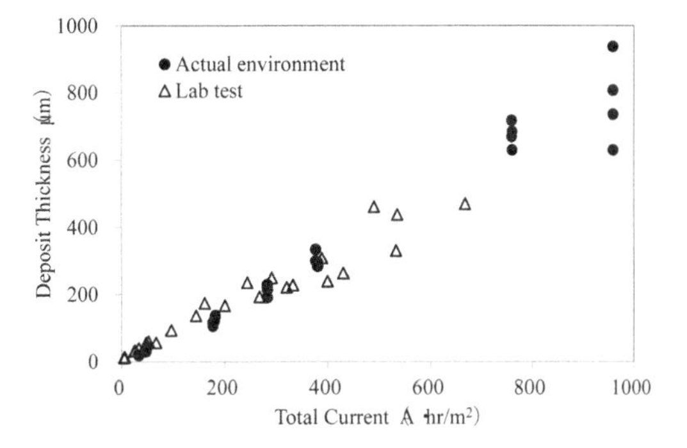


Figure 7. Relationship between the total current and the calcareous deposit thickness under the actual environment in Singapore compared with the ones in laboratory test.

Figure 7 shows the relationship between the total current and the calcareous deposit thickness in these experiments. The relationship is linear and this shows that the thickness of calcareous deposits is proportional to the total current by the external power supply at the current density condition of 1.0 A/m$^2$ and 2.0 A/m$^2$.

Here, the growth rate of coating was estimated to be 40 micro-meters per days at the current density condition of 2.0 A/m$^2$. This result under actual environment in Singapore is almost same as the one under the laboratory and the domestic harbor (Suzuki & Akamine 2009).

Moreover, we found that the chemical compositions are also same as the deposits formed in the laboratory tests by XRD analysis (Akamine & Kashiki 2002). So under the actual environment in tropical area, the forming of calcareous deposits is achieved as experiments under the laboratory and the actual environment in Japan.

### 3.2 *The evaluation of anticorrosive performance of calcareous deposits*

Next, Figure 8 shows the results of protective current density for one month. In the long-term monitoring, the protective current density to bare steel was about 0.100–0.300 A/m², while the one in the case of the test plate with 300 micro-meters coating were lower than the one of bare steel, the values were 0.020–0.030 A/m² after 10 days.

By the way, the instantaneous increases shown in the data sets of coated test piece results from the power cut in test bed, so these are negligible. This also means that the calcareous deposits decrease the protective current density and the

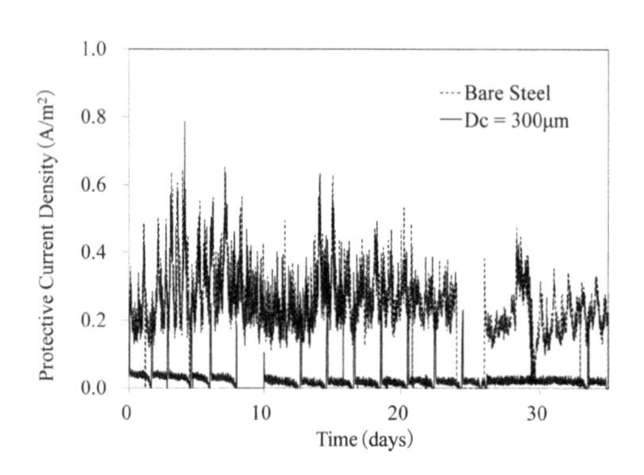


Figure 8. Results of protective current density for about one month.

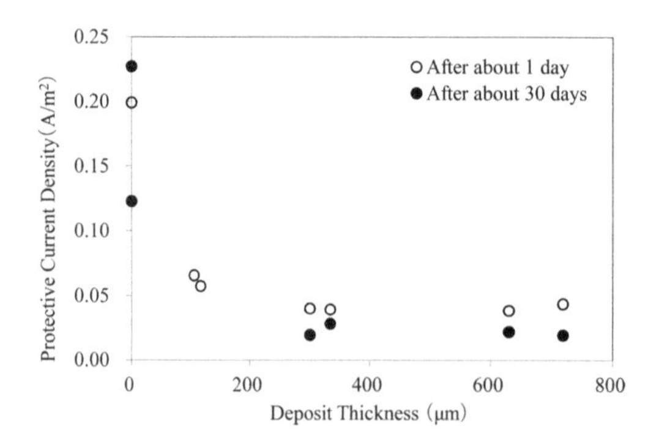


Figure 9. Relationship between the calcareous deposit thickness and protective current density.

coating has good anti-corrosive performance. At last in Figure 9, we show the relationship between the thickness of coating and protective current density in 1 day and 30 days of the determination.

This figure shows that the calcareous deposits decrease the protective current density and the decrease rate is bigger if the thickness is larger.

But the protective current densities, in the all cases of that thickness are over about 300 micro-meters, are almost equal and 0.035–0.040 A/m² after 1 day, 0.020–0.030 A/m² after 30 days respectively. These values are 1/5 of the one to the bare steel. This means that the amount of sacrificial anodes is saved to 1/5 compared to the conventional cathodic protection if the steel structure is covered with calcareous deposits.

Then, the combination process of calcareous deposit formation and cathodic protection is prospective as economic anti-corrosion method.

Moreover this figure also shows that anti-corrosive performance is almost uniform regardless of the thickness if the thickness is over about 300 micro-meters. So we estimated that the optimum thickness is about 300 micro-meters and this result matches to the ones under environment in Japan (Hamada et al. 2005, Suzuki et al. 2008 & Coastal Development Institute of Technology 2009).

## 4 CONCLUSIONS

We have developed original anti-corrosion system for repair on the marine steel structures using the calcareous electro-deposition. Here, we had some experiments in Singapore for the verification that this method was effective as the anticorrosion method under actual environment in tropical area. As a result, we verified that the calcareous deposits were formed as the same as laboratory tests. With this coating, the protective current density was decreased to 1/5 of bare steel. Then, the calcareous deposits are also effective for anti-corrosion, the combination process of calcareous deposit formation and cathodic protection is prospective as economic anti-corrosion method.


## ACKNOWLEDGEMENTS

This research was supported by Maritime Innovation and Technology (MINT) Fund from The Maritime and Port Authority of Singapore, and Jurong Shipyard Pte Ltd (JSPL)'s providing the site for experiments. We offer our appreciation to these supports.

## REFERENCES


Akamine, K. & Kashiki, I. 2002. Corrosion Protection of Steel by Calcareous Electrodeposition in Seawater (Part1) Mechanisim of Electrodeposition, *Zairyo-to-Kankyou (Corr.Eng.)* 51(11): 496–501.

Akamine, K. & Kashiki, I. 2003. Corrosion Protection of Steel by Calcareous Electrodeposition in Seawater (Part2) Mechanisim of Growth, *Zairyo-to-Kankyou (Corr.Eng.)* 52(8): 402–407.

Akamine, K. & Kashiki, I. 2004. Corrosion Protection of Steel by Calcareous Electrodeposition in Seawater (Part3)—Effects of Dilution-, *Zairyo-to-Kankyou (Corr.Eng.)* 53(7): 354–357.

Coastal Development Institute of Technology. 2009. *Corrosion Protection and Repair Manual for Port and Harbor Steel Structures.*

Cox, G.C. 1940. Anticorrosive and Antifouling Coating and Method of Application, *US Patents* 2:200.

DET NORSKE VERITAS. 2010. Cathodic Protection Design, *Recommended Practice* DNV-RP-B401.

Hamada, H., Kanesaka, K., Suzuki, Y. & Miyata, Y. 2005. An Experimental Study on the Steel Corrosion Prevention Effectiveness of Combination of Electrodeposition and Cathodic Protection, *Technical Note of the Port and Airport Research Institute.* No.1113.

Humble, R.A. 1948. Cathodic Protection of Steel in Sea Water with Magnesium Anodes, *Corrosion.* 4: 358–370.

Suzuki, Y., Akamine, K., Kanesaka, K. & Imazeki, M. 2008. Development of New Anti-Corrosion Method (IECOS) for Marine Steel Strucutures, *IHI Engineering Review* 41(2): 58–67.

Suzuki, Y. & Akamine, K. 2009. Development of New Anti-Corrosion Method (IECOS) for Marine Steel Strucutures, *Bousei-Kanri (Rust Prevention and Control Japan)* 53(9): 349–356.

Research Group for Corrosion Protection and Repair Method. 2013. *Practical Handbook about Corrosion Protection, Repair and Maintenance for Port and Harbor Steel Structures.*

*Maritime-Port Technology and Development – Ehlers et al. (Eds)*
*© 2015 Taylor & Francis Group, London, ISBN 978-1-138-02726-8*

# A risk based approach to the design of unmanned ship control systems

Ø.J. Rødseth & Å. Tjora
*Norsk Marinteknisk Forkningsinstitutt AS (MARINTEK), Norway*

ABSTRACT: An unmanned ship is a concept that causes concern related to the possible dangers it represents to itself, its cargo, other ships and to the environment. In addition, the unmanned ship is largely an unexplored concept: It is not just an extrapolation from existing ships, as it needs new sensors, new functionality and new types of shore support. This means that one has to look anew on how efficiency, safety and security are ensured for such ship systems. The MUNIN project is doing a concept study for an unmanned ship to investigate these issues among others.

As little or no previous experience exists, MUNIN has selected a risk based approach to the concept design. This is based on established methods, including formal safety assessment from IMO and an architecture framework from the shipping domain. The resulting methodology and results from its application will be reported in this paper. This includes hazard identification and risk assessment, development of risk management "hypotheses" and hypothesis testing through human interface simulators, Monte Carlo simulations, FMECA and other methods that have been adapted to the risk management approach.

The most critical aspects of this methodology are arguably hazard identification and risk assessment. Errors or omissions in these phases can lead to serious issues being overlooked or rated as not dangerous. The paper will also describe the approach used to make these phases as robust as possible.

## 1 INTRODUCTION

The MUNIN project[1] is developing a concept for an unmanned merchant ship of Handymax size. The unmanned phase will only be on the deep sea part of the voyage. Crew will be onboard for the departure from and approach to port. The concept will include new sensor systems, new technical operation and maintenance procedures, autonomous navigation functions, a Shore Control Centre (SCC) and other components (Rødseth & Burmeister 2012).

In general, autonomy can be defined as follows: *An autonomous agent is a system situated within and a part of an environment that senses that environment and acts on it, over time, in pursuit of its own agenda and so as to affect what it senses in the future* (Franklin & Graesser 1997). This definition is useful as it captures both the ability to sense and to react according to sensor information and own agenda, including the effects that own actions has on the environment and consequent changes in sensor input. However, it does not cover the different degrees of autonomy as both a simple thermostat and a human fall under this definition.

Autonomy implies that the system can do actions beyond what can be described by an analytic function of system state and sensor input and, hence, needs "intelligent" control methods such as neural networks, fuzzy logic, genetic programming or other heuristic algorithms.

This means that it may be difficult to identify and determine the importance of all critical use cases for the autonomous system and, in turn, that one may have to employ other design methods than for more conventional control system. This paper will discuss a set of methods that have been used in the MUNIN project to address this uncertainty.

## 2 INDUSTRIAL AUTONOMOUS SYSTEM

Autonomous systems are getting common in the research and defence community and are also starting to get a stronger foothold in the commercial industry. Industrial applications will be characterized by a very strong focus on benefit compared to cost, higher cost sensitivity in general, the need for highly deterministic operational characteristics and in ensuring that the new systems do not pose any danger to humans, environment or themselves. Cost will be the relevant combination of

[1]The MUNIN (Maritime unmanned ships through intelligence in networks) project is part funded under the EU 7th Framework program through the Grant Agreement number 314286. See www.unmanned-ship.org.

acquisition, operation, deployment, recovery and potential system loss costs.

While these factors will be general to all industrial autonomous systems, significant differences will be seen along the axis from "small" to "large" systems. Small systems will typically be designed for low cost and may even be disposable after use. Larger systems may have very high costs and a significant potential for causing damage. The MUNIN project, investigating a bulk carrier of about 75 000 tons deadweight, is clearly in the "large" segment.

For large systems like MUNIN, the high implementation costs and high potential for damage will translate to more specific requirements: 1) *High reliability*: It is not acceptable to lose or seriously damage the system; and 2) *High availability*: The system is part of core business processes and will need continuous availability while in operation.

For all industrial autonomous systems, a third requirement will also apply: 3) *Highly deterministic*: The system should do exactly what it is supposed to do and no more.

The latter point means that industrial autonomous systems will have limitations in the degree of autonomy they can make use of. This will be further discussed in the next section.

The successful design of an industrial autonomous system requires a combination of techniques that address the following challenges:

1. Proven safety and security levels, including proof that the system behaves deterministically within its defined boundaries and constraints.
2. Cost-efficient system solution with the appropriate levels of reliability and availability.
3. Appropriate balance between autonomy and human control for the intended tasks and for the communication links available.

This paper will mainly address the first point, but parts of the methods will also be applicable to points 2 and 3.

## 3 CHARACTERISTICS OF THE SYSTEM

This section discusses the special characteristics of industrial autonomous systems in general and large industrial systems in particular.

### 3.1 *Complexity and limited autonomy*

One consequence of the need for reliability and deterministic behaviour is that the designers need to reduce complexity to a manageable level in all parts of the system and its environment. This implies that the system should mostly rely on automated operation and only have limited "intelligence". One will normally see that the industrial autonomous system is continuously monitored from a remote control station and that it is most of the time operating in a form of automatic mode, e.g. on autopilot. Autonomy is either used to relieve the operator of tedious and expected disturbances or to handle situations where the operator cannot immediately intervene, e.g. due to communication problems. All really complex decisions are made by operator. This is illustrated in Figure 1.

"Autonomous" is here defined as freedom for the system to choose actions within certain constraints, while "Intelligent" is defined as the unrestricted freedom to act on any sensor input and system or environment state. A typical constraint for autonomous control will be that a planned track can be deviated from within a certain distance, e.g. to allow autonomous collision avoidance. Together with automatic control, direct remote control and "fail to safe" functionality, this divides autonomy into five levels. This is much less detailed than, e.g. the ALFUS (NIST 2007) taxonomy. On the other hand, this classification is more suitable to show relationship between autonomy level and determinism, which is interesting in the context of industrial autonomous systems. These will normally not accept higher levels than "Autonomous".

Although many systems, including MUNIN, could be designed as a fully remotely controlled system, there are issues related to communication availability and cost as well as workload on remote operators that favours the use of some autonomy in the system design. This also points to another problem for most autonomous system: The impact of the communication links' quality of service. One can rarely guarantee 100% availability or sufficient bandwidth at all times.

### 3.2 *Navigation, path planning and re-planning*

Object detection and recognition as well as path planning, obstacle avoidance and path re-planning are complex and central operations for mobile

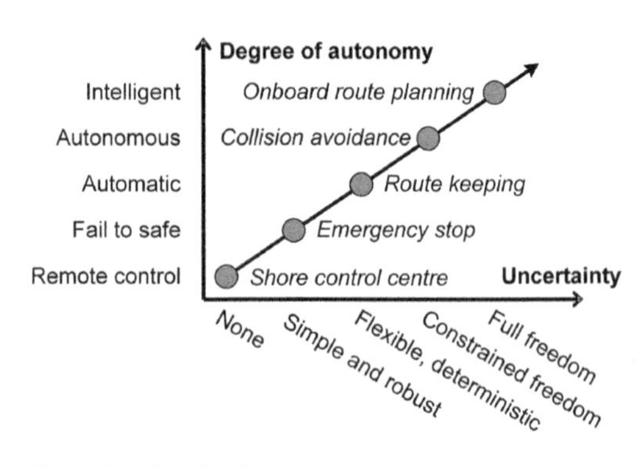


Figure 1. Levels of autonomy.

autonomous systems. These functions are also good examples of how reduction in complexity has been applied in MUNIN. Central in this is the limitation in operational area to the high seas and the use of the continuously manned Shore Control Center (SCC).

For object detection and recognition this means that objects only needs to be classified in three categories: 1) *Known ships*—typically detected by radar and recognized from Automatic Identification System (AIS) transmissions; 2) *Objects that can be ignored*—small objects of no significance that cannot cause harm; 3) *Other*—these must be inspected by the SCC operator for final classification as, e.g. 3a) *Objects that can be a danger to own or other ships*—these should be reported to maritime authorities and avoided; or 3b) *Objects related to accidents at sea*—these need to be reported to search and rescue authorities and may also in some cases be given useful assistance.

For navigation, the deep sea scenario means that the autonomous navigation system normally will at most have one or two other objects to relate to regarding path planning and avoidance. Also, the distance to objects will normally give ample time to take evasive maneuvers and in general make it possible to avoid difficult situations. More complex situations can be handled by the SCC operator through remote control. Fail to safe would, e.g. be to go dead in water.

### 3.3 *Communication quality and fail to safe*

For monitoring and control of the unmanned ship, communication is essential. As the unmanned operation is in the open sea, far from shore, communications between ship-based and shore-based systems must rely on satellite communication systems.

During normal operation, the communication from the ship to the shore will be short messages containing ship status updates for the shore control and monitoring. These messages will have relatively low requirements to the quality of the communication channel (Rødseth et al. 2013).

The main communication challenges arise when the SCC remotely controls the ship or when shore control need large amount of data (e.g. images or video from camera or radar) in order to get a full understanding of the situation. This will typically occur when the ship's autonomous controller is not able to handle the situation itself. Thus, the need for high bandwidth and high quality communication typically arises in situations that may be critical. As communication quality in general cannot be guaranteed, this also means that the system needs a "fail to safe" function as backup in such situations.

To properly design the system and determine when autonomous mode is sufficient, when remote control is required and when it is necessary to fail to safe, one need to include variable communication quality in the system design phase.

### 3.4 *Technical reliability*

On an ordinary manned voyage, there are people on board that are able to monitor and maintain the on-board equipment as well as do basic repairs if equipment should fail.

The unmanned ship will have no people on board to handle maintenance during most of the voyage. Failure of on-board equipment may lead to degradation or failure of important functions, which again will affect the ship's ability to perform the intended operations and may also result in safety problems as well as in loss of the ship.

This gives strict requirements to the technical reliability as well as the monitoring and maintenance strategies for the ship and its systems. The design phase needs to include analysis of system component reliability.

### 3.5 *Operational safety and security*

Due to a large ship's potential of causing damage as well as the value of the ship and its cargo itself, safety is of high importance for the unmanned vessel concept. The system must be designed to handle any safety issues that arise.

There are also several security challenges that must be handled. From normal shipping, issues with piracy and stowaways are known and this is also an issue for the unmanned ship. As the ship will be designed to be controlled from the shore, the possibility to hack the ship's systems must also be considered. The possibility that hostile parties disturbs or fakes communication or navigation signals is likewise an important concern.

The general challenge is that the design method must show that the risks associated with the unmanned ship are less than or at least not worse than the corresponding risks associated with traditional ships.

### 3.6 *Remote control efficiency and safety*

The responsibility of the remote or Shore Control Centre (SCC) is to monitor the unmanned ships and intervene or remotely control them if necessary. It can be assumed that a single operator will monitor several ships, but there may also be situations when focus on one of the ships is necessary (Porathe 2014). This raises several important issues for the SCC:

1. Human errors can still occur and measures must be taken to prevent this as far as possible. This has implications on technology, organisation and training.

2. Efficiency and cost is an important issue, so manning levels and organization must be optimized. This includes operations as well as technical monitoring and maintenance.
3. Cost and availability of communication is an issue as well as information overload for the operator. Right information at the right time to the right person is critical.
4. Situational awareness is a problem as the ships operate automatically or autonomously, without operator assistance most of the time. When something happens, the operator needs time to understand and respond to the situation. The response time must be a short as possible.

Analysis and measurement of these issues must also be covered in the design phase.

## 4 METHOD OVERVIEW

The design method that has emerged in the MUNIN project attempts to integrate the requirements from section 3. In addition, it has also incorporated mechanisms to cater for the limited knowledge available in the design of industrial autonomous systems. There is little previous experience to build on.

The design method needs to cover different technology areas:

1. Process design for efficient cooperation between SCC and ship.
2. Identifying and reducing risk factors to acceptable levels.
3. Efficient implementation through use of standards and interoperability enhancing mechanisms.

The methodology as it stands today is illustrated in Figure 2. The method is based on the Formal Safety Analysis (FSA) developed by and used in IMO (2007), but with additional components to also cover requirements capture (Scenario building) as well as an architectural standard framework: The Maritime Intelligent Transport System (MITS) architecture. It also includes methods to verify the most critical design decisions through hypothesis formulation and tests.

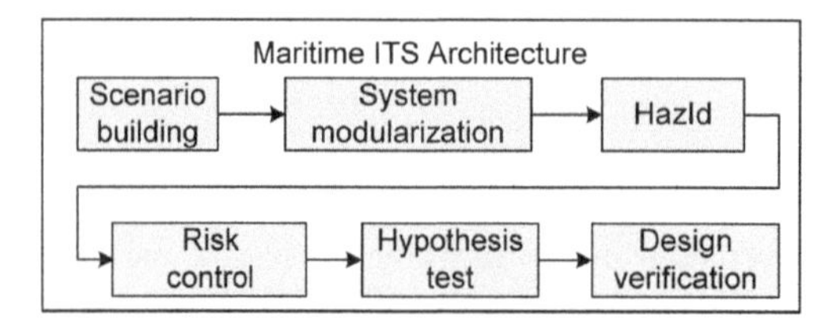


Figure 2. Overview of method.

The figure is simplified and does not show iterations or steps that are not discussed in this paper. Neither does the methodology cover software system design nor implementation as that will be handled in-house by the different subsystem manufacturers. The focus is on high level system functions and modularization, on risk assessment and on module interfaces. The individual modules will be detailed and implemented based on this overall system design.

## 5 MITS ARCHITECTURE

One of the methods employed in the MUNIN project is to use a system architecture as pattern for system development. The MITS architecture provides a set of guidelines to help define the structure of the system and the interrelationships between its parts (Rødseth 2011). It also defines a set of standards that can be used to simplify interfaces between modules in the system as well as between the autonomous systems and other remote components.

MITS is under development and ultimately aims at capturing a design pattern as well as a set of standard definitions for distributed maritime systems. One important objective of MUNIN is to provide more input to the framework. The architecture divides the system definitions into five layers as shown in Figure 3. Note that MITS does not today cover the software architecture, i.e. the internal modularization scheme and "operating system". This is currently left to the software implementers. Focus is mostly on the total system, including how the ship, other ships and shore support as well the remote control centre interacts and functions.

The domain and semantics layer supply definitions for general operations. It has had to be extended with new concepts to cover the unmanned ship, its different operational modes and in particular the new shore control centre.

MITS has provided a pattern for process descriptions through UML use cases and action diagrams. This has provided the basis for the initial descriptions of the necessary capabilities

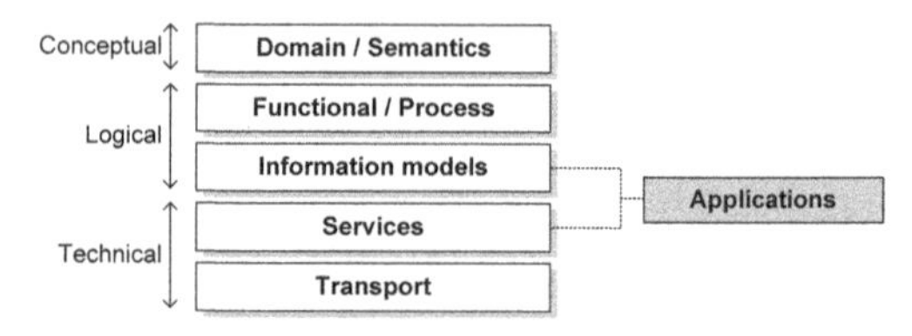


Figure 3. The MITS system architecture.

of the system and a first iteration of the system modularization.

The functional parts of MITS have provided a function breakdown that has been used as basis in the hazard identification process (HazId).

The information models have provided some of the standards for interoperability between the MUNIN components on the ship and the shore.

As MUNIN is mainly a concept study, the service and transport layers have not been used much in practical work. However, MUNIN has provided important input to the developing IEC 61162-460 (2013) standard on safe interconnection of ship data networks with off-ship entities. This standard will eventually be incorporated in MITS.

## 6 SCENARIO BUILDING AND MODULES

The very first activity in the MUNIN project was to develop scenarios for the typical and important use cases involving the unmanned ship. These included cases such as detection of flooding, unknown object detected etc. The use cases and corresponding activity diagrams where described in Unified Modelling Language (UML). The scenarios are listed in Table 1. SAR is an abbreviation for "Search and Rescue" operations at sea and ASC is the Autonomous Ship Controller (see Fig. 4).

The purpose of this exercise was mainly to create concrete descriptions of how an autonomous

Table 1. MUNIN initial scenarios*.

| |
|---|
| Autonomous ship requested for SAR |
| Collision detection and deviation by ASC |
| Communication failure |
| Flooding detection |
| GNSS (GPS/GLONASS) breakdown |
| Manoeuvring mode with malfunctions |
| Manoeuvring mode without malfunctions |
| On board system failure and problem resolution |
| Periodic status updates from vessel to SCC |
| Periodic updates of navigational data |
| Pilot unavailable, remote control to safety |
| Piracy, boarding and ship retrieval |
| Release vessel from autonomous operation |
| Release vessel to autonomous operation |
| Rope in propeller |
| Sea mode with malfunction |
| Sea mode without malfunctions |
| Small object detection |
| Weather routing |

*UML diagrams are available from http://www.mits-forum.org/munin/index.htm (June 2014).

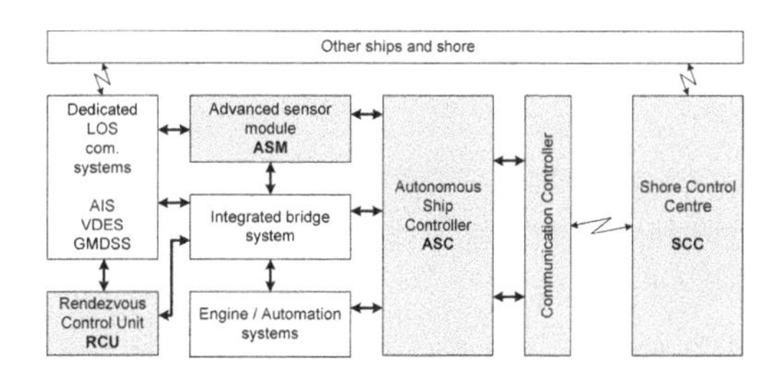


Figure 4. The MUNIN modules.

ship system should solve a set of representative problems. This, in turn, was used to develop a first iteration of a system modularization and a division of responsibilities between modules. More in general, it also represented the first user requirement phase and capabilities specification. One result of this was a more detailed specification of the unmanned voyage and the different autonomous modes that the system employs. As an example, the autonomous voyage has been limited to the deep sea passage only, partly to avoid legislative problems in costal state waters and partly to limit the overall complexity by avoiding autonomous navigation in congested waters and ports.

Figure 4 shows the physical modularisation of the MUNIN system. The unshaded rectangles are modules already present in the ship. Note that the figure shows two communication channels, one through traditional Line of Sight (LOS) systems, typically using VHF voice or digital transmission, and one through one or more digital transmission links over satellite.

The Rendezvous Control Unit (RCU) is an independent module directly interfaced to the bridge system to handle ship recovery also after complete failure in the autonomous control system. This is one of the features added to supply sufficient redundancy in the control functions. These and other definitions have also been used to update the semantic layer in the MITS Architecture.

## 7 HAZARDS AND RISK CONTROL

Initially, this paper stated that industrial autonomous system needs to be cost-effective and safe while doing operations that no existing system can perform alone. Designers of such systems do not have much previous experience to build on.

To address this problem the MUNIN project has applied a risk based design process. This means controlling the risk elements while developing effective solutions for the problem at hand and providing documented evidence that the risk level will be acceptable. The method employed has been adopted from parts of the Formal Safety Assessment method from IMO (2007).

The method is illustrated in Figure 5 with a flow from hazard identification, via risk control options, cost-benefit analysis to decisions made. The shaded rectangles show the elements of FSA that has been used in MUNIN.

The hazard identification (HazId) has been done with the help of a functional decomposition from the MITS framework, systematically breaking down ship function into main groups and sub-groups (Rødseth 2014). The groups are listed in Table 2.

The groups have been combined with other definitions from the MITS semantic layer covering ship operational mode, operational restrictions and voyage stages to create a framework for the hazard identification.

Through a workshop, a total of 65 main hazards were identified. Each of the hazards was then classified according to consequence if the event should happen and the probability that it would happen.

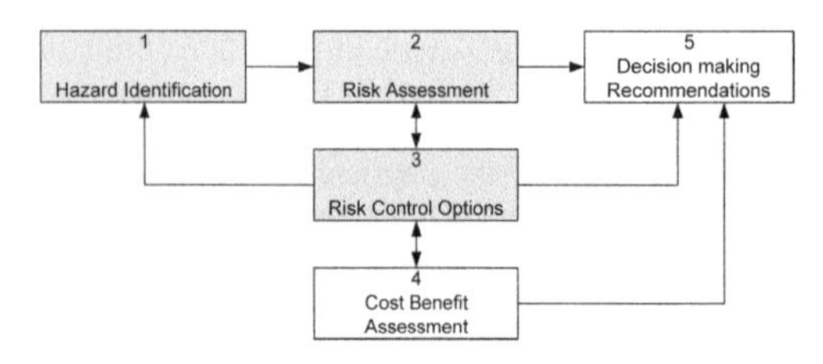


Figure 5. FSA main steps.

Table 2. Function groups used in MUNIN HazId.

| Group | Description |
|---|---|
| Voyage | High level voyage planning, execution and monitoring |
| Sailing | Manoeuvring, avoidance, communication |
| Observations | Environment, objects, ship |
| Safety, emergencies | Other ships, own ship, environment |
| Security | Antipiracy, ISPS, access control and lock-down |
| Crew, passenger | Not applicable to unmanned ship |
| Cargo, stability, strength | Ship stability, hull integrity, cargo monitoring |
| Technical | Power generation and distribution, emissions to air/water |
| Special functions | Not applicable to bulk carriers |
| Administration | Log keeping, operational, communication, reporting |

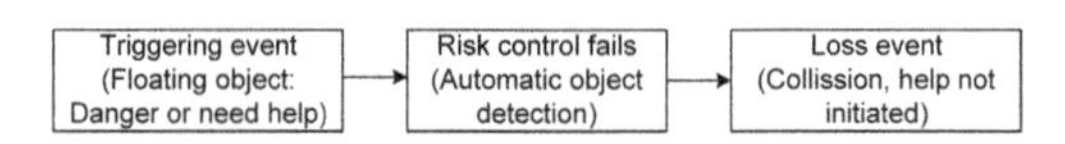


Figure 6. Risk assessment framework.

The typical hazard will consist of some triggering event, e.g. the presence of a floating object that can present a danger to the ship or that should be reported as it may contain persons in need of help, e.g. a raft or a lifeboat. The loss event will occur if the risk control method also fails, i.e. the system fails to detect or recognize the object, as illustrated in Figure 6. The risk control methods where identified partly in the scenario building and partly during the HazId. The probability of a loss event and its consequence is determined from both these factors.

The hazards are sorted according to their risk level which is the probability multiplied with the consequence. They are further classified into unacceptable, i.e. high consequence and high probability, acceptable, i.e. low consequence or low probability and "As Low As Reasonably Practicable" (ALARP). Hazards in the not acceptable regions must be eliminated or reduced and those in the ALARP regions should be examined for optimal risk reduction.

Normally, the risk assessment is based on "expert knowledge", i.e. best effort semi-quantitative assessment by persons experienced in the question at hand. To further analyse the risk levels, two more steps are made:

A. The technology experts check if there are operational or technical measures that can be taken to reduce the risk level. As an example, it was decided to not use heavy fuel oil on the autonomous ship as that would require human intervention to clean filters and maintain the fuel system in general.
B. When no further technical or operational risk control can be found, a test hypothesis is formulated to check the residual risk level. For object detection this will typically be that "All objects that may pose a danger to the ship shall be detected, regardless of weather or other environmental factors". For the remote control centre a similar hypothesis would state how many people should be the maximum manning for different situations, including normal monitoring.

The next step is to test these hypotheses by using a selection of tools available to the partners. This will be discussed in the next section.

## 8 HYPOTHESIS TESTS

The formulated hypotheses were generally of four types and will be tested accordingly.

At time of writing, the hypothesis tests have just begun and it cannot yet be reported on the experience from the approach.

### 8.1 *Technical capability*

Technical capability includes, e.g. object detection, autonomous collision avoidance or technical condition monitoring. Hypotheses related to these functions have to be tested by using the corresponding software and hardware modules with as close to real world data as possible as input and by that verify the capabilities of the software. As an example, the object detection system will use recorded data from several field trials under realistic and known conditions. This type of test may also include Monte Carlo type simulations if sufficient variance in module input cannot be got by real world data alone. The latter applies, e.g. to test of collision avoidance routines.

### 8.2 *Operator capability*

Operator capability may include expectations to minimum or maximum number of operators for given situations or the maximum response time in specific cases of emergency. This needs to be tested in a simulator environment that can as close as possible emulate the environment the operators will work in with the support systems they use. Not all aspects of the environment need to be modelled; this depends on the hypothesis at hand. This may be the most complex hypothesis types to test as the building of the simulated environment may be very demanding.

Operator capability should also include normal operational situations to establish the baseline for remote control centre manning.

### 8.3 *Infrastructure capability*

Infrastructure capability is, at least in the MUNIN project, mostly related to communication quality of service, e.g. communication bandwidth, response times and availability. Tests will be done by doing a workflow analysis for situations that needs infrastructure support. This will determine situation based communication requirements which will be combined with a Monte Carlo simulation to see how the statistical combination of situations can add up to increase overall requirements.

Some preliminary analysis has been made, indicating that available communication bandwidth is sufficient (Rødseth et al. 2013).

Table 3. Duration of voyage for different ship speeds.

| Speed (knots) | Duration (days-hours) |
|---|---|
| 15 | 11 d 10 h |
| 10 | 17 d 3 h |
| 7.5 | 22 d 20 h |
| 5 | 34 d 6 h |

### 8.4 *Technical system availability*

A typical example of a technical system availability problem is guaranteeing the availability of propulsion power throughout the whole voyage. The corresponding hypothesis will be tested through analytic methods, most commonly through a Failure Mode Effects and Criticality Analysis (FMECA). This is based on expected reliability of the components and sub-systems used in implementing the critical functions as well as the interdependencies between them.

The establishment of a sufficiently high availability is not only dependent on the technical components. The risk control method should also include the organisational elements involved in monitoring and maintenance functions (Rødseth H. Mo B, 2014). The FMECA should also include the effects of these factors.

## 9 DESIGN VERIFICATION AND CBA

The design verification will be based on an analysis of results from the hypothesis testing. If the hypotheses can be proven with sufficient statistical validity, one can with a high degree of confidence conclude that the autonomous system should work as expected and be tolerant to the dangerous situations that have been identified. The HazId is obviously a critical factor here, as any hazards that were missed in that phase will represent an unaccounted for risk for the system.

The cost-benefit analysis will compare costs of a normal ship with costs of the unmanned ship for different voyage scenarios. Operational costs have to include variable cost elements, such as fuel use—including higher costs if distillates have to be used, cost of any crew onboard, crew in the SCC, communication costs etc. Capital costs will be dependent on the equipment fitted, lifetime, additional maintenance investments etc. In addition, one may also want to look at benefits related to lower environmental foot-print due to lower speeds and less fuel used and possibly also effects of using cleaner fuels. The premise of the safety and security analysis is that the unmanned ship shall be at least as safe and secure as the manned ship. If improvements in safety or security can be quantified, these may also be added to the benefits. However, these will most likely be societal benefits which may be of lower importance for the individual ship owner and maybe not so relevant for the CBA.

Investigations so far shows that the main driving force for unmanned ships is probably not reduced crew costs. With additional ship system costs, communication costs and the costs of manning the shore control center etc., operational costs for the unmanned ship will most likely be similar to or even higher than for the unmanned ship. However, if ship owners are aiming for reduction in fuel costs by slow and ultra-slow steaming, they are probably looking at problems getting crew for these long voyages. A voyage between Rotterdam and Ponta Da Madeira in Brazil is about 4108 nautical miles and voyage duration for different speeds is listed in the table. Staying more than 30 days in open sea with limited Internet connectivity is not something that may create a lot of interest among highly qualified and sought after ship crew.

On the other hand, it is also clear that crew costs will be an issue at these long voyages so reductions here will also help to decrease overall costs. A more extensive discussion on the operational cost issues can be found in Rødseth & Burmeister (2012).

## 10 EXPERIENCES GAINED

The MUNIN project has attracted considerable attention although it is only a concept study. The automation of merchant shipping is obviously considered a relevant possibility while the professional community also clearly sees the technical, legal and economical obstacles. This has led the project to approach the proof of concept phase with much more care than originally envisaged so that it can supply high quality documentation of the project's results and conclusions.

Also, activities in other areas of industrial autonomous systems, particularly in aquaculture and in the oil and gas industry, show the same mix of high interest and expectations with doubts and lack of willingness to invest in autonomous systems for critical operations.

The methodology emerging from MUNIN is promising in that it provides a more formalized framework around the assessment of system reliability, cost and effectiveness. Better and more trustworthy documentation of benefits of autonomous systems will be an important and perhaps crucial factor in increasing the commercial interest in industrial autonomous systems.

So far, the project has completed the scenario analysis, the system modularization, the HazId and the hypothesis formulation. All the hypothesis have been collected in a set of 10 test scenarios where the different test methods discussed in section 8 will be applied to the different sub-hypothesis within each test scenario. The final test scenarios are listed in Table 4.

Table 4. Final test scenarios.

| | |
|---|---|
| 1 | Normal monitoring |
| 2 | Object sensing |
| 3 | Precise manoeuvres |
| 4 | High sea navigation |
| 5 | Heavy Weather |
| 6 | Engine problem |
| 7 | Crew change |
| 8 | Ship efficiency optimizing |
| 9 | Weather routing |
| 10 | Pirate attack |

## 11 CONCLUSIONS

This paper has presented parts of a design and analysis method for industrial autonomous systems, i.e. autonomous systems with high demands on availability, reliability, cost-effectiveness, security and safety.

The most complex operations performed by an autonomous ship are arguably object detection and recognition as well as path planning, obstacle avoidance and path re-planning. An important aspect of the method presented in this paper is to minimize the complexity of these operations as far as possible, *before* the operational requirements are determined. In the case presented here, this is done by limiting autonomous operation to the deep sea passage, using a shore control centre for continuous monitoring and defining "fail to safe" routines as a final fall-back. The method also emphasizes a systematic approach to risk management and testing with the purpose of providing as far as possible quantitative assessments of system performance. This is critical for industrial autonomous systems: Documented reliability and cost-benefit.

The method as described here has mainly covered the initial system analysis phase as well as hazard identification, risk control method development and verification. The main emphasis is on a structured and documented approach to risk control.

As the method has been developing as the project has progressed, it is too early to judge the success of the method, except for the general opinion that it has provided a very useful structure to the analysis and design activities and has been a useful framework for the developed results. A more comprehensive report will be made when the project is finished in the summer of 2015.

## REFERENCES


Franklin, S., Graesser, A. 1997. Is it an Agent, or just a Program?: A Taxonomy for Autonomous Agents. In Intelligent agents III agent theories, architectures, and languages (pp. 21–35). Springer Berlin Heidelberg.

IEC 61162-460. 2013, Maritime navigation and radiocommunication equipment and systems—Digital interfaces—Part 460: Multiple talker and multiple listeners—Ethernet interconnection—Safety and security. Committee Draft July 2013.

IMO 2007. MSC 83/INF.2, Formal Safety Assessment: Consolidated text of the Guidelines for Formal Safety Assessment (FSA) for use in the IMO rule-making process. May 14, 2007

NIST Special Publication 1011-II-1.0 2007. Autonomy Levels for Unmanned Systems (ALFUS) Framework. Volume II: Framework Models, Version 1.0, October 2007.

Porathe, T. 2014. Remote Monitoring and Control of Unmanned Vessels—The MUNIN Shore Control Centre. In Proceedings of the 13th International Conference on Computer Applications and Information Technology in the Maritime Industries (COMPIT '14) (pp. 460–467).

Rødseth, H.; Mo, B. 2014. Maintenance Management for Unmanned Shipping. In proceedings of 13th conference on computer and IT applications in the maritime industries—Compit '14, Redworth, UK, May 12–14 2014.

Rødseth, Ø.J.; Tjora, Å. 2014. A System Architecture for an Unmanned Ship, in proceedings of 13th conference on computer and IT applications in the maritime industries—Compit '14, Redworth, UK, May 12–14 2014.

Rødseth Ø.J. Ed. 2014, MUNIN Deliverable D4.5 Architecture specification, 2014-02-08. Available from www.unmanned-ship.org (June 2014).

Rødseth, Ø.J., Kvamstad, B., Porathe, T., & Burmeister, H.-C. 2013. Communication architecture for an unmanned merchant ship. In Proceedings of IEEE Oceans 2013. Bergen, Norway.

Rødseth Ø.J., Burmeister, H.-C. 2012. Developments toward the unmanned ship, in proceedings of International Symposium Information on Ships—ISIS 2012, Hamburg, Germany, August 30–31, 2012.

Rødseth Ø.J. 2011. A Maritime ITS Architecture for e-Navigation and e-Maritime: Supporting Environment Friendly Ship Transport, in Proceedings of IEEE ITSC 2011, Washington, USA, 2011.

*Maritime-Port Technology and Development – Ehlers et al. (Eds)*
*© 2015 Taylor & Francis Group, London, ISBN 978-1-138-02726-8*

# Offshore upstream logistics for operations in arctic environment

A.-S. Milaković, S. Ehlers & M.H. Westvik
*Department of Marine Technology, NTNU, Trondheim, Norway*

P. Schütz
*DNV GL, Oslo, Norway*

ABSTRACT: As oil and gas companies move their offshore operations further north and into the Arctic, numerous challenges are faced. Therefore, well-established operational models that are used in, e.g. the North Sea, may have to be revised in order to make them applicable for operations in the Arctic. Based on information gathered through literature review, individual meetings with stakeholders, as well as through participation in workshops, this paper seeks to identify what kind of adjustments are necessary for the upstream logistics system of an offshore oil and gas field in the Arctic to become economically and environmentally sustainable. We describe a typical offshore logistics system and point out Arctic-specific challenges. Further, the influence of arctic conditions on the offshore supply chain is discussed. Finally, we present a stakeholder interdependence matrix, showing where the different stakeholders' interests overlap and where they might possibly be in conflict. The findings of this paper can serve as a basis for further research and should be solidified and extended through in-depth interviews with stakeholders. Some of the major questions that need to be answered, in order to achieve a complete overview of stakeholders' perspective on offshore upstream logistics in arctic environment, are also mentioned in this paper.

## 1 INTRODUCTION

Over the lifetime of an offshore oil field, supply operations and supporting logistics are two of the key operational segments required to have a high level of functionality in order to make offshore operations both economically and technically sustainable. Upstream logistics, i.e. the delivery of all products and services necessary for operations to and from the offshore field, is a major part of the logistics system. The logistics system has to meet two main requirements:

i. Efficiency in order to minimize the costs of product and service delivery (Aas et al. 2009), and
ii. Robustness to maximize productivity and reduce the risks of operational delays caused by delayed deliveries of goods to and/or from the offshore field (Berle et al. 2011).

Delivery of products and services to offshore fields is commonly accomplished by using Offshore Supply Vessels (OSVs) operating between an Onshore Supply Base (OSB) and the offshore field. These supply systems are well established for current operations in, e.g. the North Sea, but may have to be further developed and adjusted for operations that will take place in remote and harsh environments of arctic regions.

As the global climate rapidly changes and the arctic ice cap steadily diminishes, along with the world's ever increasing energy needs, oil and gas exploration and production is moving further north. Motivation for that is the fact that an estimated 22% of the world's undiscovered oil resources are located in the Arctic, 84% of which is projected to be offshore (USGS 2008). Therefore, waters of the High North, along with all the potential benefits, hazards and vulnerabilities that they entail, are becoming a new frontier for the petroleum industry. When moving their operations to the Arctic, oil and gas companies, together with other stakeholders, are facing significant challenges amongst which are: low temperatures, icing, remoteness, darkness, sea ice (in some parts), polar lows, fog, vulnerability of the ecosystem, geopolitical issues etc.

The aim of this paper is to assess which adjustments of the offshore logistics systems need to be done in order to overcome challenges posed by the arctic environment. Additionally, interdependencies between various stakeholders in the offshore logistics operations that may affect the necessary adjustments are presented and discussed.

## 2 EXISTING LITERATURE

Researchers have been addressing the problem of planning offshore field development for more than 40 years (see e.g. Frair & Devine 1975).

The upstream logistics system however has only recently started to attract the interest of the academic community.

Examples of contributions have mainly focused on routing problems for OSVs: Fagerholt & Lindstad (2000) develop an optimal routing policy for a set of supply vessels serving several offshore installations from one onshore supply base. They focus on the case where some or all installations operate with opening hours for deliveries. Aas et al. (2007) formulate a vehicle routing problem with pickups and deliveries to solve an OSV routing case from the Norwegian Sea. Halvorsen-Weare et al. (2012) consider a similar problem. They first generate all candidate voyages before determining the supply vessel fleet and assigning vessels to voyages. The impact of weather condition on sailing and loading times, as well as the consequences for the optimal routing policies, are studied in Halvorsen-Weare & Fagerholt (2011).

Contributions from other areas than Operations Research can also be found: Aas et al. (2008) discuss the outsourcing of logistics activities and use a case from the Norwegian oil and gas industry to illustrate their analysis. Aas et al. (2009) provide a good overview over planning upstream logistics. They pay special attention to the role the OSV plays in upstream logistics. Kaiser (2010) develops a methodological framework for quantifying the spatial relationship between offshore activities and the onshore supply base. The framework is applied to the planning and routing of OSVs and crew boats in the Gulf of Mexico for the purpose of servicing local oil and gas operations. Leite (2012) studies how to improve the efficiency of deck cargo transportation to offshore units off the coast of Brazil.

Existing literature on upstream logistics in the offshore industry has traditionally focused on operations in non-arctic environments. A notable exception is the early paper by Juurmaa & Wilkman (2002) that provides a short overview over the challenges of operating in ice-covered waters and the requirements to OSVs. Since then, several publications have started to address the necessity of modifying the existing operational systems for offshore oil and gas activities when moving to the Arctic. CARD (2012) presents an Arctic Development Roadmap focusing on key issues in research & development that need to be addressed in order to fill the technological and methodological gaps related to offshore oil and gas development in the Arctic. In a similar way, a joint Russian and Norwegian initiative (INTSOK 2013) depicts the need for development of new technologies and cooperation between Norway and Russia to facilitate the oil and gas developments in the High North, focusing on infrastructural, logistical and regulatory issues. Borch et al. (2012) present an overview of arctic challenges and identify the gap (from technological and business point of view) that needs to be overcome in order to move the operations from the North Sea to the Arctic. Main focus is put on the design aspects for offshore service vessels operating in the Arctic. Jimenez-Puente (2013) uses a similar approach and identifies arctic hazards to the oil and gas industry with focus on building Arctic-compliant onshore terminals.

We extend the existing literature in two ways: First, we place offshore supply operations in an arctic environment and discuss the consequences. Second, we introduce a stakeholder interdependence matrix to identify both mutual interests that may promote collaboration, and possible conflicts of interest that could reduce industrial activities in the Arctic.

## 3 OFFSHORE LOGISTICS SYSTEMS

### 3.1 *Offshore supply operations*

The offshore oil and gas support network presents a supply chain whose purpose is to supply offshore operations with all necessary products and services. The supply chain and the logistics that accompany it need to be highly reliable and robust, as potential stopping of operations due to failure of this system would induce substantial financial losses for the operator.

Figure 1 provides an overview of the offshore logistics system, including infrastructure, equipment, and typical operations, as well as the main stakeholders involved in each segment of the operations. The logistics system can e.g. be described based on location of activities. Based on Kaiser (2010), we identify three sectors:

i. Inbound sector,
ii. Port sector, and
iii. Offshore sector.

Various suppliers are delivering cargo to OSB where it is stored and prepared to be moved further through the supply chain. The inbound sector covers the infrastructure for transporting cargo destined for offshore installations to the OSB. This includes the networks for all different means of transportation of products and services to the port (highways, railways, waterways, air) and is a generalization of the inland transportation sector used by Kaiser (2010).

The port sector consists of the Onshore Supply Base (OSB), including all associated infrastructure. The OSB is usually run by one of the stakeholders in the supply chain—the supply base operator. Main purpose of the onshore supply base is to create a connecting point between the inland and

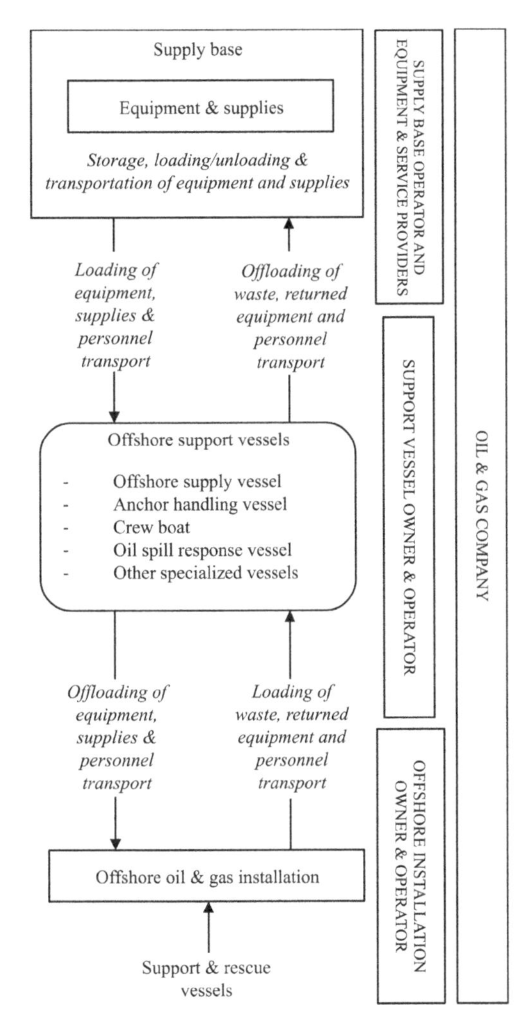


Figure 1. Overview over an offshore upstream logistics supply chain.

offshore sectors. Apart from cargo handling, other activities take place at the OSB, such as warehousing, fabrication, repair, and vessel stowage. Also, the OSB needs to have means of attending to waste produced at offshore installations.

The offshore sector consists of offshore installations and various types of vessels providing support for offshore activities. Such vessels are (see e.g. Aas et al. 2009): OSVs, Anchor Handling Vessels (AHVs), crew boats, rescue/standby vessels, etc. Cargo is loaded at the OSB and transported to offshore installations using OSVs. Once the outbound cargo is delivered at the offshore installation, the OSV takes the backhaul cargo on board (e.g. used equipment, empty containers, waste, etc.) and returns to the OSB. Two main types of cargo can be identified (Leite 2012): deck cargo (containerized cargo and larger pieces of equipment) and bulk cargo (dry and liquid cargo).

Logistics operations can also be distinguished in supply logistics and in-field logistics (INTSOK 2013). Supply logistics are related to operations of vessels running from the onshore base to one or more offshore installations, delivering products and services, while in-field logistics deal with vessels that are in permanent service of an offshore oil or gas field (e.g. support and rescue vessels, firefighting vessels, storage vessels, etc.).

Another important concept is to describe the logistics in terms of the different stages of field development as each of the phases has significantly different logistics requirements. Kaiser (2010) distinguishes four different stages during the life-cycle of an offshore installation:

i. Exploration,
ii. Development,
iii. Production, and
iv. Abandonment.

The exploration phase consists of geophysical surveying and exploratory drilling stages. While the former is rather self-sufficient and usually does not require much logistic support (except for crew change), the latter requires significant onshore support. Drilling operations require substantial amounts of material (drilling equipment, drilling fluids, etc.) to be transported to and from the drilling site by the OSVs. Naturally, the amount of equipment and material transported depends largely on the type of drilling rig used, as some have longer autonomy in terms of larger storage capacity and therefore require less onshore support. The need for logistic support during this phase is characterized by high volatility in the demand for materials and equipment (Kaiser 2010). Apart from OSVs providing supply services, AHVs are also engaged in operations of relocating of drilling rigs.

The development phase involves design, fabrication and installation of the rig. Production wells are drilled during this phase and thus, the requirements for the logistic support are similar to those during the exploration phase.

A need for high regularity in logistic support characterizes the production phase of offshore field development. Installations are usually visited by OSVs based on a predetermined schedule. Higher volatility in demand for logistic services arises during times of drilling of additional wells or infrastructure expansion.

During the abandonment phase, all installed structures are removed, wells are plugged and debris is cleared. In terms of requirements for logistic support, this phase is very similar to the development phase.

Table 1 summarizes the demand characteristics of the different phases of offshore field development.

### 3.2 *Operating in the arctic environment*

Arctic-specific challenges that are considered to be most significant for offshore operations are presented in Table 2. It is necessary to identify gaps that need to be overcome in order to make offshore operations in the Arctic environmentally and economically sustainable.

An important aspect of understanding and planning operations in the Arctic is the fact that not all arctic regions involve the same challenges nor is the severity of those the same in all regions. INTSOK (2013) therefore divides the Norwegian and Russian Arctic in different areas based on the regional severity of arctic challenges, suggesting a gradual development over time. Similarly, Statoil splits the Arctic in three areas (Karlsen 2014):

i. Workable Arctic—oil and gas activities possible with today's technologies and operational models (e.g. Snøhvit gas field, Goliat oil field),
ii. Stretch Arctic—incremental technological and operational development necessary, and
iii. Extreme Arctic—radical technological and operational development necessary.

It is a common opinion in the Norwegian oil and gas industry that the offshore development in the Arctic will have to follow a stepwise transition and will move in direction from Barents Sea SW to NE and further east (see e.g. INTSOK 2013; DNV GL 2014).

Table 1. Demand characteristics for logistic services for the different phases during offshore field life-cycle.

| Phase | Predictability of demand | Vessels used |
|---|---|---|
| Exploration | LOW | OSV, AHV, Crew boats |
| Development | LOW | OSV, AHV, Crew boats |
| Production | HIGH | OSV, Crew boats, Standby/rescue vessels |
| Abandonment | LOW | AHV, Crew boats |

Table 2. Arctic challenges for offshore operations.

| Challenge | Possible effects | Season |
|---|---|---|
| Low temperatures | Hampering SAR operations | Winter |
| Icing | Reduced operability, freezing of mechanisms; Shutdown of communication and evacuation systems | |
| Sea ice | Presents serious challenges to vessels and installations; Sea ice, drifting icebergs | |
| Darkness | Reduced visibility | |
| Fog | Reduced visibility | Summer |
| Remoteness | Large distances; Lack of infrastructure (power grid, roads and railways to OSBs, SAR, communication systems) | All year |
| Polar lows | Safety risk and challenge to operations; Difficult predictability | |
| Delicate eco-systems | Risk of pollution | |
| Geopolitical issues | Uncertainty regarding future of political relations in the Arctic | |

### 3.3 *Impact of arctic conditions on different sectors of the offshore supply chain*

Offshore installations are usually supplied from an OSB in relative proximity to the installation, e.g. installations in the North Sea are supplied amongst others from Aberdeen or Stavanger, while the OSB in Kristiansund serves the Haltenbank installations in the Norwegian Sea. The Goliat field in the Barents Sea will be supplied from Hammerfest. The oil and gas industry's move north will therefore have direct and indirect implications for all sectors (namely inbound, port and offshore sector) of the logistics system with respect to how they have to adapt to the arctic environment.

The *inbound sector* comprises the land-based infrastructure. Cargo is transported to the OSB using different modes of transport: road, rail, water or air. Limited infrastructure and hinterland connections for ports in the High North will therefore pose a significant logistical challenge to efficient transportation solutions. The necessary investments in infrastructure development (e.g. transportation and communication networks) in the High North will be very high. Environmental conditions, snow and ice, low temperatures, and darkness during winter months, present an additional risk to the reliability of the logistics chain. One of the ways to mitigate these challenges is to increase ship-borne transportation for inbound cargo. However, this will require additional investments in an ice-classed and winterized fleet. Also, a ship-based transportation system depends upon sufficient cargo handling capacity at the ports.

The *port sector* serves as link between inland sector and offshore sector. The limitations of the inbound sector therefore affect the port sector as well. The inability to efficiently move cargo is a significant hindrance to offshore operations. With the exception of Norway, port infrastructure along the arctic coastline is in general poorly developed and has to be upgraded considerably in order to handle both inbound and outbound cargo volumes. But not only cargo-handling will have to be improved, increased offshore activity will also require sufficient Search and Rescue (SAR) capabilities in the ports. On the other hand, future Arctic-compliant rigs may have higher autonomy and will be able to store larger quantities of equipment and consumables, comparing to rigs that are used now-days. This might reduce the need for supplies from the OSB (based on an interview with a rig operator). Further, it is expected that a large portion of field developments in the Barents Sea will be subsea developments, contributing to a lower activity level for the base (compared to other areas) during the production phase. This uncertainty regarding future capacity requirements may delay investments in port infrastructure. Another important challenge for the development of the port sector is the limited availability of qualified labour in the High North (based on an interview with a supply base operator).

The *offshore sector* is the sector most exposed to arctic conditions and a strong impact is therefore expected. Firstly, in order to operate year-round in the area, common opinion is there will be a need for highly specialized Arctic-compliant installations and vessels (OpLog 2014). These will be built specifically for the High North and the winterization will come with a high price tag. The arctic design requirements can also reduce the second-hand value of mobile units, which can be a challenge for rig owners or supply ship owners to commit to the necessary investments without long term contracts.

Secondly, the transfer of supplies and personnel from the OSV to the offshore installation might be hampered by environmental conditions. Icing can prevent a supply vessel from sailing into certain area and disrupt the supply schedule. Relative motions between installation and OSV due to ice loads for example, can interfere with the physical connection between the OSV and an installation. As transportation of cargo is one of the main tasks of the logistics system, disruptions in service present a major challenge.

Thirdly, the long distances in the Arctic stretch the supply lines between port and installation. As the operations move further north, distance between supply base and installation will increase significantly, comparing to the North Sea and the Norwegian Sea. Any unexpected failures of equipment may have serious consequences as the necessary time for replacing it, even from the nearest base, may cause delays in production. An oil spill representative put it like this: "If you need it in the Arctic, you better bring it with you." (DNV GL 2014). An alternative of mitigating this risk is to increase the redundancy by using multi-purpose vessels that can perform several tasks, e.g. anchor handling, ice management, or oil spill response. From a supply point of view, technology development such as 3-D printing, ability to make more types of e.g. mud from a more limited set of base bulk products, can reduce the effect of the uncertainty as well.

The distances also affect the crewing of installations. Common crew rotation of 2 weeks on and 2 weeks off for installations in the North Sea or Norwegian Sea is based on helicopter transport between onshore airfields and offshore installations. Direct helicopter transportation in the Arctic however, may only be feasible in southern parts of the Barents Sea. In case of vessel-based crew transportation, additional robustness will have to be added to the crew rotation to account for uncertainties e.g. in sailing time. Also, workers may need to be compensated for the additional time spent on board a crew vessel.

Table 3 provides a short overview over some of the expected impacts of moving further north (see e.g. INTSOK 2013; Karlsen 2014; OpLog 2014).

An important topic related to arctic operations is emergency evacuation in case of a disaster as well

Table 3. Arctic development for different sectors in offshore supply chain.

| Sector | Workable Arctic | Stretch Arctic | Extreme Arctic |
|---|---|---|---|
| Inbound | Infrastructure sufficiently developed | Lack of land-based infrastructure. Supply network needs to be developed | |
| Port | Sufficiently developed | Existing and new bases can be used and/or upgraded, with improvement of supply fleet | Forward supply bases will be needed |
| Offshore | Current fleet of ice-classed vessels sufficient | Need for novel, multi-purpose, ice-going vessels with various functions | |

as Search and Rescue operations. The oil and gas industry's offshore service vessels must play an important role in the SAR contingency plans. Vessels in the area will have the role of first responders and as such need to be equipped with the necessary capabilities. This reinforces the need for winterized, multi-purpose vessels. According to INTSOK (2013), the following functions need to be provided:

- SAR capabilities,
- Emergency centre for operations,
- Hospital facilities,
- Fire-fighting equipment,
- Oil spill recovery equipment,
- Ice breaking capabilities,
- Ice management capabilities, etc.

The concept of forward supply bases has been suggested as a possible solution to address some of the issues related both to upstream logistics and Search and Rescue preparedness (see e.g. Haugen 2013; INTSOK 2013). Located between OSB and installation, these supply bases can serve as intermediate hubs for crew change, cargo handling, intermediate storage, etc. Adding hospital facilities and/or helicopter refuelling capabilities can help alleviate the challenges of long distances and limited helicopter range.

## 4 STAKEHOLDERS INTERDEPENDENCE

In order to overcome arctic challenges and specific conditions, which are discussed in the previous chapters, increased level of cooperation between the stakeholders will be required. Thus, a stakeholders interdependence study to cope with the logistics management issues for the Arctic is necessary.

The offshore logistics supply chain is a complex environment involving numerous stakeholders each having their own interests and goals. The purpose of this preliminary analysis is to look into interdependencies between various stakeholders, and to examine, and also raise awareness, where interests overlap or possible conflicts may arise. We consider the following main stakeholders in the offshore logistics supply chain:

i. Oil and gas company,
ii. Rig owner and operator,
iii. Supply vessel and operator,
iv. Supply base operator,
v. Equipment and service providers,
vi. Working personnel, and
vii. Society at large (local population, environment, government institutions).

The oil and gas company is the stakeholder that dominates the entire supply chain and has a managing role. Other stakeholders are specialized in their own field of activity but may have an interest in other parts of the supply chain as they all depend on each other, e.g. a drilling rig operator is dependent on regular visits by the supply vessels and deliveries of necessary supplies. Society at large is here used as a paramount stakeholder, representing amongst others: local population, local and regional government institutions regulating the framework for operations, as well as environmental groups' concerns regarding oil and gas operations in delicate eco-systems. Other minor stakeholders, such as companies subcontracted by larger stakeholders for performing various smaller tasks (e.g. diving companies, ROV operators, catering companies etc.) have been excluded here as focus is put only on main stakeholders in the supply chain.

The necessary technological, infrastructural and operational improvements for successfully moving offshore operations to the Arctic require collaboration of the different stakeholders. The stakeholders themselves agree that in order to make arctic operations feasible, there is a need for higher cooperation between actors included in the process (OpLog 2014). However, as Arctic-compliant offshore installations, and novel, multi-purpose vessels as well as infrastructure come with a high price tag, there will be interesting dynamics amongst stakeholders concerning the question of financing, controlling, and operating such expensive assets.

The stakeholder interdependence matrix in Figure 2 shows some important areas of mutual interest or possible conflicts between the main stakeholders. Areas of mutual interest will promote collaboration between the stakeholders and support the move to the Arctic, whereas the possible points of conflicting interest can delay the development of offshore logistics for the High North.

An example of such an area of possible collaboration between stakeholders is development of new business models that could include private-public partnerships between e.g. oil and gas company and government institutions. This may be a way to address the need for investment in highly expensive infrastructure projects. Also, cooperation between stakeholders is expected related to technological and operational improvements such as the development of Arctic-compliant rigs, multi-purpose service vessels, forward based supply bases, logistics systems etc. On the other hand, possible points of conflicting interests between stakeholders could be regarding increased risk of pollution, scheduling of supply in case of servicing

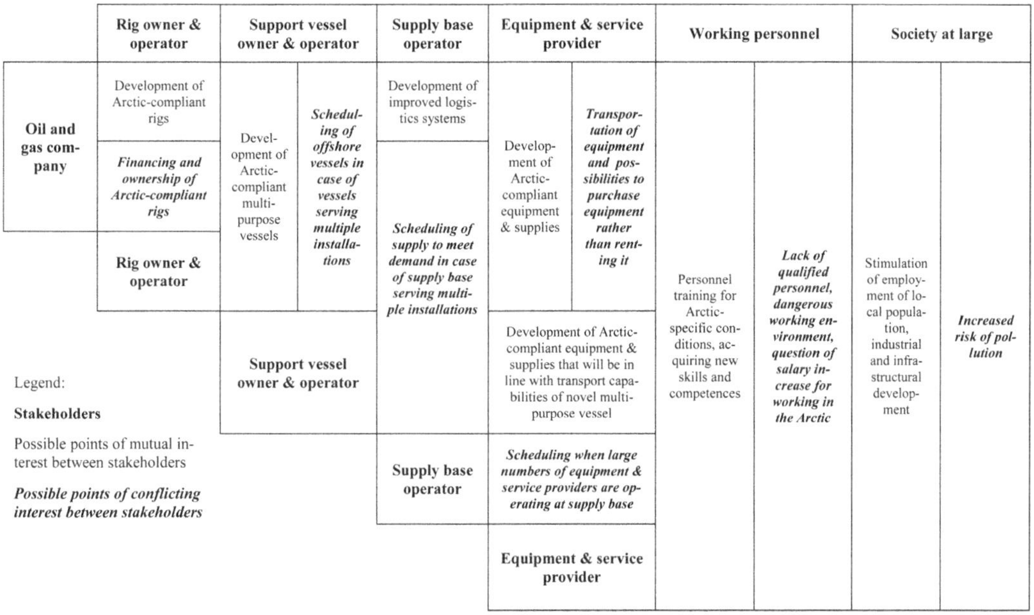

| | Rig owner & operator | Support vessel owner & operator | Supply base operator | Equipment & service provider | Working personnel | Society at large |
|---|---|---|---|---|---|---|
| **Oil and gas company** | Development of Arctic-compliant rigs<br>***Financing and ownership of Arctic-compliant rigs*** | Development of Arctic-compliant multi-purpose vessels<br>***Scheduling of offshore vessels in case of vessels serving multiple installations*** | Development of improved logistics systems<br>***Scheduling of supply to meet demand in case of supply base serving multiple installations*** | Development of Arctic-compliant equipment & supplies<br>***Transportation of equipment and possibilities to purchase equipment rather than renting it*** | Personnel training for Arctic-specific conditions, acquiring new skills and competences<br>***Lack of qualified personnel, dangerous working environment, question of salary increase for working in the Arctic*** | Stimulation of employment of local population, industrial and infrastructural development<br>***Increased risk of pollution*** |
| **Rig owner & operator** | | | | | | |
| **Support vessel owner & operator** | | | | Development of Arctic-compliant equipment & supplies that will be in line with transport capabilities of novel multi-purpose vessel | | |
| **Supply base operator** | | | | ***Scheduling when large numbers of equipment & service providers are operating at supply base*** | | |
| **Equipment & service provider** | | | | | | |

Legend:

**Stakeholders**

Possible points of mutual interest between stakeholders

***Possible points of conflicting interest between stakeholders***

Figure 2. Stakeholders interdependence matrix.

multiple installations, as well as the financing and ownership of assets.

## 5 SUMMARY

The upstream logistics system for offshore oil and gas operations has been described. Further, Arctic-specific challenges have been addressed and their impact on logistic operations has been assessed. Finally, main stakeholders participating in the supply chain have been pointed out and a stakeholders interdependence matrix has been formulated. This matrix lists possible points of mutual interest as well as possible conflict that may affect collaboration between stakeholders and the development of business models for upstream logistics in the High North.

Further work should include forming of a questionnaire for empirical research on stakeholders' view of the expected impact of arctic challenges to their operations. This work can solidify and possibly extend the interdependence matrix. Some of the main questions that should be answered through interviews with the stakeholders concern description of relations between individual stakeholders. Also, it is important to clearly define where do stakeholders' responsibilities and activities start and where do they end. Further, most important performance indicators (KPIs) as well as risk factors should be collected for each of the stakeholders. All questions should be answered while keeping in mind that the operations will be taking place in the arctic environment.

## REFERENCES


Aas, B., Gribkovskaia, I., Halskau, Ø. and Shlopak, A. 2007. Routing of supply vessels to petroleum installations. *International Journal of Physical Distribution & Logistics Management* 37(2): 164–179.

Aas, B., Buvik, A. and Cakic, D.J. 2008. Outsourcing of logistics activities in a complex supply chain: A case study from the Norwegian oil and gas industry. *International Journal of Procurement Management* 1(3): 280–296.

Aas, B., Halskau, Ø. and Wallace, S.W. 2009. The role of supply vessels in offshore logistics. *Maritime Economics & Logistics* 11: 302–325.

Berle, Ø., Rice, J.B. and Asbjørnslett, B.E. 2011. Failure modes in the maritime transportation system: A functional approach to throughput vulnerability. *Maritime Policy & Management* 38:6, 605–632.

Borch, O.J., Westvik M.H., Ehlers, S. and Berg, T.E. 2012. Sustainable Arctic field and maritime operations. In: *Proceedings of the Arctic Technology Conference, OTC 23752, 3-5 December, Houston, USA.*

CARD 2012. Arctic Development Roadmap, *C-CORE Report R-11-275001-CARD v2, January 2012.*

DNV GL 2014. The Arctic—the next risk frontier. *DNV GL, Høvik, Norway.*

Fagerholt, K. and Lindstad, H. 2000. Optimal policies for maintaining a supply service in the Norwegian Sea. *Omega* 28(3): 269–275.

Frair, L. and Devine, M. 1975. Economic Optimization of Offshore Petroleum Development. *Management Science* 21(12): 1370–1379.

Halvorsen-Weare, E.E. and Fagerholt, K. 2011. Robust supply vessel planning. In: Pahl, J., Reiners, T. and Voß, S (eds.), *Lecture Notes in Computer Science, Vol. 6701: Network Optimization, Springer, Berlin.* pp. 559–573.

Halvorsen-Weare, E.E., Fagerholt, K., Nonås, L.M. and Asbjørnslett, B.E. 2012. Optimal fleet composition and periodic routing of offshore supply vessels. *European Journal of Operational Research* 223(2): 508–517.

Haugen, S.F. 2013. Offshore Supply Base—A Concept for the Barents Sea. *Department of Marine Technology, Norwegian University of Science and Technology, MSc thesis, Trondheim, Norway.*

INTSOK 2013. RU-NO Barents Project, Logistics and Transport-Report, 15. November 2013.

Jimenez-Puente, I. 2013. Offshore field development in cold climate—with emphasis on terminals. *Department of Civil and Transport Engineering, Norwegian University of Science and Technology, MSc Thesis, Trondheim, Norway.*

Juurmaa, K. and Wilkman, G. 2002. Supply operations in ice conditions. *Proceedings of the 17th International Symposium on Okhotsk Sea & Sea ice, 24-28 February, Mombetsu, Japan.*

Kaiser, M.J. 2010. An integrated systems framework for service vessel forecasting in the Gulf of Mexico. *Energy* 35(7): 2777–2795.

Karlsen, M. 2014. Statoil's perspective on Arctic operational challenges offshore. *Arctic Frontiers, 18-23 January, Tromsø, Norway.*

Leite, R.P. 2012. Maritime transport of deck cargo to Petrobras fields in Campos Basin: An empirical analysis, identification and quantification of improvement points. *Departamento de Engenharia Industrial, Pontificia Universidade Catolica do Rio de Janeiro, MSc Thesis, Rio de Janeiro, Brazil.*

OpLog 2014. Presentations from the OpLog Workshop on Offshore service vessel operations in the High North, 6–7 May, Bodø, Norway.

USGS 2008, Circum-Arctic Resource Appraisal: Estimates of Undiscovered Oil and Gas North of the Arctic Circle. *U.S. Geological Survey Fact Sheet 2008–3049.*

*Maritime-Port Technology and Development – Ehlers et al. (Eds)*
*© 2015 Taylor & Francis Group, London, ISBN 978-1-138-02726-8*

# An evaluation of evacuation systems for arctic waters

O.E. Staalesen & S. Ehlers
*Department of Marine Technology, Norwegian University of Science and Technology, Trondheim, Norway*

ABSTRACT: The increased exploration for new petroleum resources in the Barents Sea presents new challenges with regards to evacuation systems that can bring the crew to safety under arctic conditions. These systems will need to perform in waters where the presence of ice and subzero temperatures pose challenges that are difficult to fulfill with current solutions. This paper seeks to conduct an investigation on the requirements for an arctic evacuation system. Critical scenarios and hazardous conditions are identified by application of a Preliminary Hazard Analysis (PHA), as well as a Failure Mode Effect and Criticality Analysis (FMECA). Supplementary design requirements to existing regulations are proposed, and relevant concepts are presented and evaluated according to these by ordered weight averaging. Results indicate that there are many considerations that need to be taken into account with consideration to designing an arctic survival vehicle. Launch systems, hull structure and maneuverability must be designed with consideration to the extreme environmental conditions. Further, the self-sustainability of the vessel needs to be improved in order to ensure crew comfort for long durations while awaiting rescue.

## 1 INTRODUCTION

The exploration for oil and gas has until now been performed in relatively accessible areas both on- and offshore where resources have been fairly easy to extract with existing technology. However, as the global demand for oil and gas continues to increase operators have seen the need to venture into the Arctic regions of the world in the search for new petroleum resources. Of these regions, the Barents Sea has proven itself to be an area of interest with a high possibility of finding new profitable sources of hydrocarbons, proven by exploratory drilling conducted by major contenders in the oil industry.

Following the increased activity in the high north a new class of oil platforms and ships have been given life to, especially designed for operation in the extreme weather and subzero temperatures of the Arctic. Nonetheless, new challenges have arisen with the Arctic exploration, where maintaining safe operations for the vessels and crew when faced with the long distances and the arctic environment has proven challenging. As a consequence of these matters, evacuation systems that can provide a viable solution for evacuation in arctic waters must be installed, but at the present-day there is a lack of systems that are specially designed for Arctic operations. This, as most systems are designed for open water conditions where the presence of ice and extended survival capability is not accounted for. Thus, the need for an arctic evacuation system that can be launched and operated in ice infested waters has made itself present, sparking the development of new evacuation systems especially designed for safe evacuation under the exposure of the extreme arctic environment. Reports worthwhile noting in this regard are the upcoming polar code, at least for the shipping side, DNV's Barents 2020 HSE study, CARD's arctic roadmap and the INTSOK's RUNO activities.

To assist in this development, this study has sought to look into the requirements for an arctic evacuation system applicable for operation in the Barents Sea. Critical scenarios and hazardous conditions will be evaluated with the intent of determining areas of importance for the design of such a system, where current concept will be evaluated and weighed in accordance to their compliance with these specifications. The study will be based on environmental criteria for a single location in the Barents Sea, respectively North East of Bjørnøya at coordinates N75°00′00″ E25°00′00′.

## 2 EMERGENCY PREPAREDNESS AND EVACUATION

### 2.1 *Escape, evacuation and rescue*

On petroleum installations the presence of explosive and flammable substances create a significant potential risk for fire and explosions if safe operational procedures are not followed. Therefore, much effort is put into preventing events of a harmful nature to human health and the environment, but regardless of this critical situations may arise. If this occurs, the chosen solution is often to terminate operations and relocate the crew to a safe location. This operation is generally separated into three different phases according to Bercha et al. (2004), respectively referred to as Evacuation, Escape, and Rescue (EER). The definition of the terms is as follows:

**Escape** The crew aborts their activities and move to a safe location on the installation. The crew prepares for leaving the installation by helicopter, lifeboat or bridge-links in an orderly fashion.

**Evacuation** Conditions are deemed to be not safe for the crew to remain onboard the installation. The crew is evacuated to a safe location directly by means of helicopter, bridge-walk or lifeboat. The preferred method of evacuation is chosen to be the one that poses the minimum risk for the crew, often deemed to be evacuation by helicopter if weather allows.

**Rescue** The process of recovering evacuees following an evacuation or escape from the installation and moving them to a safe location such as a ship. The phase can be performed directly by allowing a life raft to be retrieved by a standby vessel, transfer via helicopter, or transfer via MOB boats.

It should be noted that the focus for the study lies on the first two phases, respectively evacuation and escape. The general procedure is outlined in the flowchart below based on the work of Skogdalen et al. (2012) and Nedrevåg (2011).

### 2.2 *Means of evacuation*

Helicopter is often preferred as the main mean of evacuations as the whole procedure can be conducted as a transport operation. As this is a common procedure offshore the risk for the health and safety of the crew is assumed to be minimal. However, evacuation by helicopter is dependent on weather conditions. Thus, with consideration to the operational range and distances to the installations in the Barents Sea helicopters may not be a viable solution for evacuation. Subsequently, the final mean of evacuation will be by lifeboat. Open lifeboats, partially enclosed lifeboats, and inflatable life rafts are normally applied for vessels or installations where a high protection against the environment is not required. For offshore vessels and petroleum installations totally enclosed lifeboats (TEMPSC, see Fig. 1) and Free Fall Lifeboats (FFL) are the preferred evacuation systems that can comply with regulations.

In general TEMPSC are more suitable for merchant vessels, tanker, and offshore related vessels as they supply full protection against the environment while maintaining a lower weight and space requirement in relation to FFLs. Thus, they are beneficial for structures where the deadweight is of importance.

Figure 1. TEMPSC.

Figure 2. FFL on oil rig.

For offshore installation and vessels where the personnel needs to move away from the installation to a safe distance quickly, FFLs are the common means of evacuation. These lifeboats are generally certified for launch heights up to 35 m. The FFLs can be stored in davits hanging by wire and quick release, but most commonly they are stored on sloping skids and secured by a retaining mechanism. With regards to quick evacuation the skid launched FFLs are deemed as the safest as they transform the vertical momentum into forward momentum upon water entry, allowing the vessel to move to a safe location away from the installation quickly (PSA, 2005), see also Figure 2.

## 3 CLIMATE AND CONDITIONS IN THE BARENTS SEA

Many of the challenges faced in the Arctic by the maritime industry can be directly related to the operational climate. Wind, polar lows, sea ice, and temperatures need to be accounted for in order to run continuous operations in the north. Even though the Barents Sea climate is fairly stable throughout the year compared to the North Sea, normal operations will have to proceed under conditions that are viewed as extreme in other parts of the world. General environmental parameters applied for the study are thus based on the following information.

### 3.1 *Sea conditions*

With regards to sea conditions in the Barents Sea, data from the NORSOK (2007) standards indicate a maximum wave height, $H_s$, of 15 m in the southwest, reducing to 14 m towards the North East. Storms may additionally contribute to violent sea and wave conditions that can hinder safe evacuation. An arctic lifeboat will therefore have to perform under these conditions as evacuation is rarely performed in calm seas. Data from the Norwegian Meteorological Institute (2013) present additional information of a significant wave height $H_s > 5$ [m] for more than 4.60% of the time in the east, and 6.61% in the southwest.

### 3.2 *Temperature*

Temperatures in the Arctic present additional challenges as extreme subzero temperatures may induce atmospheric icing on the vessel as well as influence the performance of static and moving components in technical systems. Fluid based systems may be prone to freezing or increased viscosity, while mechanical failure may occur as a consequence of thermal fatigue or embrittlement. Temperatures may vary depending on year and location, but NORSOK N-003 (2007) indicate that the chosen location may expect temperatures in the range of 20°C in the summer to negative 30°C in the winter. The ocean temperature is found to vary between +7.5°C to +5°C in the summer and +0°C to −2°C in the winter. As the freezing point of seawater is −1.7°C surface ice can be expected. The water temperature in the area close to the ice sheet generally remains stable around 0°C year round.

### 3.3 *Wind*

Based on meteorological data from Bjørnøya (NMI, 2013), being the closest relevant position, the yearly average wind speed 10 meters above ground level is 12.19 m/s. In the winter months the average is found to be approximately 2 m/s higher and 2 m/s lower in the summer. However, the wind speed can reach significantly higher levels due to polar lows, sometimes referred to as arctic hurricanes, which can develop rapidly in the Arctic. These systems occur in the autumn and winter season with a frequency of 2–4 times per month and can result in wind speeds in the range of 17–30 m/s according to Jacobsen (2012). Together with heavy snow falls they are known to cause white-outs and cover equipment in snow.

### 3.4 *Icing*

Atmospheric icing is a known problem for vessels operating in cold climates, occurring from a combination of low temperatures and snow-, sleet-, or rainfall. Ice can subsequently build up on exposed surfaces if the surface and surrounding materials hold temperatures below freezing and the material remains in a liquid state. For lifeboats this will in general not pose a large threat, but the build-up of ice can affect the operation of the launching equipment and delay evacuation.

Another issue is sea spray icing which may occur when the vessel moves through the water. If actions are not taken this may de-stabilize the vessel. According to Guest (2005) sea spray icing can occur if the following criteria are fulfilled:

1. Wind speed above 5 [m/s] for small vessels.
2. Air temperature below freezing.
3. Water temperature lower than 7°C.

Thus, icing can occur even though the seawater temperature is above the freezing point, and may cause a rapid build-up of ice on the vessel in combination with an arctic hurricane, i.e. polar low.

### 3.5 *Sea ice and icebergs*

First year ice may be expected in the Barents Sea and the probability of encountering ice must therefore be accounted for. According to NORSOK N-003 (2007) the risk of encountering ice increases as one moves further north. The ice generally starts to form in the early winter as pancake ice, floating on the surface and growing to larger flakes if left undisturbed. For the considered location NORSOK estimates a probability larger than $10^{-2}$ of encountering first year ice.

For operation of a lifeboat the type and concentration of ice is of interest. A dense ice cover may cause problems for the launching and maneuvering of the lifeboat. According to information retrieved from the U.S National Ice Center (USNICE) a low ice coverage is expected, but this does not exclude the presence of pancake-ice on the surface. Hence, according to NORSOK and USNICE the risk of pancake-ice is present and shall be accounted for.

Another issue that must be accounted for is the risk of icebergs. With regards to the operation of lifeboats, large icebergs do not pose a great risk, but smaller icebergs known as growlers can hinder an efficient launch of the lifeboats in an emergency situation. According to NORSOK N-003 (2007) there is a risk of icebergs at the chosen location, a matter, which will not be accounted for in this study.

### 3.6 *Polar night, distances and low visibility*

The polar night experience in the high north will affect operations normally conducted during daytime. This will pose challenges with regards to emergency and rescue operations, but as procedures exist for night time operations this is not considered a critical aspect. However, the darkness will require consideration in the planning of operations to maintain the operational standards. Additionally, new equipment may need to be installed to monitor ice and detect icebergs, as visual detection will be unreliable in the polar night. Similar challenges are caused due to the low visibility in the summer months due to fog.

Furthermore, the large distances between the installations and land will require a higher level of self-sufficiency as it may take days for help to arrive in the case of an emergency in combination with poor weather. Consequently, new guidelines for the operational limit of the installation and evacuation equipment may be required to ensure that safety is maintained. Moreover, evacuation procedures and limits for evacuation may have to be adapted to comply with the large distances and dark conditions.

## 4 HAZARD IDENTIFICATION AND CRITICALITY ANALYSIS

For a lifeboat operating in the Barents Sea challenges may arise that conventional life systems are not exposed to. It is therefore of interest to discover the critical events that may occur and the severity of these occurrences. These factors form the foundation of an evaluation of the areas that will need improvement for an arctic evacuation system.

Identification of critical scenarios and hazardous conditions will be identified by application of a Preliminary Hazard Analysis (PHA) as well as a Failure Mode Effect and Criticality Analysis (FMECA). The work is based on the previous work of Nedrevåg (2011) and methods discussed by Kristiansen (2001).

### 4.1 *Hazardous scenarios*

The applied analysis is based on the use of a conventional FFL as applied in the offshore industry today. The boat is assumed to be launched by free fall from a launching system containing movable components which should remain ice free. It is further assumed that there is a need for an alternative launching system for situations where ice is present around the installation, so the corrective measured are assumed valid for this application as well, see Table 1. After launch the lifeboat will need to navigate through pancake ice to a safe location and await rescue.

Based on the PHA the following corrective measures should be considered:

- Prevent freezing of moving components.
- Prevent icing on lifeboats during and before operation.
- Prevent damage to the propulsion system.
- Improve maneuverability during operation.
- Facilitate for delayed rescue of evacuees.

### 4.2 *Launching and operation*

In the case of a critical event the launch and operation has been divided into five different phases. The phases are chosen to relate hazard identification and corrective measured to a specific part of the evacuation. This is due to the fact that a hazard may only be critical at certain stages of the evacuation. The phases are as follows:

**Phase 1** A critical situation has occurred. Lifeboats to be made operational.
**Phase 2** The decision to evacuate is made and the lifeboat is lowered to the surface.
**Phase 3** Lifeboat reaches the surface and moves away from the installation.
**Phase 4** Evacuation is complete. Evacuees to be kept alive and in good health until they can be transferred to a safe location.
**Phase 5** Rescue operation initiated. Evacuees need to be transferred to a safe refugee.

### 4.3 *Criticality analysis*

From the PHA critical scenarios and corrective measures are identified. However, the probability of an event occurring is not equal and the severity of the events remains unanswered. Therefore, an FMECA analysis is conducted to determine which events that are the most critical for the evacuation process.

By evaluating each phase based on "what if" questions and considering the probability and severity of an event, one can estimate the criticality of an event from the product of the probability and severity factors. These assist in rating the importance of the critical scenarios in relation to each other. The rating of probability applied is based on the following scale and presented in Table 2:

**One** Low probability of occurrence.
**Two** Probable, can occur under certain circumstances.

Table 1. PHA and corrective measures.

| Hazard condition | Event | Consequence | Corrective measure |
|---|---|---|---|
| Low temperature | Moving components freeze | Evacuation cannot be completed. Fatalities. | Heating of area where lifeboat is stored. |
| Low temperature | Engine fluids freeze. | Engine will not start. Delay of evacuation. Fatalities. | Anti-freeze in engine fluids. External heating of fluid tanks. |
| Low temperature | Temperature below ductile brittle transition for materials. | Materials fail and lifeboat falls to surface. Fatalities. | Evacuation before certain temperature. Increased requirements with regards to materials used. |
| Wind | Lifeboat motion is uncontrolled due to strong winds. | Loss of control. Damage to vessel. Fatalities | Guide system to reduce to horizontal movement. |
| Wind | White out. Low visibility. | Prevents safe navigation away from installation. | Radar or navigation system installed in lifeboats. |
| Wind | Lifeboat cannot resist wind in initial operation. | Collision with installation. Damage to vessel. Fatalities. | Sufficient engine power. |
| Wind | Strong winds prevent rescue operation. | Rescue is delayed. | Alternative retrieval methods for rescue operations. |
| Atmospheric icing | Components are covered by snow or ice. | Evacuation cannot be performed. Delay. | External heating to prevent ice buildup. |
| Atmospheric icing | Ice increases weight of lifeboat. | Material failure. Delay. Lifeboat falls to surface. Fatalities. | External heating to prevent ice buildup. |
| Sea spray icing | View of driver is blocked by ice. | Collision with platform. Hinders safe evacuation. | Windows need to be heated. |
| Sea spray icing | Build up of ice on lifeboat. | Loss of stability. Fatalities. | Low center of gravity. Heating of hull or frequent removal of ice. Design of lifeboat surfaces to prevent ice buildup. |
| Ice | Ice is present around the installation. | Free fall launch impossible. Evacuation cannot be completed. Delayed Evacuation. Fatalities. | Evacuation before ice concentration is too high. Alternative launching system. |
| Ice | Ice is present around the installation. | Maneuvering reduced. Failure to navigate away from installation. Delayed evacuation. Fatalities. | Increased engine power. Hull design that is designed for ice. Evacuation before ice concentration is too high. |
| Ice | Ice is present around the installation. | Ice interaction with propeller. Damage to propeller. | Propeller need to be protected against ice. Design of alternative propulsion system for ice conditions. |
| Ice | Ice is present around the installation. Large pieces. | Maneuvering reduced. Delayed evacuation. Fatalities. | Hull shape modified so vessel can get onto the ice and propel itself on the ice sheet. |
| Distance | Response time for rescue vessels is longer than 24 hours. | Hypothermia. Run out of fuel. Fatalities. | Heating source in vessel. Increased fuel tanks and supplies. |

Table 2. Rating of critical events.

| What if.. | Severity | Probability | Criticality |
|---|---|---|---|
| **Phase 1: Pre-launch** | | | |
| Moving components freeze due to low temperature | 3 | 2 | 6 |
| Engine fluids freeze or engine will not start | 2 | 1 | 2 |
| Temperature below specifications for materials used in load carrying components | 3 | 1 | 3 |
| Components are covered by snow or ice | 2 | 2 | 4 |
| Lifeboat is covered by snow or ice | 2 | 2 | 4 |
| Maintenance not performed properly/defect equipment | 2 | 1 | 2 |
| **Phase 2: Launch** | | | |
| Ice present in launching area | 3 | 2 | 6 |
| Boat is to heavy during launch due to ice | 3 | 1 | 3 |
| Strong winds rock the lifeboat when alternative launching system is used | 2 | 2 | 4 |
| Ice present in launching area when alternative launch system is used | 1 | 2 | 2 |
| Large ice pieces are present in launching area when alternative launch system is used | 2 | 1 | 2 |
| **Phase 3: Initial operation** | | | |
| Surface ice is present hindering maneuvering | 3 | 2 | 6 |
| Strong winds make it difficult to maneuver the boat | 2 | 1 | 2 |
| Low visibility or white-out due to wind and snow | 1 | 2 | 2 |
| Sea spray icing occurs during initial phase | 2 | 1 | 2 |
| Large pieces of surface ice is present around the installation | 1 | 1 | 1 |
| **Phase 4: Operation** | | | |
| Fuel capacity is to low | 3 | 1 | 3 |
| There are large pieces of ice present on the surface | 2 | 1 | 2 |
| The surface is covered by pancake ice | 2 | 2 | 4 |
| Low visibility or white-out due to wind and snow or icing on windshields | 1 | 2 | 2 |
| Sea spray icing occurs on the hull | 3 | 2 | 6 |
| Ice impacts with the propeller and causes damage | 2 | 2 | 4 |
| Wind and waves make it difficult to maneuver efficiently | 2 | 2 | 4 |
| **Phase 5: Rescue** | | | |
| Response time is more than 24 hours | 2 | 2 | 4 |
| Weather conditions prevent rescue vessel from safely transferring the occupants | 2 | 2 | 4 |
| Weather conditions prevent pick-up from helicopter | 1 | 2 | 2 |

**Three** Highly probable, can occur under normal circumstances.

For severity the rating is based on the following scale:

**One** Minor consequences, will most likely not lead to fatality or injury of personnel.

**Two** Medium consequences, can lead to damage of vessel or equipment. In rare cases fatality or injury.

**Three** Major, can lead to loss of life or serious injury.

The results indicate that there are common denominators to the most critical situations marked in red and yellow, which are:

- Moving components freeze of are covered by snow in the storage area.
- Presence of ice in the launch zone.
- Surface ice in the launch zone hinders efficient maneuvering.
- Ice-build up on hull surfaces.

Thus, the conducted analyses indicate that measure should be taken to prevent icing to the lifeboat and launch equipment both pre- and post-launch. Further, actions have to be taken to improve performance in ice-covered waters, and lifeboats will need to be adapted for extended survival above 24 hours. A secondary launch system is necessary for evacuation in ice-covered waters and to reduce the risk of damage during launch.

## 5 DESIGN SPECIFICATIONS

Present day rules and requirements are in general not constructed to take arctic requirements into consideration. Nonetheless, an arctic survival vessel will have to comply with existing legislations and recommendations for lifeboats, respectively International conventions (SOLAS), classification societies (ABS/DNV-GL), and national legislations (PSA). It should be noted that the national legislations are normally compliant with the rules set by the classification societies, such as the *DNV-OS-E406 Offshore Standard* which provides design requirements for free fall lifeboats. From 2015 all lifeboats on the Norwegian Continental Shelf must provide a safety level equivalent of the regulations in *DNV-OS-E406*, making it a national legislation regarding lifeboat safety.

Operation under arctic conditions will pose great challenges with regards to designing a lifeboat that can function adequately. In general the lifeboat will need to be designed specifically for the conditions at the location to ensure decent functionality and performance. The existing regulations, such as *DNV-OS-E406,* take this into account by referring to design criteria based on weather conditions on site. However, it has been found that the regulation focus more on structural integrity and loads inflicted rather than operational criteria. It has therefore been concluded that the regulations are more suitable for open water design and do not account for the challenges posed by ice, extreme winds, and low temperatures. The requirements should further specify, in detail, the operational criteria and equipment consideration that should be made to accommodate for extended survival and crew welfare.

Due to the matters mentioned above it has been concluded that additional requirements should be amended as a supplement to existing regulations, so as to ensure a vessel more suitable for the conditions on-site in the Barents Sea. Hence, the specifications listed should assist in improving the lifeboat operation in partly ice-covered waters and the extreme climatic conditions it will encounter. Reference is given to the previous work of Nedrevåg (2011). The main areas of interest are taken to be temperature, wind, icing, and prolonged survival.

**Temperature**

a. The lifeboat must be stored in a manner that prevents accumulation of ice on the vessel surface and launch equipment. Indoor storage is recommended, but must not hinder or delay evacuation in any way.
b. Engine must be ready for startup without any pre-operation. Can be achieved by pre-heating or storage in a heated environment.
c. Fluids vital to engine operation should under no circumstance freeze. Storage in a warm environment, anti-freeze, or direct heating of fluid tanks are viable alternatives.
d. An auxiliary independent heating system should be installed. The system should be capable of heating the interior independently of engine operation.
e. The launch mechanism should be operable when the operator is wearing thick clothes and gloves. It should be not rely on the use of electric energy. If indoor storage is applied, the hangar doors must open and clear the launch energy without the use of electric energy.

**Wind**

a. Launching must be possible under all expected wind conditions. Wind speed from polar lows, $U_{10} \leq 30$ [m/s], should not hinder launch.
b. If the vessel is lowered by means of winches the horizontal movement of the lifeboat should be restricted to prevent structural damage and risk for the crew. Guide lines may be applicable.
c. The distance between the lifeboat should allow for the vessel to clear the platform zone independently. The lifeboat must be able to resist and navigate against wind condition in the initial phase of operation.

**Atmospheric icing**

a. Materials should be painted with low friction paint, and indirect or direct heating of components should be applied to prevent accumulation of ice. Alternatively, established routines for ice removal and monitoring should be applied.
b. Ice removal equipment must be available in the storage area if heating systems fail.
c. Latches, propulsion equipment, windows etc. should be designed in such a manner that ice—accumulation is prevented.

**Sea spray icing**

a. Ice removal equipment should be available in the interior compartment.
b. A sufficient amount of hatches in the superstructure must be available to allow for ice-removal.
c. The superstructure should be designed in a manner to prevent sea spray icing, allowing accumulated ice to fall off.
d. Latches, windows and surfaces should be coated with a low friction material to prevent ice-accumulation.
e. The operator's windows should be heated.
f. The launching system should be considered if placed in a location where sea spray icing may be expected.

**Ice maneuvering**

a. Operation in open drift ice equivalent to 6/10 of the surface should be possible.
b. Breaking of thin ice layers in the range of 30 [cm] while proceeding at 3 knots should be possible.
c. If thick ice layers or large concentrations are encountered the vessel should be able to climb onto the ice.
d. The hull should be designed to reduce crushing loads from compressive ice and lift it out of the water.

e. The helmsman should be provided with sufficient view of the area in the vicinity of the bow to facilitate navigation through ice-covered waters.
f. The propeller and nozzle must be designed for the loads they may experience, and the blade thickness must be sufficient to endure repeated impacts from ice. Nozzle protection should be applied to direct ice away from the propeller.
g. The bottom hull shape should direct broken ice along the sides, preventing ice from entering the propulsion system.
h. Design of an alternative propulsion system should be considered. The system should comply with recommendations listed above, i.e. a pair of Archimedes screws.
i. The engine must be allowed to provide sufficient over-torque in case of ice-propeller interaction.

**Prolonged operation**
a. The lifeboat shall be equipped with drinking water and food for a total of 72 hours. Alternatively, a water purification system may be considered, but this must be compliant with SOLAS.
b. The interior heating must be able to operate continuously for 72 hours independent of the main engine.
c. The fuel capacity should be sufficient for the lifeboat to proceed at full speed for 3 hours, followed by 45 hours at 60% of maximum power (extension of article A1201 in *DNV-OS-E406*.

### 5.1 *Secondary launch system*

The primary launch method for the lifeboat should be by free fall, but a secondary launch method should be implemented to allow for safe deployment if ice is present in the area surrounding the platform. Such a system should be of a manner that allows for a reduction of vertical speed when the vessel meets the sea surface at a level positions at a distance approximately one or two ship lengths from the installation. The system should be hydraulically operated, where the energy for deployment is stored in accumulators to allow for independent operation.

There exist systems that might fulfill this task with minor modifications. The Preferred Orientation and Displacement System (PROD) is a launch system developed to move a TEMPSC outwards to a preferred orientation and displace it away from the installation. The system applies a davit launch survival craft in cooperation with a large boom to clear the platform installation. When the TEMPSC starts to descend it induces tension in a tag line, causing to boom to extent out—and bend downwards. This moves the lifeboat away from the installation while the boom flexes downwards until the vessel reaches the surface. At this stage the tag line continues to pull on the bow, pulling the lifeboat forward. When the lifeboat is clear of the launch zone the tag line releases (Wrigth et al., 2002).

The PROD system is currently in use on a few platforms and has proven to be functional in open waters. It is believed that the presence of ice could interfere with the use of the systems tag line, but that this could be solved by redesign. The system should be applicable for a wide range of arctic survival vehicles.

## 6 EXISTING AND PROPOSED CONCEPTS

With an increase in the exploration for oil and gas in the arctic areas it is expected that the market demand for arctic evacuation systems will increase, leading forth new developments and production of arctic survival vehicles. Some concepts and systems already exist, and will be briefly presented here.

### 6.1 *ARKTOS™*

ARKTOS, see Figure 3, is a specialized amphibious survival vehicle developed in the 1980's by Watercraft International for use in the Beaufort Sea (Wright et al., 2002). The system is especially designed for crossing the transition between open water according to ARKTOS Developments Ltd. (2010).

The system consists of two closed hull units linked together through a hydraulic link with belts mounted at the vehicles sides to allow for climbing onto the ice. For propulsion, the system is equipped with water jets driven by a diesel engine. Maneuverability is achieved by manipulating the hydraulic link.

Currently the ARKTOS is approved by ABS and is found to be suitable for applications where it can be driven off or lowered by large davits from bottom-bearing installations. The system can

Figure 3. ARKTOS (ARKTOS Developments Ltd., 2010).

accommodate 52 persons and has an independent air supply to allow for movement through burning oil slicks and $H_2S$ saturated environments.

### 6.2 *AMV lifeboat*

The AMV lifeboat, see Figure 4, is a conceptual design for a free fall lifeboat suitable for arctic operation developed by Team Innovation Trondheim A/S and patented by Foo et al. (2009). The configuration is that of an enclosed hull structure propelled by a pair of rotatable Archimedes' screws, allowing it to move on water, ice, and snow. Additionally, the screws allow for climbing over ice and up-righting itself if it is turned over. The screws further allow for maneuvering by adjustment of the rotational speed of the screws, enabling sideways motion if necessary. De-icing is of the hull structure is performed by a closed heating system, drawing heat away from the engine and circulating it around the hull superstructure. With additional attachment points implemented in the hull, the vessel is adapted for normal davit launch and drive of capability in addition to free fall. The concept provides a feasible and promising solution for an arctic evacuation vehicle.

### 6.3 *Seascape*

Seascape, see Figure 5, was developed in the 1980's as an alternative lifeboat system, being stored on a large pivoting steel arm. Upon deployment the steel arm extends out, lowering the lifeboat down to the surface. When the arm is fully extended the lifeboat is gravity lowered from the arm by the means of a winch on the platform. The solution allows for the vessel to be placed 20–30 [m] away from the platform and has recently been modified for use in ice-conditions. This was accomplished by relocating the pivoting arm to the main deck system so ice would not cause structural damage. The system is designed to be applied together with a large high powered TEMPSC designed to maneuver and function well in ice. Tests revealed that the lifeboat could operate in ice concentrations up to 8/10, where it was able to ram its way through a 24 hour frozen channel. The system does however not allow the vessel to proceed onto the ice which is a large drawback.

Figure 4. AMV lifeboat (Team Innovation Trondheim, 2010).

Figure 5. Seascape system (O'Brien, 2004).

### 6.4 *Schat-Hardinger FFL 1200*

The Schat-Harding FFL 1200, see Figure 6, is the only commercially available lifeboat that is designed in accordance with the *DNV-OS-E406* offshore standard, and complies with regulations on the Norwegian Continental Shelf. It is not specifically designed for arctic conditions, but may provide a viable alternative for installations in the Barents Sea.

### 6.5 *Arctic FFL*

The Arctic FFL, see Figure 7, is a design concept proposed by Nedrevåg (2011) for a lifeboat specifically designed for operation in areas where there is a high likely-hood of encountering large ice concentrations. The vessel is intended to comply with *DNV-OS-E406*, sharing the same dimensions as conventional FFLs to allow for a capacity of 70 persons. The hull is to be constructed From Reinforced Polyester (FRP) to ensure adequate strength and reduced weight. Furthermore, the design intends to reduce ice accumulation by inclining all hull surfaces, creating a vertical down-force from the weight of the ice which generates a shear force parallel to the hull. The ice will thus break off by its own weight.

The lifeboat will operate as FFL or by davit launch, being stored indoors in a heated environment. To facilitate maneuvering in ice a V-shaped hull and a flat keel will allow improved progress in low ice concentrations, leading the ice away from the path and reducing crushing loads, see Figure 8. A low stem angle is applied at the bow

Figure 6. FFL1200 (Schat-Harding, 2013).

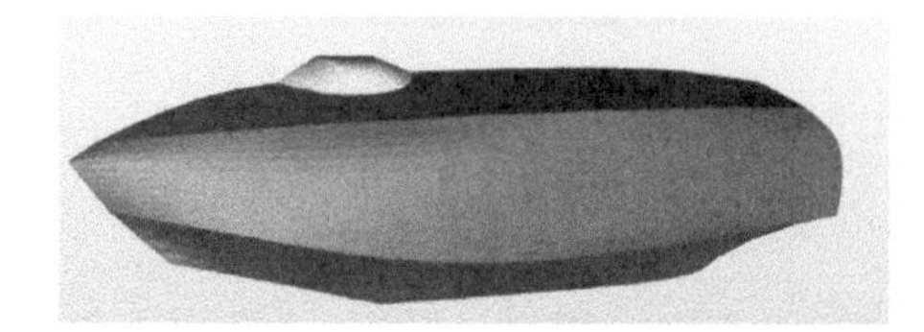

Figure 7. Arctic FFL hull design (Nedrevåg, 2011).

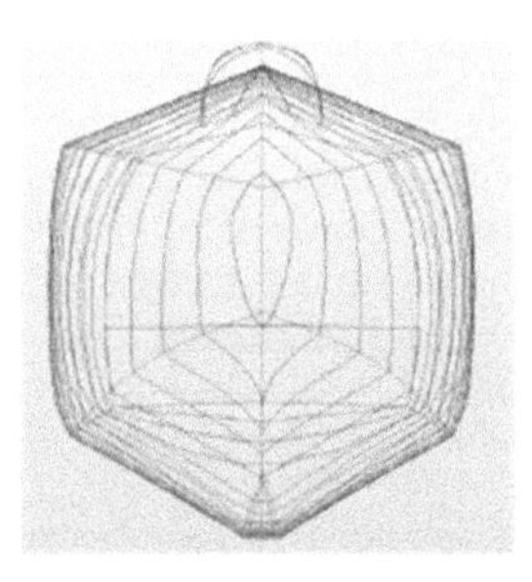

Figure 8. A-FFL body plan (Nedrevåg, 2011).

to improve icebreaking capabilities and assist in allowing the vessel to climb onto the ice. Furthermore, the system is propelled by two Archimedes' screws located at the flat keel to allow for the vessel to climb onto the ice if necessary. This propulsion system will assist in reducing the ice-interaction problems and the blocking that is encountered with conventional nozzle propellers. It should be noted that two stabilizing fins need to be added to the hull to prevent the lifeboat from falling over when on the ice.

To improve visibility the coning position will be located at the front, allowing for easier navigation through high concentrations of ice. The lifeboat is designed with the intension of using diesel engines to generate energy for the propulsion system. Overall, the system is highly suitable and provides a solution for many of the challenges encountered when operating in ice infested waters.

## 7 EVALUATION OF LIFEBOAT SYSTEMS

To evaluate the systems in relation to the requirements for an arctic survival vehicle a hierarchic weighted evaluation is applied based on all categories and articles mentioned in Section 5. Each category is weighted based on a best judgment with a scale of 1–5 within each category. The value zero is applied if the article is not applicable. In those cases where information is not supplied an estimate of the systems compliance will be made. The calculation method applied for evaluation of each category and the total score is thus stated as:

$$S_n = \frac{\sum_{i=1}^{m}\left(S_{n,m}a_{n,m}w_{n,m}\right)}{\sum_{i=1}^{m}\left(a_{n,m}w_{n,m}\right)}$$

$$S_T = \frac{\sum_{i=1}^{n}\left(S_n a_n w_n\right)}{\sum_{i=1}^{n}\left(a_n w_n\right)}$$

where $S_T$ is the total score, $w$ is the weight factor, a is the applicability factor, and the subscripts $n$ and $m$ indicate respectively category and article identification.

### 7.1 *Results*

The independent scores for each category with their associated weights are presented in Table 3, along with the total score for each concept. The table allows for comparison of the concepts. The following sections provide a discussion on the results of each system.

#### 7.1.1 *ARKTOS™*

With the ARKTOS being specifically designed for operation in extreme arctic environments, but shall be stored in a protected environment. However, when in operation the craft scores poorly with regards to wind and sea spray icing due to its many appendices and flat surfaces. It is further believed to have large disadvantages for operation in the large waves created by polar lows. Due to its ability of proceeding in ice and climbing onto the ice cover it scores fairly well in the ice maneuvering category. The vessel does however have a major drawback in the prolonged operation category due to its 12 hour operational limit.

Nonetheless, ARKTOS differs from the other concepts as it has proven its capabilities, found to perform well in areas of high ice concentration and continuous ice. Being designed with a low freeboard and drive-off application it is in general found to be unsuitable for the Barents Sea, but does however present an impressive overall score in comparison to the other crafts.

Table 3. Summary of evaluation.

| | | Score $S_n$ | | | | |
|---|---|---|---|---|---|---|
| Category $n$ | Weight $\omega_n$ | ARKTOS | AMV | Sea-scape | S-H FFL1200 | Arctic FFL |
| Temperature | 0.2 | 4.75 | 3.40 | 3.30 | 3.25 | 5.00 |
| Wind | 0.1 | 1.50 | 2.50 | 4.25 | 3.50 | 4.50 |
| Atmospheric icing | 0.1 | 4.20 | 3.60 | 2.10 | 4.10 | 5.00 |
| Sea spray icing | 0.2 | 2.65 | 3.80 | 2.85 | 3.20 | 4.10 |
| Ice maneuvering | 0.3 | 3.34 | 4.02 | 3.22 | 2.62 | 4.14 |
| Prolonged operation | 0.1 | 1.00 | 3.00 | 3.00 | 3.00 | 5.00 |
| Total | 1.00 | 3.15 | 3.57 | 3.13 | 3.67 | 4.51 |

#### 7.1.2 *AMV lifeboat*

Being on a conceptual stage little is known about the true performance of the survival vehicle in ice. The evaluation is therefore largely based on estimation of performance and fulfillment of requirements. As the vehicle is specifically designed for operation as an arctic free fall lifeboat it scores well in categories regarding temperature, atmospheric-, and sea spray icing. However, these categories are highly dependent on the storage of the vessel, and this has influence the temperature category as storage requirements and subzero starting abilities are not known. Moreover, the craft has a heated hull that will assist in minimizing icing and the need for manual ice removal. Some large appendices do however contribute to a reduction in score.

With its large Archimedes' screws the vessel is believed to handle well in and on the ice, but uncertainties and lack of information influence the final score. A good score for prolonged survival has been granted as the vessel should comply with these requirements. Nonetheless, the preliminary design lacks the hydrodynamic properties that will result in a good performance for high wind and wave conditions, giving it a poor score in this category.

It is believed that a redesign of the hull will make the AMV lifeboat a viable alternative for the Barents Sea, improving the performance in waves and FFL capabilities. It is a strong contender against ARKTOS.

#### 7.1.3 *Seascape*

The Seascape system is believed to perform exceptionally in strong wind conditions due to its launching system and the vessel design. Being designed intentionally for arctic conditions it is believed to comply well with all criteria posed. However, the outdoor storage exposes the craft to icing and reduces the overall score of the vessel. For maneuvering the vessel has proven to perform well, but lacks the ability to climb onto the ice sheet which is a major drawback. Additionally, the vessel can only proceed in ice by ramming and not breaking, with a bow design that is believed to hinder the operators view. The large flat deck with many edges and appendices are in terms believed to expose it to atmospheric and sea spray icing.

The overall score places it behind the ARKTOS and AMV, being the main contesters for this concept. The placement is influenced by the poor resistance to icing and storage method, but it has proven its functionality during ice trials. Nonetheless, the launching system is believed to be the most valuable part of this concept, even though it will have to compete with the relatively shorter launch time of the FFL systems.

#### 7.1.4 *Schat-Harding FFL1200*

The Schat-Harding FFL1200 receives a fair score in comparison to the other systems. This is surprising as it is not designed for specific use in arctic conditions, but rather to comply with the *DNV-OS-E406* offshore standard for the Norwegian Continental Shelf. It should be noted that limited information was available, influencing the score as performance has been judged on requirements listed in the *DNV-OS-E406* standard. However, the lifeboat performs fairly well with regards to atmospheric icing as it has a round smooth hull that facilitates manual ice removal, with additional heating of the davit winches. As temperature scores are largely dependent on storage and subzero startup capabilities, the overall score has been reduced by limitations in these areas.

With regards to operation, the system gets an adequate score for sea spray icing, but few latches and rail appendices affect the score. The lifeboat is assumed to handle poorly in ice as it is not designed for ice breaking and cannot enter the ice cover. It is however designed with a propulsion system that is protected against debris, but without an auxiliary system. When it comes to survival it is equipped for a minimum of 24 hours of independent operation, giving a fair score in this category.

The FFL 1200 is currently the only known option known that is approved for operations in the Barents Sea. Nonetheless, it is believed that modifications should be made to ensure improved operation in arctic waters.

#### 7.1.5 *Arctic FFL*

The Arctic FFL is specifically designed to operate as a FFL, with the additional capability of performing in ice covered waters with a concentration of 8/10. Specific consideration has been taken to design a hull that can break and maneuver in ice with the ability to proceed onto solid ice covers. The hull is designed with few appendices to minimize atmospheric and sea spray ice accumulation. The vessel is equipped so that it can await rescue for up to three days.

The care taken to specifically design a FFL suitable for the conditions in the high north reflects itself on the score it receives. The vessel outperforms the other survival crafts in every category. Indoor storage and dual launching methods allow the vessel to fulfill all criteria in the temperature category, with a launch system that allows for deployment during extreme weather and quick clearance of the installation. The lifeboat is believed to operate well in ice cover waters with the only drawback being the lack of an alternative propulsion system. It is further questioned how the vehicle will enter and remain stable when it gets onto the ice cover.

The Arctic FFL is currently the best alternative for application in the Barents Sea, but as it has never been tested and the design is still conceptual there are many uncertainties regarding actual performance. It should however be noted that the application of the design criteria could significantly improve the performance of other lifeboat systems for arctic environments.

### 7.2 *Concludatory remarks on the evaluation*

The evaluation of the five concepts reveals that most of the scores are in the intermediate range. The Schat-Hardinger FFL1200 and the AMV lifeboat received fairly even scores even though there are large differences between the designs. Equally, the ARKTOS and Seascape system proved equal, with the Seascape system having a slightly lower score. The highest score was appointed to the Arctic FFL which proves that a free fall lifeboat specifically designed for ice-going purposes outmatch the contesters.

It should be noted that all crafts designed for arctic conditions received fair scores within all categories in comparison to those that were not designed with the arctic in mind. This enforces the importance of considering the actual environmental conditions that the craft will encounter and design accordingly.

## 8 CONCLUSION

As we move further north in the search for new oil and gas resources we are faced with new challenges with regards to conducting safe operations. The arctic presents demands and requirements that far supersede those we work with today, as can be seen by the extensive modifications done to winterize and adapt drilling rigs and ships for operation in arctic waters. Extreme subzero temperatures, polar lows, and the presence of ice requires modification and development of new designs if we are to operate under these conditions.

The main criterion for conducting any operation is to ensure that neither people nor the environment is exposed to any harm. With regards to evacuation this requires an evacuation system that can launch safely and survive the brutal condition of the northern seas.

This study has focused on the conditions encountered in the northern part of the Barents Sea, seeking to determine hazardous and critical conditions for the operation of a lifeboat. Many challenges are posed by the conditions on site, but it has been found that the most important considerations must be made with regards to:

- Prevent the freezing of moving components and equipment vital to launch.
- Design hull structures and storage to prevent accumulation of ice pre- and post-launch.
- Design propulsion systems that can handle ice-interaction.
- Improve maneuverability and performance in water with high ice concentrations.
- Facilitate and equip survival crafts for delayed rescue and long waits.
- Design a launch system that can safely transfer the vessel to the sea surface and away from the installation when ice is present.

Based on these considerations, supplementary design criteria for existing regulations have been recommended. The most important measures are found to be:

- Improvement of storage to prevent ice accumulation and freezing of components.
- Improvement of maneuverability in waters with high ice concentrations.
- Ensure that the launch system can operate in low temperatures and when ice is present in the water around the installation.
- Prevention of sea spray icing on hull surface.
- Sufficient fuel, food, and water to allow for long rescue times.

Several concepts where tested in accordance with the corrective measures. Of these, the ones that were specifically designed to operate in the arctic provided the best solutions and performance. It was found that the Arctic FFL, a conceptual design by Nedrevåg (2011), provided the best solution according to the given design criteria. As the concept shows a high compliance with *DNV-OS-E406* and the additional requirements posed by the environment in the Barents Sea it would be interesting to see it developed further and tested.

## REFERENCES


ARKTOS Development Ltd. *Arktos amphibious vehicle*, (2010). Accessed December 21, 2013, from http://www.arktoscraft.com.

Bercha, F.G., Cervošek, M., & Abel, W. (2004). Reliability assessment of arctic EER systems. In the *17th International Symposium on Ice, IAHR*, St. Petersburg, Russia.

Det Norske Veritas. (2010). *Offshore Standard DNV-OS-E406: Design of free fall lifeboats*. Høvik, Norway.

Foo, K.S., Quah, C.K., Li, K., Scharffscher, M., Scharffscher, P., & Loset, S. (2009, September). *Vehicle*. US Patent App. 12/559,808.

Guest, P. (2005). Vessel icing. *Mariners Weather Log*, 49(3).

International Maritime Organization. (1980). *Solas 1974: Brief history-list of amendments to date and where to find them*. Accessed December 20, 2013, from http://www.imo.org/about/conventions/listofconventions/pages/international-convention-for-the-safety-of-life-at-sea-(solas),-1974.aspx.

Jacobsen, S.R. (2012). *Evacuation and Rescue in the Barents Sea* (Master's thesis, University of Stavanger). Stavanger: University of Stavanger.

Kristiansen, S. (2001). *Risk Analysis and Safety Management of Maritime Transport*. Trondheim: Department of Marine systems Design, Faculty of Marine Technology, Norwegian University of Science and Technology.

Nedrevåg, K. (2011). *Requirements and Concepts for Arctic Evacuation* (Master's thesis, Norwegian University of Science and Technology). Trondheim: Norwegian University of Science and Technology.

NORSOK N-003. (2007). *Actions and action effects* (2nd edition). Oslo: Standards Norway.

NORSOK Z-013. (2001). *Risk and emergency preparedness analysis* (2nd edition). Oslo: Standards Norway.

Norwegian Meteorological Institute (NMI) (2013). *Meteorologiske og oseanografiske parametre frakyst-, hav- og landposisjoner*. Accessed October 29, 2013, from http://www.regjeringen.no/pages/14137473/Vedlegg_12_6.pdf.

O'Brien, D.P. (2004). Seascape system of evacuation. In the *17th International Symposium on Ice, IAHR* (pages 21–25), St. Petersburg, Russia.

Petroleum Safety Authority Norway. (2005). *Products*. Accessed December 18, 2013, from http://www.ptil.no/news/free-fall-lifeboats-safest-means-of-evacuation-at-sea-article2275-878.html.

Schat-Harding. (2013). *Products*. Accessed December 18, 2013, from http://www.schat-harding.com/products/.

Skogdalen, J.E., Khorsandi, J., & Vinnem, J.E. (2012). Evacuation, Escape, and Rescue Experiences from Offshore Accidents Including the Deepwater Horizon. *Journal of Loss Prevention in the Process Industries, 25*(1), 148–151.

Team-Innovation Trondheim. (2010). *AMV lifeboat*. Accessed October 21, 2013, from http://www.teaminnovationtrondheim.com.

U.S National ice Center (2013). *Ice extent data*. Accessed October 21, 2013, from http://www.natice.noaa.gov/.

Wright, B., Timco, G., Dunderdale, P., & Smith, M. (2002, October). Evaluation of Emergency Evacuation Systems in Ice-covered Waters. *PERD/CHC Report 11–39*, pp. 1–106. Canada: National Research Council Canada.

*Maritime-Port Technology and Development – Ehlers et al. (Eds)*
*© 2015 Taylor & Francis Group, London, ISBN 978-1-138-02726-8*

# An approach towards the design of robust arctic maritime transport systems

M. Bergström, S. Ehlers & S.O. Erikstad
*Department of Marine Technology, Norwegian University of Science and Technology (NTNU), Trondheim, Norway*

ABSTRACT: This paper describes a simulation-based approach towards the design of robust arctic maritime transport systems that are adaptable to uncertain future ice conditions. It makes it possible to simulate the performance of the transport system for various future ice scenarios and to compare various ice mitigation strategies in terms of cost. A case study is carried out to demonstrate how the approach could be applied in practice. The outcome from the case study indicates that the approach can provide valuable insights into the economics of an arctic maritime transport system and that its components can easily be modified or replaced for improved accuracy.

*Keywords*: arctic ship; arctic maritime transport system; ice class; icebreaker assistance; simulation

## 1 INTRODUCTION

Shipping in the Arctic is predicted to grow both in volume and diversity over the coming years. This prediction is due to the large oil, gas, and mineral discoveries found in the arctic region, as well as due to the increased interest in the Northern Sea Route.

When designing an arctic maritime transport system, here defined as a system consisting of any number of vessels transporting cargo between two or more ports through partially ice-covered waters, several arctic specific challenges, such as uncertain future ice conditions, need to be considered. To make such a transport system robust, in the sense that it is adaptable to such uncertain future ice conditions, it is necessary to consider a range of various possible future ice conditions along the intended route and to define an ice mitigation strategy that is able to deal with each of those conditions. To this aim, a simulation based approach is developed that can be used to simulate the performance of an arctic maritime transport system for various future ice scenarios and to compare various ice mitigation strategies for those scenarios in terms of cost.

In the current approach, the ice-vessel interaction is limited to the ice resistance, i.e., only the power demand of the vessel is considered. As a simplification, the ice is assumed to be level ice, i.e., possible ridges, ice channels, etc. are not considered. In addition, the ice thickness is assumed to remain constant between consecutive waypoints along the route.

A case study is carried out to demonstrate how the developed approach could be applied in practice. The results of the case study indicate that it can provide valuable insights into the economics of an arctic maritime transport system and that it can be developed further as it's components can easily be modified or replaced for improved accuracy.

The developed approach can be considered a further development of an approach towards mission-based design of arctic maritime transport systems developed by (Bergström, et al., 2014), which in turn was partly based on an approach developed by (Erceg, et al., 2013). Other related work include (Valkonen, et al., 2013) and (Riska, et al., 2001).

## 2 DESCRIPTION OF THE APPROACH

The transport task is defined by the route, the transport demand, and the period of time the transport will be taking place (for instance within the period 2016–2025). The transport route is determined by waypoints (coordinates along the route). Waypoint and date (voyage) specific ice thickness estimates are obtained from ice scenarios determined based on the prevailing ice conditions and various possible future development trends determined by the user.

In case of independent operation in ice, i.e., operation without icebreaker support, the speed of the vessel is calculated using a so-called *h-v* curve described by (Juva, et al., 2002) that determines the speed of a ship as a function of the ice thickness.

The speed of a vessel being escorted by an icebreaker is assumed to correspond to an assumed average speed of the icebreaker. Icebreaker assistance is assumed to be required from the first to the last waypoint along the route where the ice thickness exceeds a specific value determined by the user based on the ice class and the propulsion power of the ship. A flowchart describing the developed approach is presented in Figure 1.

Possible ice mitigation strategies, i.e., strategies for how to deal with sea ice include for instance the following:

1. Use of ships with a low ice class that are able to operate independently in thin ice only and use of icebreaker assistance when the ice conditions exceeds the class capabilities.
2. Use of ships with a high ice class and propulsion power to reduce/minimize the amount of icebreaker assistance required.
3. Avoidance of difficult ice conditions by limiting the operation to periods with little or no ice.

Costs related to various ice mitigation strategies are calculated based on estimates for the following cost items:

- Daily cost for icebreaker assistance.
- Additional investment and operating cost related to a higher ice class and propulsion power.
- Additional fuel costs due to additional ice resistance.

The total voyage specific sailing times are calculated based on the leg distances and the corresponding leg specific speeds. Calculated sailing times for the time span simulated are then imported into a SimEvents (a discrete event simulation tool developed by MathWorks) simulation model. By using the simulation tool, it is then possible to simulate stochastic transit time including stochastic factors such as time spent waiting for icebreaker assistance, loading and unloading times, variations in the transit time caused by weather etc. Additional parameters can be included as needed.

The simulation model can then be used to simulate how the transit times vary during the time span simulated due to varying ice conditions, to simulate the total accumulated amount of cargo transported from location A to location B, and to simulate the required number of days of icebreaker assistance.

## 3 CASE STUDY

### 3.1 *Transport task*

The case study deals with the maritime transport of Liquefied Natural Gas (LNG) from the port of Sabetta (Russia), which is still under construction, to the port of Narvik (Norway), from where the LNG is assumed to be transported onwards. The transport from Sabetta to Narvik is carried out by ice-strengthened LNG carriers, while the onwards transport from Narvik to the large transhipment terminals in central Europe and Asia is carried out by more cost efficient LNG carriers without ice class. The route, which is approximately 1489 NM, is presented in Figure 2.

The average LNG production rate in Sabetta is assumed to be 100,000 $m^3$/day (Total S.A., 2014). Thus, to avoid production stops the average

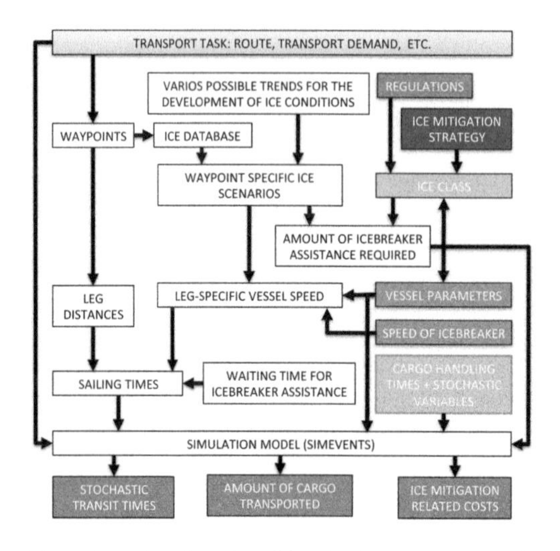


Figure 1. Flow chart describing the developed approach.

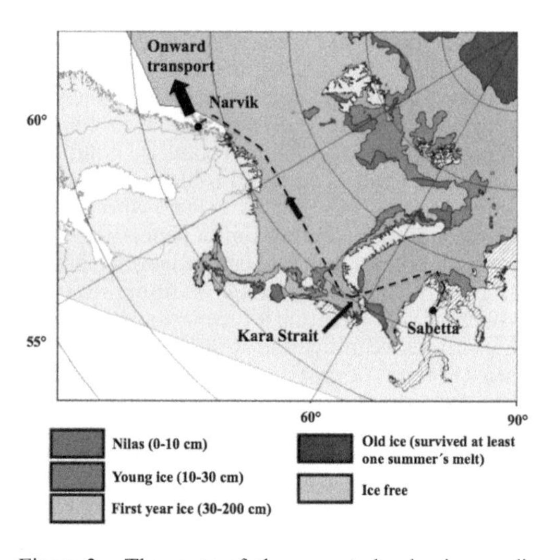


Figure 2. The route of the case study plus ice conditions along the route in mid-march 2014 as determined by (AARI, 2014).

transport capacity of the system needs to be at least 100,000 m$^3$/day × 365 days/year = 36,500,000 m$^3$/year. The objective of the case study is therefore to design a transport system with sufficient capacity to avoid production stops resulting in very significant economic losses.

The assumed transport task can be seen as an alternative to the plan to use Arc 7 classified 170,000 m$^3$ LNG carriers to transport the LNG directly from Sabetta to the large transhipment terminals in central Europe and Asia (Renton, M., 2013,). Such heavy ice-strengthened ships are, in open waters, general significantly less cost-effective than lighter non-ice-strengthened ships. Thus, it could be more economical to limit the use of ice-strengthened vessels to the part of the distance where ice strengthening is needed, and carry out the onward transport using normal ships. The planning of the transport system is assumed to be in the conceptual design phase. Operation is assumed to start at January 1 2016 and to continue for at least 10 years. The time span simulated is therefore 01.01.2016–31.12.2025.

### 3.2 *Determination of ice scenarios*

The starting point, i.e., the assumed prevailing ice conditions, was determined by modifying ice data obtained from a numerical climate model developed by SINTEF called SINMOD (Slagstad, et al., 2005) (SINTEF, 2014) to correspond to ice data from satellite imagery from year 2012 and 2013 provided by (AARI, 2014). Based on the assumed prevailing conditions, four possible future ice scenarios were then generated for the time span simulated based on four assumed ice thickness development trends presented in Figure 3. The trends, which include one trend of increasing ice thicknesses, two trends of decreasing ice thicknesses, and one trend of more or less unchanged ice thickness, were determined based on coefficients generated at random between pre-determined intervals. Ice scenario specific average ice thicknesses along the distance Kara Strait-Sabetta, where first-year ice occurs, are shown in Figure 4.

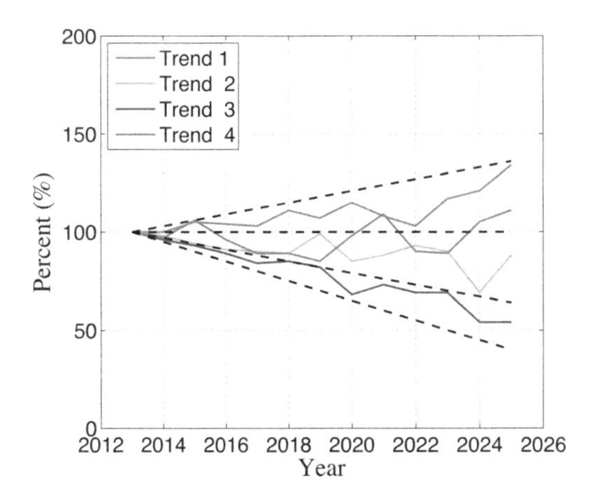


Figure 3. Applied ice thickness development trends.

### 3.3 *Ice conditions along the route*

The route goes through the Kara and the Pechora Sea, both of which according to satellite imagery based ice maps provided by (AARI, 2014) are normally covered by first year ice in the winter. An ice map, that was determined based on one of the ice maps from (AARI, 2014) showing the ice conditions along the route in mid-march 2014, is presented in Figure 2.

An example of applied date specific ice forecast for the route is shown in Figure 5, which shows the predicted ice thicknesses along the route for 31.03.2026 in accordance with ice scenario 1. On that date, as shown in Figure 3, the predicted maximum ice thickness along the route is around 2.0 m. This is assumed to be the maximum ice thickness that can occur along the route during the simulated period of time.

The sailing time is determined based on the date of departure, i.e., based on the ice conditions that

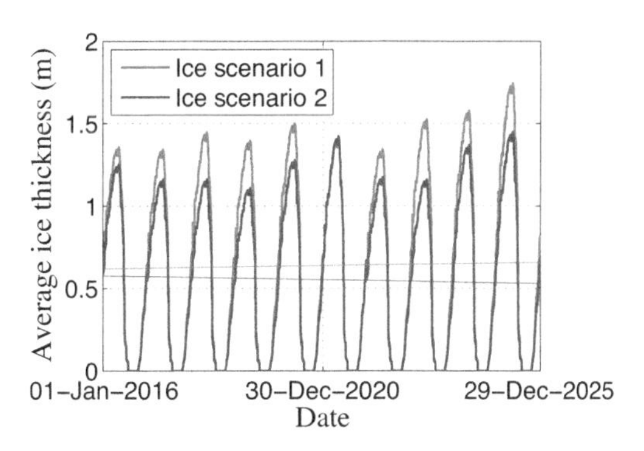


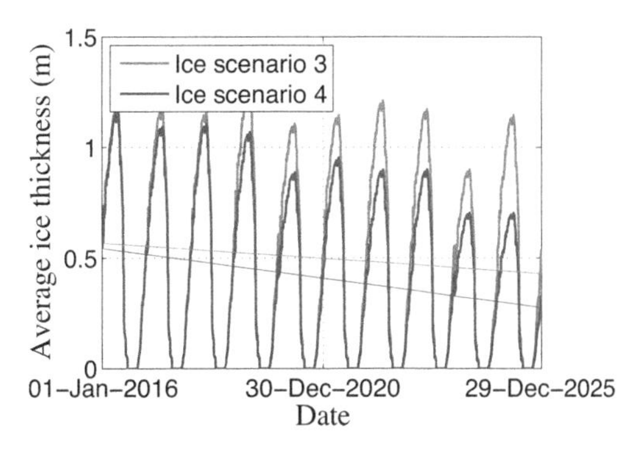


Figure 4. Determined ice scenarios.

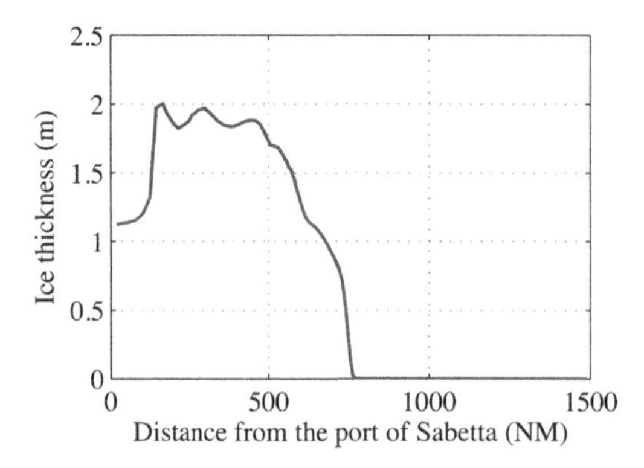


Figure 5. Ice thickness along the route at 31.03.2026 in accordance with ice scenario 1.

occur along the route as the ship leaves the harbour. This means that the ice thicknesses estimated for the various legs are assumed to remain constant during a voyage. In addition, the ice thickness is assumed to be homogenous between waypoints, which in the case study are between 7 and 22 nautical miles (nm) apart along the part of the route where ice occur.

### 3.4 *Ice mitigation strategies considered*

Three different ice mitigation strategies were considered:

1. Use of Polar Class (PC) 7 classed ships that are able to operate independently in up to 0.7 m thick ice. Use of icebreaker assistance when the ice thickness exceeds 0.7 m.
2. Use of PC 5 classed ships that are able to operate independently in up to 1.2 m thick ice. Use of icebreaker assistance when the ice thickness exceeds 1.2 m.
3. Use of PC 4 classed ships that are able to operate independently in up to 1.7 m thick ice. Use of icebreaker assistance when the ice thickness exceeds 1.7 m.

Periods with little or no ice are expected to be very short along the present route. Thus, one of the in section 2 mentioned possible ice mitigation strategies, to avoid difficult ice conditions by limiting the operation to periods with little or no ice, was excluded while it was considered infeasible.

A single icebreaker is assumed to cost USD 50,000 per day. Convoys are not considered, i.e., the icebreaker costs are not divided on multiple ships.

### 3.5 *Estimation of vessel parameters and costs*

The assumed vessel parameters are presented in Table 1. The main dimensions of the vessel were determined based on a LNG carrier of the fleet

Table 1. Assumed ship parameters.

| | |
|---|---|
| Length w.l. | 280 m |
| Breadth | 45.8 m |
| Draft | 12 m |
| Cargo capacity | 172,000 m$^3$ |
| Tonnage | 110,920 GT |
| Speed o.w. | 19.5 kn |
| Ice class | PC 7/PC 5/PC 4 |
| Propulsion power at 0.85% MCR | 30,000 kW/57,000 kW/ 90,000 kW |
| Specific fuel consumption (HFO) | 180 g/kWh |
| Initial investment | USD 242 M/USD 264 M/ USD 297 M |
| Annual operating costs related to a higher ice class | USD 0/USD 3.96 M/ USD 9.9 M |

of Knutsen OAS Shipping (Knutsen OAS Shipping AS, 2014). The initial investment costs were determined assuming that a PC 7 vessel costs 10% more than a standard vessel without ice class that is assumed to cost USD 220 M. The corresponding additional investment cost for PC 5 and PC 4 vessels are assumed to be 20% and 30%, respectively.

The additional operating costs related to PC 5 and PC 4 were determined assuming that the annual operating costs correspond to around 3% of the initial investment.

The required propulsion power for each ice class were determined so that the ship at 85% MCR is able to operate with a speed of around 3 kn in the maximum ice thickness for independent operation specified for the ice class in question. The 15% sea margin can be utilized, for instance, in case the vessel gets stuck in an ice ridge.

### 3.6 *Transit times*

Regardless of ice scenario and ice mitigation strategy, the transit times vary significantly between seasons. Simulated transit times for ice scenario 2 and PC 7 vessels are shown as example in Figure 6.

In this case, the total duration of a return trip varies between 16 days (2 × 8 days) during peak ice conditions and 6.5 days (2 × 3.25 days) during periods with no ice. The average return trip is around 10.2 days (2 × 5.1 days) and the median is around 9.2 days (4.6 × 2 days). Please note that the above-mentioned transit times are examples only. All transit times applied in the simulations are voyage and date specific, i.e., unique.

### 3.7 *Determination of transport capacity*

The ensure a sufficient transport capacity also in the worst assumed ice conditions, i.e., ice scenario 1,

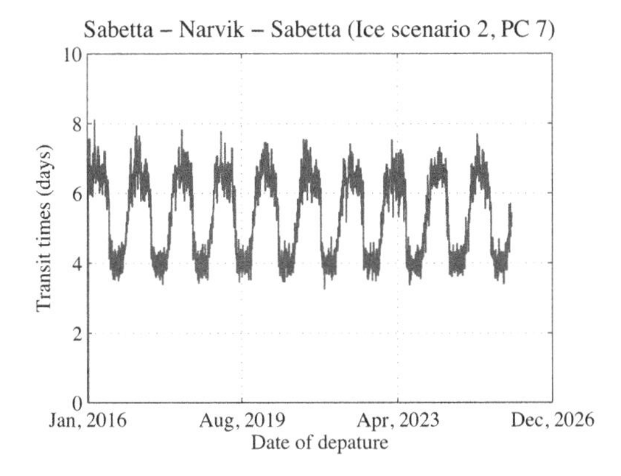


Figure 6. Example of simulated transit times.

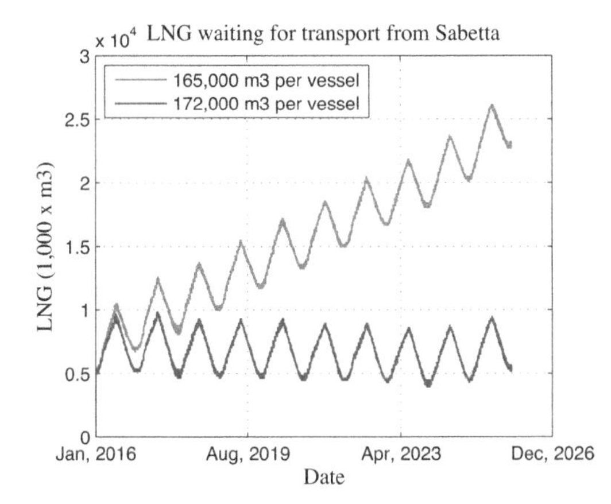


Figure 7. The amount of LNG waiting to be transported from Sabetta for various vessel capacities (Ice scenario 1, PC 5).

six vessels each with a capacity of 172,000 m$^3$ are, regardless of the polar class of the vessels, needed to meet the transport demand. If the cargo capacity of the vessels is reduced to for instance 165,000 m$^3$, the amount on LNG waiting to be transported from Sabetta will start to increase. This is demonstrated in Figure 7, which in case of ice scenario 1 and use of PC 5 vessels, shows the amount of LNG waiting to be transported from Sabetta for various vessel capacities. As the storage capacity in Sabetta is limited, an increasing amount of LNG waiting for onward transport will eventually enforce a production stoppage. Therefore, assuming the costs related to such a production stoppage are very significant, it was decided that the vessels need to have a capacity of at least 172,000 m$^3$.

The drawback of having a transport capacity that is adjusted to the worst assumed ice conditions is that there inevitable will be some overcapacity in less severe ice scenarios. However, the amount of overcapacity depends on the selected ice mitigation strategy. Thus, the various ice mitigation strategies are in the following investigated to find out which of them is the least sensitive to uncertain future ice scenarios, i.e., which of them represents the most robust solution.

In case of ice scenario 2, in which there is a trend towards decreasing ice thickness, the overcapacity is limited to around 1% for both PC 5 and PC 7. However, for PC 4, the overcapacity is around 7%. In case of ice scenario 3, with the least amount of ice, i.e., the overcapacities for PC 7 and PC 5 are around 4% and 5% respectively while the overcapacity for PC 4 is up to 13%. In case of ice scenario 4, in which the ice thickness does neither significantly increase nor decrease, the overcapacity for both PC 7 and PC 5 is around 1% while the overcapacity for PC 4 is around 5%. Transport capacity utilization per ship for the various ice scenarios and ice mitigation strategies is presented in Table 2.

### 3.8 *Determination of the number of days of icebreaker assistance required*

Icebreaker assistance is assumed to be required when the ice thickness exceeds the maximum ice thickness for independent operation specified for each ice mitigation strategy. Since the present LNG carriers are 45.8 m wide, two icebreakers will be required to escort them. The time spent waiting for icebreaker assistance is drawn from a normal distri-

Table 2. Capacity utilization per ship for various ice scenarios and ice mitigation strategies.

| Ice class | Number of ships | Capacity utilization per ship | |
|---|---|---|---|
| | | Cubic meters | Percent |
| *Ice scenario 1* | | | |
| PC 7 | 6 | 172000 m$^3$ | 100% |
| PC 5 | 6 | 172000 m$^3$ | 100% |
| PC 4 | 6 | 172000 m$^3$ | 100% |
| *Ice scenario 2* | | | |
| PC 7 | 6 | 170000 m$^3$ | 99% |
| PC 5 | 6 | 170000 m$^3$ | 99% |
| PC 4 | 6 | 160000 m$^3$ | 93% |
| *Ice scenario 3* | | | |
| PC 7 | 6 | 165000 m$^3$ | 96% |
| PC 5 | 6 | 163000 m$^3$ | 95% |
| PC 4 | 6 | 150000 m$^3$ | 87% |
| *Ice scenario 4* | | | |
| PC 7 | 6 | 171000 m$^3$ | 99% |
| PC 5 | 6 | 171000 m$^3$ | 99% |
| PC 4 | 6 | 163000 m$^3$ | 95% |

bution with a mean value of 2 hours and a standard deviation of 1 hour. The relatively low waiting time was determined on the assumption that the icebreaker service in the area would be adjusted to the demands of the assumed regular service route. The icebreakers are assumed to assist the vessels from the first to the last waypoint along the route where the ice thickness exceeds the determined maximum value for independent operation. The average speed of the icebreakers and the assisted vessel is assumed to be 8 kn. Figure 8 shows an example of how the speed of a ship that operates in up to 1.2 m thick ice is affected by icebreaker assistance when the ice thickness exceeds 1.2 m.

The number of days of icebreaker assistance required for the whole fleet of 6 vessels for various ice scenarios and ice mitigation strategies is shown in Table 3.

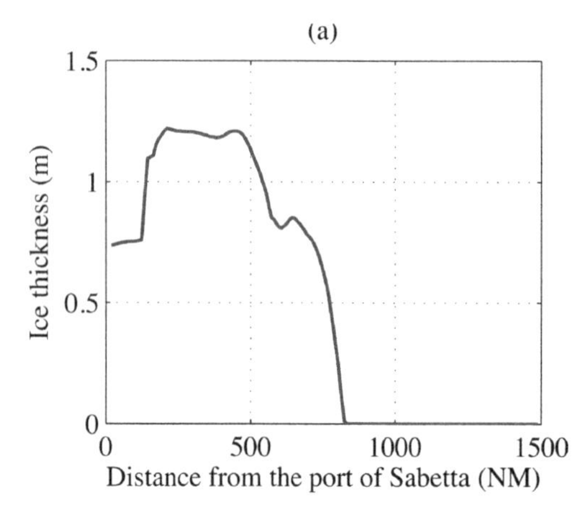


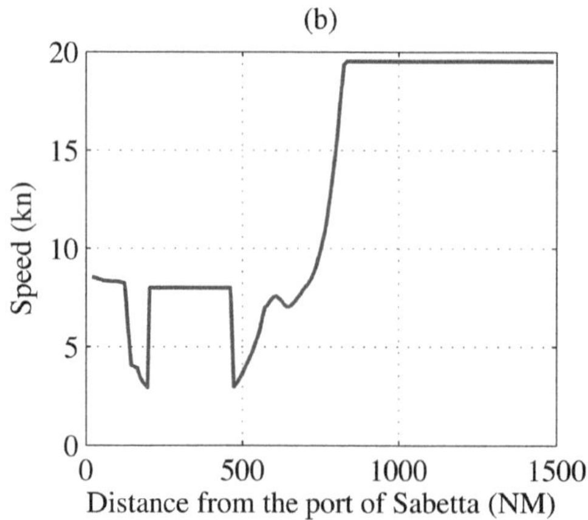


Figure 8. Example of how the speed of a ship that operates independently in up to 1.2 m thick ice is affected by icebreaker assistance when the ice thickness exceeds 1.2 m: (a) The ice thickness along the route; (b) The corresponding speed of the vessel.

Table 3. Required icebreaker assistance in days for various ice scenarios and ice classes (for the whole fleet of LNG carriers).

| | Ice scenario 1 | | | Ice scenario 2 | | |
|---|---|---|---|---|---|---|
| Year | PC 7 | PC 5 | PC 4 | PC 7 | PC 5 | PC 4 |
| 2016 | 521 | 198 | 0 | 438 | 110 | 0 |
| 2017 | 523 | 185 | 0 | 433 | 104 | 0 |
| 2018 | 560 | 238 | 5 | 424 | 100 | 0 |
| 2019 | 541 | 220 | 1 | 492 | 175 | 0 |
| 2020 | 582 | 257 | 8 | 403 | 61 | 0 |
| 2021 | 554 | 231 | 2 | 428 | 97 | 0 |
| 2022 | 523 | 191 | 0 | 456 | 125 | 0 |
| 2023 | 597 | 273 | 21 | 441 | 111 | 0 |
| 2024 | 610 | 278 | 58 | 296 | 0 | 0 |
| 2025 | 636 | 316 | 133 | 421 | 97 | 0 |
| Total | 5647 | 2385 | 227 | 4233 | 979 | 0 |
| | Ice scenario 3 | | | Ice scenario 4 | | |
| Year | PC 7 | PC 5 | PC 4 | PC 7 | PC 5 | PC 4 |
| 2016 | 427 | 101 | 0 | 469 | 153 | 0 |
| 2017 | 395 | 46 | 0 | 424 | 98 | 0 |
| 2018 | 405 | 58 | 0 | 424 | 99 | 0 |
| 2019 | 378 | 22 | 0 | 409 | 61 | 0 |
| 2020 | 289 | 0 | 0 | 488 | 164 | 0 |
| 2021 | 317 | 0 | 0 | 557 | 231 | 3 |
| 2022 | 298 | 0 | 0 | 439 | 107 | 0 |
| 2023 | 301 | 0 | 0 | 437 | 104 | 0 |
| 2024 | 130 | 0 | 0 | 526 | 194 | 0 |
| 2025 | 126 | 0 | 0 | 561 | 241 | 4 |
| Total | 3065 | 227 | 0 | 4734 | 1453 | 8 |

### 3.9 *Fuel costs related to the choice of ice mitigation strategy*

Operation in ice-covered water requires large amount of propulsion power to overcome the resistance between the ice and the ship's hull. A ship built to operate independently in up to 1.7 m of ice requires therefore significantly more propulsion power than a ship built to operate independently in maximum 0.7 m of ice. This is shown in Table 1 that presents propulsion power requirements for vessel with various ice-going capabilities or polar classes.

A larger power requirement results in both higher investment costs and significantly higher fuel consumption as the fuel consumption can be considered directly related to the power demand. Thus, the additional fuel costs related to the PC 5 and PC 4 ships in the present study need to be considered. To this aim, the number of days when the PC 5 and PC 4 vessels need their additional power was determined as shown in Table 4.

The PC 5 classified ship is assumed to need its additional power of 57,000 kW − 30,000 kW = 27,000 kW when the ice thickness

Table 4. Number of days when the additional power of the PC 5 and PC 4 vessels is needed for various ice scenarios (for the whole fleet of LNG carriers).

IS = Ice Scenario

| Year | IS 1 | IS 2 | IS 3 | IS 4 |
|---|---|---|---|---|
| *PC 5: Number of days when 0.7 m ice < thickness < 1.2 m* | | | | |
| 2016 | 323 | 328 | 326 | 316 |
| 2017 | 338 | 330 | 349 | 325 |
| 2018 | 322 | 324 | 347 | 325 |
| 2019 | 321 | 316 | 356 | 348 |
| 2020 | 326 | 342 | 289 | 324 |
| 2021 | 323 | 332 | 317 | 325 |
| 2022 | 332 | 331 | 298 | 332 |
| 2023 | 324 | 330 | 301 | 333 |
| 2024 | 332 | 296 | 130 | 332 |
| 2025 | 321 | 324 | 126 | 319 |
| *PC 4: Number of days when 1.2 m ice < thickness < 1.7 m* | | | | |
| 2016 | 198 | 110 | 101 | 153 |
| 2017 | 185 | 104 | 46 | 98 |
| 2018 | 233 | 100 | 58 | 99 |
| 2019 | 219 | 175 | 22 | 61 |
| 2020 | 249 | 61 | 0 | 164 |
| 2021 | 229 | 97 | 0 | 228 |
| 2022 | 191 | 125 | 0 | 107 |
| 2023 | 251 | 111 | 0 | 104 |
| 2024 | 221 | 0 | 0 | 194 |
| 2025 | 183 | 97 | 0 | 237 |

is larger than 0,7 m and smaller than 1.2 m. The PC 4 vessel is assumed to need the same amount of additional power as the PC 5 ship as long as the ice thickness is less than 1.2 m. When the ice thickness is larger than 1.2 m and less than 1.7 m, the PC 4 ship is assumed to need an additional power of 90,000 kW – 30,000 kW = 60,000 kW. With icebreaker assistance both the PC 5 and the PC 4 vessels are both assumed to have the same power requirement as the PC 7 vessels. Assuming use of HFO as fuel, an average HFO price of USD 750 per ton, and a specific fuel consumption of 180 g/kWh, the additional fuel cost for a fleet of PC 5 vessels amount to 6 × USD 87,000 per day = USD 524,000 per day (when the additional power is required). The corresponding figure for a fleet of PC 4 vessels is 6 × USD 194,000 = USD 1,166,000.

In the above fuel cost calculation, only the use of HFO as fuel is considered. It should be mentioned that LNG carriers are typically fitted with a so-called dual-fuel engine that can run on either natural gas or HFO. However, currently most LNG carriers use HFO as fuel as it for the moment is cheaper than natural gas. Thus, use of natural gas as fuel with not be further discussed in the present paper.

Table 5. NPC of ice mitigation costs for various ice scenarios and ice mitigation strategies.

| | PC 7 | PC 5 | PC 4 |
|---|---|---|---|
| *IS 1* | | | |
| IB support (days) | 5,647 | 2,385 | 227 |
| Addl. fuel cons. (t) | 0 | 380,000 | 940,000 |
| NPC (USD) | 3.7E+08 | 5.1E+08 | 8.8E+08 |
| *IS 2* | | | |
| IB support (days) | 4,233 | 979 | 0 |
| Addl. fuel cons. (t) | 0 | 380,000 | 633,000 |
| NPC (USD) | 2.9E+08 | 4.2E+08 | 7.2E+08 |
| *IS 3* | | | |
| IB support (days) | 3,065 | 227 | 0 |
| Addl. fuel cons. (t) | 0 | 331,000 | 390,000 |
| NPC (USD) | 2.2E+08 | 3.5E+08 | 6.1E+08 |
| *IS 4* | | | |
| IB support (days) | 4,734 | 1,453 | 8 |
| Addl. fuel cons. (t) | 0 | 383,000 | 757,000 |
| NPC (USD) | 3.1E+08 | 4.4E+08 | 7.7E+08 |

### 3.10 *Comparison of ice mitigation related costs for the various ice mitigation strategies*

To enable a holistic comparison of the various ice mitigation strategies, the Net Present Cost (NPC) of all their related costs were calculated. All costs except the additional investment costs related to the PC 5 and PC 4 vessels were discounted using an assumed interest of 8%. The obtained NPC values are presented in Table 5.

The figures presented in Table 5 indicate clearly that ice mitigation strategy 1 with PC 7 vessels is the most economical alternative for all ice scenarios. However, the outcome is quite sensitive to the assumed costs for icebreaker assistance. Assuming that the two icebreaker required to escort one of the LNG carriers would cost USD 80,000 × 2 = USD 160,000 or more per day instead of the USD 50,000 × 2 = USD 100,000, ice mitigation strategy 2 with PC 5 built ships would be more economical.

## 4 CONCLUSIONS

The present study resulted in an approach towards the design of robust arctic maritime transport systems that are able to deal with various possible future ice scenarios. It makes it possible to assess how a complex arctic maritime transport system, consisting of a single or multiple vessels, with or without icebreaker assistance, is able to cope with various possible future ice scenarios.

A case study was carried out to demonstrate how the approach could be applied in practice.

The outcome from the case study indicates clearly that it, for the investigated route, is more economical to use vessels with a low or medium level ice going capabilities in combination with icebreaker assistance instead of vessels with high ice going capability and a minimum demand for icebreaker assistance. In other words, the results indicate that costs related to higher ice going capabilities are high in comparison with the costs for icebreaker assistance. Especially in case of decreasing ice conditions, the transport system with PC 4 vessels performed poorly while the utilization of the vessels ice going capabilities was limited to the start of the 10-year period, and resulted only in additional capital costs and operating costs towards the end of the period. In reality the PC 4 vessels would most likely perform even worse in comparison with the vessels with lower ice classes as their additional weight would significantly harm their fuel consumption in all ice conditions including open water.

The presented approach can be further developed as its components can easily be modified or replaced for improved accuracy. Components that should be improved include for instance the method for calculation of differences in fuel costs between ships with various ice going capabilities as well as the applied ice data, which should be extended to include openings, ridges, etc.

## ACKNOWLEDGEMENTS

The financial support of MAROFF Competence building project funded by the Research Council of Norway on "Holistic risk-based design for sustainable arctic sea transport" is greatly acknowledged.

## ABBREVIATIONS

| | |
|---|---|
| LNG | Liquefied Natural Gas (LNG) |
| IB | Icebreaker |
| IS | Ice Scenario Liquefied |
| NM | Nautical Mile |
| NPC | Net Present Costs |
| PC | Polar Class |
| USD | United States Dollars |

## REFERENCES

AARI (Russian Arctic and Antarctic Research Institute). 2014. Retrieved from: http://www.aari.ru/projects/ecimo/.

Bergström, M., Ehlers, S., Erikstad, S.O., Erceg, S., Bambulyak, A. 2014. Development of an approach towards mission-based design of arctic maritime transport systems. Proc. 33nd International Conference on Ocean, Offshore and Arctic Engineering OMAE 2014, June 8–13, 2014, San Francisco, USA.

Erceg, S., Ehlers, S. Ellingsen, I.H., Slagstad, D. von Bock und Polach, R., Erikstad, S.O. 2013. Ship performance assessment of Arctic transport routes, Proc. 32nd International Conference on Ocean, Offshore and Arctic Engineering OMAE 2013, June 9–14, 2013, Nantes, France.

Juva, M., Riska, K. 2002. On the power requirement in the Finnish-Swedish ice class rules. Winter navigation Research Board, Res. Rpt. No. 53, Helsinki, Finland.

Knutsen O.A.S., Shipping AS. 2014. Retrieved from: http://knutsenoas.com/shipping/lng-carriers/barcelona/.

Renton, M. 2013. Yamal LNG puts Russian gas on the map, Loyd's List Intelligence. Retrieved from: http://info.lloydslistintelligence.com/yamal-lng-puts-russian-gas-on-the-map/.

Riska, K., Patey, M., Kishi, S., Kamesaki, K. 2001. Influence of Ice Conditions on Ship Transit Times in Ice, Proc. 16th International Conference On Port and Ocean Engineering under Arctic Conditions (POAC'01), August 12–17, 2001, Ottawa, Canada.

SINTEF. 2014. SINMOD—A physical—chemical—biological model system. Retrieved from: http://www.sinmod.no/images/SINMOD.pdf.

Slagstad, D., McClimans, T.A. 2005. Modeling the ecosystem dynamics of the Barents Sea including the marginal ice zone: I. Physical and chemical oceanography, Journal of Marine Systems, 58, 1–18.

Total S.A. 2014. Yamal LNG. Retrieved from: http://www.total.com/en/energies-expertise/oil-gas/exploration-production/projects-achievements/lng/yamal-lng.

Valkonen, J., Løvoll, G., Strandmyr, E.M., Walter, E.L. 2013. Cossarc-Concept Selection for Shipping in the Arctic. Proc. 32nd International Conference on Ocean, Offshore and Arctic Engineering OMAE 2013, June 9–14, 2013, Nantes, France.

*Maritime-Port Technology and Development – Ehlers et al. (Eds)*
*© 2015 Taylor & Francis Group, London, ISBN 978-1-138-02726-8*

# Defining operational criteria for offshore vessels

T.E. Berg, Ø. Selvik & B.O. Berge
*Department of Ship Technology, Norwegian Marine Technology Research Institute (MARINTEK), Trondheim, Norway*

ABSTRACT: Operational criteria for marine operations are traditionally specified in terms of maximum permissible significant wave heights. Motions of ships and offshore structures depend on their dynamic characteristics and the nature of the environmental excitation forces. These parameters should be reflected in how operating limits are defined. Particularly for vessels operating in harsh environments, where the quality of weather forecasts tends to be low, such as in open Arctic waters, operational criteria should be related to vessel motions and environmental conditions described by wave, current and wind parameters. This paper reviews previous work on parameters influencing operational performance for ships. A brief look at specification of operational limits for marine operations are given as background knowledge for a discussion of how operational limits could be developed for safe and efficient marine operations in open Arctic waters.

## 1 INTRODUCTION

For most marine operations the project owner estimates the time needed to complete the actual operation and some upper environmental condition threshold below which the operation can take place. A well-known example is the wave conditions specified for starting the offshore loading hook-up process on the Norwegian continental shelf. Statoil, for instance, specifies that the significant wave height should be less than 4.5 m when starting the hook-up and that the loading process has to stop when $H_s > 5.5$ m. MARINTEK's considers that significant wave height alone is not a suitable way of defining operational limits. For any offshore operation, safety depends on the actual metocean situation (wind, waves and current), the vessel's dynamic characteristics and the tools involved in the operation. For launch and recovery from an offshore vessel, the relative motion between the sea surface and the object to be launched/retrieved and forces when it enters the splash zone might better define safe operational limits.

The first part of this paper offers a brief review of previous work related to defining operational parameters for different types of vessels and discusses existing operational limits for a range of offshore operations. The second part presents our view on how future operational limits should be specified and how model tests in combination with calculations and simulations can be used to investigate operational characteristics during the design phase. The final part describes operability studies for an intervention vessel designed for all-year operation in the southwestern part of the Barents Sea.

## 2 REVIEW OF PREVIOUS WORK ON OPERATIONAL PARAMETERS

### 2.1 *NORDFORSK project*

In 1987 a group of Nordic research organizations completed the "Seakeeping Performance of Ships" project (NORDFORSK, 1987). One of the aims of the project was to develop a Nordic standard for seakeeping criteria and standard methods of evaluating the motion characteristics to be proposed. The intention was to enable different ship designs to be compared, and thus improve seakeeping performance. In its work on specification of seakeeping criteria the group was looking for practical evaluation methods or indexing systems that could measure the ability of a vessel to perform its functions under a wide range of environmental conditions. Part of that effort involved developing criteria for acceptable levels of ship motions related to specific ship subsystems and operations such as:

- Ship hull
- Propulsion machinery
- Ship equipment
- Cargo
- Personnel efficiency
- Passenger comfort
- Special operations involving
    - Helicopter
    - Sonar
    - Crane

Criteria were specified with regard to:

- Slamming
- Deck wetness

- Vertical acceleration
- Horizontal acceleration
- Roll angle
- Pitch angle
- Vertical motion
- Vertical velocity
- Relative motion

As an example, Table 1 lists limits to the ability to perform various types of work in a safe and efficient manner.

### 2.2 *STANAG*

The NATO Standardization Agreement (STANAG) 4154 (NATO, 2000) describes procedures for assessing seakeeping in the ship design process. It is based on a systems approach that starts with establishing a set of naval missions and support activities. The interactions between missions, activities, specific systems and tasks are then investigated. The document defines three groups of criteria for ship motions:

- Motion response criteria
- Derived response criteria
- Other criteria

Motion response criteria are based on ship motions. These could include Root Mean Square (RMS) values for roll, pitch and yaw angles, vertical and lateral displacements, vertical velocity (absolute and relative with respect to wave surface), longitudinal, transverse and vertical accelerations.

Examples of derived response criteria are Motion-Induced Interruptions (MII), Motion Sickness Incidence (MSI), propeller emergence index, slamming index and deck wetness index.

Other criteria could be effects of relative wind velocity for helicopter landing/takeoff, involuntary speed loss in waves, maximum and RMS forces in mooring or towing lines.

The document shows operational envelopes for different environmental parameters and mission tasks such as fixed-wing and helicopter takeoff and landing and replenishment at sea.

Table 1. Criteria for safe performance of work tasks (NORDFORSK, 1987).

| Work task description | Root mean square criterion | | |
|---|---|---|---|
| | Vertical acc. | Lateral acc. | Roll |
| Light manual work | 0.20 g | 0.10 g | 6° |
| Heavy manual work | 0.15 g | 0.07 g | 4° |
| Intellectual work | 0.10 g | 0.05 g | 3° |
| Transit passengers | 0.05 g | 0.04 g | 2.5° |
| Cruise liner | 0.02 g | 0.03 g | 2° |

The NATO Naval Armaments Group Subgroup 61 on Virtual Ships is developing standards for modelling and simulation applied to ship and maritime systems acquisition. The Virtual Ships STANAG (STANAG 4684) is based on the High Level Architecture (HLA) for simulation; it is oriented to HLA federations performing physics-based calculations, in which fidelity is generally given higher priority than runtime performance. Example federations have been applied to landing of aircraft on moving ships, launch and recovery of smaller ocean vessels from a mother ship, and replenishment at sea with coupled ship motions.

### 2.3 *Crew performance in heavy seas*

Stevens and Parsons (2002) discuss crew reduction. A requirement in any analysis of possible reductions is that all crew members should be able to perform their duties under severe environmental conditions. Table 2 presents a comparison of operability criteria for the US Navy and US Coast Guard (USCG). The mission profile of the USCG cutter means that it has a significantly higher value for MII and a different limit for MSI.

### 2.4 *DNV GL's recommendations for marine operations*

In 2011, DNV published a general document on marine operations (DNV, 2011) that specifies (Sec. B 600) that limiting operational environmental criteria ($OP_{LIM}$) must be established and clearly described in the marine operation manual. This document states that the $OP_{LIM}$ shall not be set higher than the minimum values of:

a. The environmental design criteria.
b. Maximum wind and waves for safe working- (e.g. at vessel deck) or transfer conditions for personnel.

Table 2. Comparison of operability criteria (RMS is Root Mean Square; SSA is Significant Single Amplitude. SSA = 2xRMS), (Stevens & Parsons, 2002).

| | NATO STANAG 4154 U.S. Navy | U.S. coast guard cutter certification plan |
|---|---|---|
| Motion Sickness Incidence (MSI) | 20% of crew in 4 hours | 5% in 30 min exposure |
| Motion-Induced Interruption (MII) | 1 tip per minute | 2.1 tips per minute |
| Roll amplitude | 4.0° RMS | 8.0° SSA |
| Pitch amplitude | 1.5° RMS | 3.0° SSA |
| Vertical acceleration | 0.2 g RMS | 0.4 g SSA |
| Lateral acceleration | 0.1 g RMS | 0.3 g SSA |

c. Equipment—(e.g. ROV and cranes) specified weather restrictions.
d. Limiting conditions for position-keeping systems.
e. Any limitations identified, e.g. in HAZID/ HAZOP, based on operational experience with the vessel(s), equipment, etc. involved.
f. Limiting weather conditions for executing specific contingency plans.

Uncertainty in both monitoring and forecasting of the environmental conditions also needs to be considered. It is recommended that this should be done by defining forecast (and, if applicable, monitored at the start of operations) operational criteria—$OP_{WF}$, defined as $OP_{WF} = \alpha \times OP_{LIM}$. Tables are given for α-values for both wind and waves. For waves the α-factor depends on the significant wave height, the planned operation time and the quality of the weather forecast.

DNV published a major revision of its Recommended Practice (RP) for modelling and analysis of marine operations in 2011. New amendments have been made since then (DNV, 2012). As an example of operational criteria, in Sec. 4.4 DNV present the accepted criteria for lifting through the wave zone in the form of:

- Characteristic total force
- Slack sling criteria

Chapter 8 in DNV's RP describes how to apply weather criteria in availability analysis for weather-restricted operations. In Sec. 8.5.2 an example of operability calculation is given for moonpool-based operations. Different limits for maximum significant vertical amplitude of water in the moonpool are specified for a given vessel. Figure 1 shows the relation between significant wave height, wave period and maximum significant vertical amplitude limit for water motion in the moonpool.

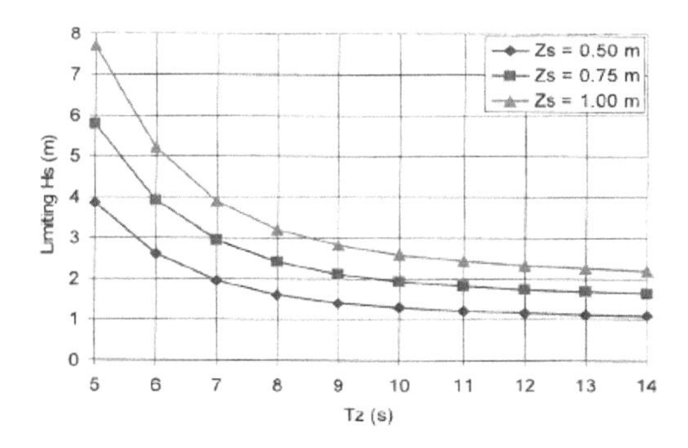


Figure 1. Influence of wave parameters on water motion in moonpool (ZS is maximum significant vertical amplitude limit for water in the moonpool, TZ is zero-up-crossing period), DNV (2012).

Table 3. Calculated operability for moonpool-based operations. The limiting parameter is vertical water motion in the moonpool (DNV, 2012).

| Criterion | Downtime $P(Z_S > Z_{S0})$ % | Operability $P(Z_S < Z_{S0})$ % |
|---|---|---|
| $Z_{s0} = 0.5$ m | 35.0 | 65.0 |
| $Z_{s0} = 0.75$ m | 19.4 | 80.6 |
| $Z_{s0} = 1.0$ m | 6.6 | 93.4 |

Using wave statistics for a given working site, the operability is found as shown in Table 3.

## 2.5 *Company-specific operational limits—equipment manufacturers' specifications*

In order to maintain the integrity of subsea modules during installation, manufacturers often define critical parameters for the individual phases of marine operations. As a part of a previous R&D project MARINTEK invited equipment suppliers to discuss how to specify operational limits for critical subsea installation and maintenance tasks. In most cases operational limits were established on the basis of experience from similar types of operations. In very few cases did the providers take into account the dynamics of the actual vessel on which the equipment was to be installed. Suppliers often ended up with operational limits based only on the significant wave height. Some examples of operational limits are presented in the following sub-sections.

### 2.5.1 *Operational limits for short-time operations*

A review of task characteristics and operational limits was among the initial activities of a project to study an intervention vessel for Arctic operations. Some tasks and experience-based operational limits in the form of maximum significant wave height ($H_S$) are listed in Table 4.

Discussions with equipment manufacturers made it clear that $H_s$ were used as the only operational limit for most intervention tasks. In some specific cases the significant wave height criteria could be extended to include a wave period requirement. In connection with crane operations where the lifted body would be entering subsea structures, additional criteria for angles of docking cones entering guideposts will normally be given. Vertical speed during the docking phase is another criterion for subsea lifting operations.

### 2.5.2 *Launch and recovery of Hugin AUV*

The standard launching system for the Hugin AUV is the Stinger ramp system (Fig. 2). The Stinger is

Table 4. Examples of operational limits for intervention tasks.

| Task | Duration | Significant wave height limit ($H_S$) | Comments |
|---|---|---|---|
| Launch ROV | 15 min | 5 m | To 300 m water depth |
| Launch running tool | 30 min | 2.5 m | To 300 m water depth |
| Land/connect tool on module | 30 min | 2.5 m | Heave compensation assumed |
| Retrieve running tool to vessel deck and sea fastening | 30 min | 3 m | From 300 m |
| Retrieve ROV onboard vessel and prepare for transit | 30 min | 3 m | From 300 m |

Figure 2. Stinger ramp system for launching of Hugin (courtesy of Fugro GeoServices).

installed on an open deck or inside a container. The system is stated to be operable in Sea State 4.

## 3 FUTURE WORK ON THE DEVELOPMENT OF OPERATIONAL CRITERIA FOR OFFSHORE VESSELS

### 3.1 *MARINTEK's position*

As mentioned in sec. 2 of the paper there exists multiple suggestions for definition of operational limits for offshore vessels. In future studies of the operability envelope of offshore vessels, MARINTEK intends to employ methods that take the quality of the vessel into account. This means that the dynamic characteristics of the vessel and the equipment used should be a basic part of studies to develop location-specific operational limits. Such limits will be based on location-specific scatter diagrams for waves, as changes in wave periods may move the energy-dense part of the wave spectrum closer to the resonance frequencies of offshore vessels. Thus both significant wave height and frequency-peak wave period should be included in future operational criteria.

The individual phases of a vessel's missions should be investigated separately. As the transit leg of the operating cycle of an offshore vessel increases for new fields far from the coast line, the performance of the vessel during transit sailing should be checked with respect to criteria specified by NORDFORSK and STANAG 4154. However, the primary focus should still be on the ability to perform on-site services for offshore drilling and production units as well as on construction, installation and intervention work connected to sub-sea oil and gas systems.

This paper does not look into operational limits due to visibility. Low visibility is an important factor due to summer fog and heavy snow showers in the winter season. A brief description of wind influence on station-keeping ability will be given.

### 3.2 *Background for MARINTEK's position*

The motion of a vessel is given by the dynamic characteristics of the vessel itself, its speed and heading and excitation forces from the environment (waves, wind, current and ice). In general the wave-related dynamics are given in form of Response Amplitude Operators (RAOs) shown in Figure 3. For short wave periods the motion response is dominated by the inertia of the vessel. For long waves it is stiffness-dominated, and the vessel will ride on the waves. The resonance peak position is defined by the natural frequency of the vessel, while the peak value depends on the damping of the vessel. The RAO curve is linked to the vessel's size and hull shape.

Wave characteristics are often given in the form of scatter diagrams based on observations or by specific sea spectrum functions. Compared to a standard North Sea spectrum, the spectrum peak period for the Barents Sea is lower and thus closer to some of the natural frequencies of an offshore vessel.

The vessel's response spectrum can be calculated from the RAOs and the wave spectrum. Using different criteria it is possible to calculate limits of

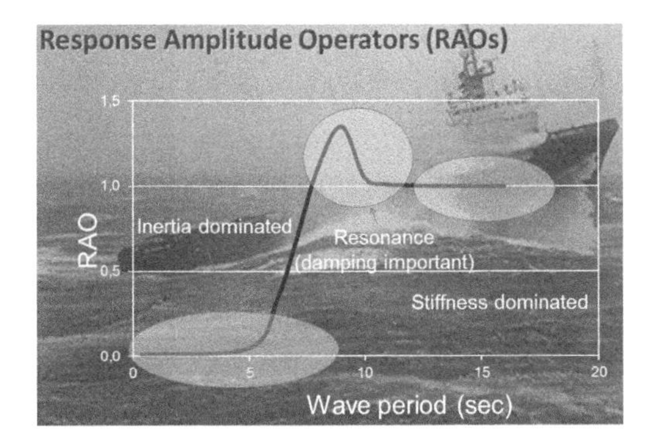


Figure 3. Response Amplitude Operators (RAOs).

operability. These criteria should be based on values of critical parameters related to specific operations and phases of the operation such as:

- Motion-Sickness Incidence during transit (MSI)
- Position-keeping at work site
- Crane operations
- Lowering objects through the splash zone
- Entering guide post on a subsea installation
- ROV and AUV launch and recovery

### 3.3 *Development of operational criteria for offshore vessels*

To reduce operating costs by increasing operability it is necessary to relate operational limits to motions (absolute and relative motions, velocities and accelerations) at the locations where tasks are performed (crane and winch positions, moonpools, working decks, etc.). For some operations loads in lifting lines or mooring lines should be included in the list of operational limits. This is a significant step forward from the current widespread use of significant wave height as the only parameter for specifying operational limits for offshore operations. The new approach will increase the value of location-specific design of vessels for operation in sensitive sea areas such as the Barents Sea.

## 4 BARENTS SEA AND SPECIFICATION OF OPERATING LIMITS—A CASE STUDY

### 4.1 *Case study description*

As part of a MAROFF R&D project supported by the Research Council of Norway (and industrial partners Statoil and STX OSV (now VARD)), MARINTEK studied the design of a construction and intervention vessel for Arctic operations. The objective of the project was to extend the season for installation and maintenance of subsea oil and gas structures in waters with seasonal ice. More information about this project can be found on the project's web site (http://www.sintef.com/home/MARINTEK/Projects/Maritime/CIV-Arctic/).

As a business case the Olga Field east of Svalbard was chosen (Fig. 4). The shorebase for the vessel was at Hammerfest. For this operation the mission profile was divided into three phases:

- At logistics base: 20%
- In transit: 20%
- At worksite: 60%

The case vessel developed in the project is illustrated in Figure 5. Even though the vessel is based on the double acting principle ice management is

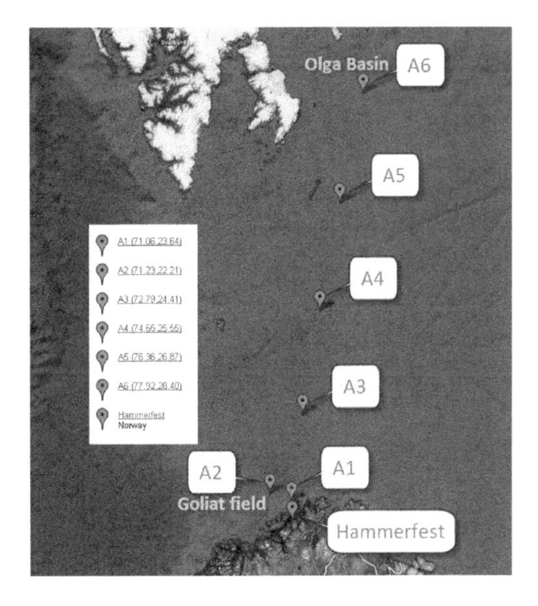


Figure 4. Olga Basin location and defined waypoints for investigation of transit operability.

Figure 5. Case vessel design.

required for all types of operations in ice covered waters. This paper presents some study outcomes, transit operability based on bow acceleration and station-keeping performance in open water at the predefined work site.

### 4.2 *Operability in transit*

The critical parameters for operability during the transit to/from the work site were defined as:

- After-ship slamming
- Vertical accelerations
- Horizontal accelerations (especially in following seas)
- Roll motion (in following seas)

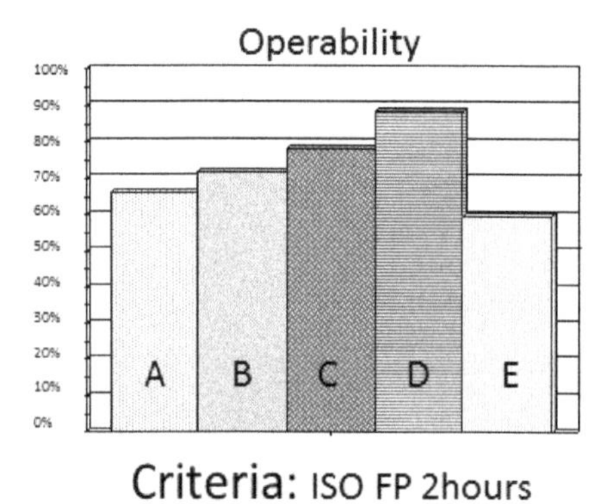


Figure 6. Differences in operability, based on bow vertical acceleration at waypoint A1 for different ship length.

Another important parameter is fuel consumption, both as an economic factor, but equally important in terms of its environmental footprint.

This paper looks at one of the criteria, bow acceleration in a seaway during transit to/from the work site in the Olga Basin. Hull form data is used as input for MARINTEK's hydrodynamic work bench ShipX (MARINTEK, 2012a) to calculate the response amplitude operators for different ship motions. From the combined heave and pitch motions the vertical accelerations at the bow can calculated based on a given wave spectrum or available wave scatter diagrams. Here the new NORA10 wind and wave hindcast model has been applied. This model has a data archive for a 10 km grid in the North Sea, the Norwegian Sea and the Barents Sea (Norwegian Deepwater Programme, 2014). Calculated vertical accelerations are compared with specified operational limits for the bow motion to find the operability percentage with respect to the vertical bow acceleration criterion.

Figure 6 shows an example of operability investigation for bow vertical acceleration at one of the waypoints. Calculations are done for a transit speed of 15 knots and in head seas. The baseline model vessel (ship A) has a length of approximately 120 m. This figure illustrates the influence of vessel size (ship B 140 m, ship C 160 m, ship D 180 m and ship E 100 m). Operability ranges from 60 to 88%, from the smallest to the largest vessel, based on available wave statistics. The operability for the transit voyage (based on the bow acceleration criterion) differs at each waypoint (which are

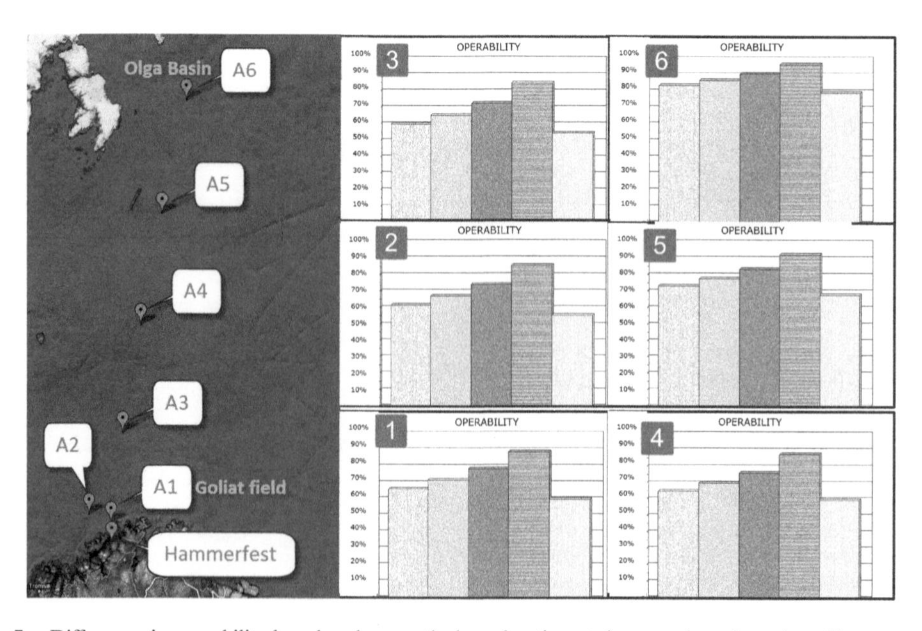


Figure 7. Differences in operability based on bow vertical acceleration at six waypoints along the sailing route.

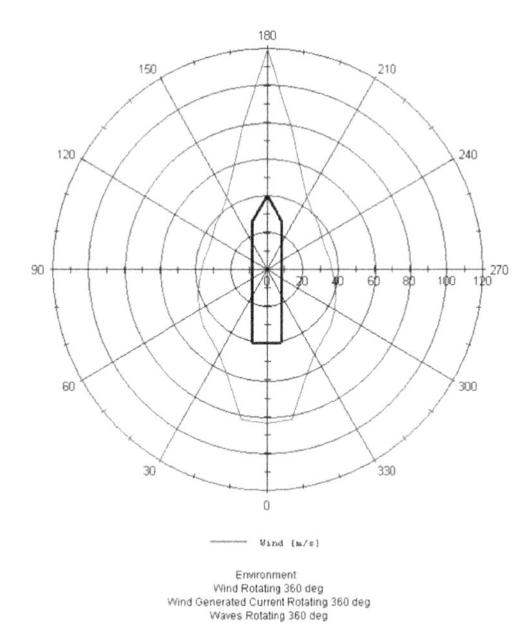


Figure 8. Capability plot for station-keeping in wind.

defined in Fig. 4). This is illustrated in Figure 7. As can be expected, the longest vessel has the highest operability with respect to the bow acceleration criterion.

### 4.3 *Station-keeping capability*

For DP based operations the wind forces on the superstructure are an important station-keeping parameter. Figure 8 is a capability plot for an offshore vessel. This plot was generated using the ShipX—Station Keeping simulation package (MARINTEK, 2012b). Equivalent capability plots will have to be developed for managed ice operations by vessels that will be operating in the marginal ice zone. Offshore vessels should not operate on DP without ice management support. More information on DP operations in ice has been produced in another MAROFF project "Arctic DP: Safe and Green Dynamic Positioning Operations of Offshore Vessels in an Arctic Environment", see http://www.marin.ntnu.no/arctic-dp/.

## 5 RECOMMENDATIONS FOR FURTHER WORK ON THE DEFINITION OF OPERATIONAL LIMITS FOR OFFSHORE VESSELS

In the course of discussions with offshore vessel designers and ship officers, MARINTEK believes that there is a need to further develop methods for specifying work task-specific operating limits. Such limits should be based on the actual dynamic characteristics of the vessel, environmental conditions (waves, wind, current, ice and visibility), the type of equipment to be used and critical phases of the operation (splash zone entry, guidepost entry to subsea structures etc.). It is important that experienced ship masters and equipment specialists are involved in development of operational limits based on an understanding of the physics involved in ship motions and loads in safety critical equipment used when completing complex marine operations. MARINTEK recommends a close cooperation between seagoing personnel performing marine operations, ship designers, equipment manufacturers and researchers to develop an overall operational criterion to be applied when comparing different ship designs for location specific missions.

## ACKNOWLEDGEMENTS

The authors thank project partners Statoil, STX OSV, VTT and Aker Arctic for their support for the MAROFF R&D project "Construction and Intervention Vessels for Arctic Operations" (CIVARCTIC). The project was funded by the MAROFF research programme of the Research Council of Norway.

## REFERENCES

DNV 2011: Marine Operations, General. DNV-OS-H101, Høvik, 2011.

DNV 2012: Modelling and Analysis of Marine Operations. DNV-RP-H 103. Amended December 2012. Høvik, 2012.

MARINTEK 2012a: Fact sheet: ShipX Integrated Ship Design Tool—Hydrodynamic Workbench ShipX. http://www.sintef.no/upload/MARINTEK/PDF-filer/Sofware/ShipX.pdf.

MARINTEK 2012b: Fact sheet: ShipX Station Keeping. http://www.sintef.no/upload/MARINTEK/PDF-filer/FactSheets/ShipX_StationKeeping.pdf.

NATO 2000: STANAG 4154 (Edition 3)—Common Procedures for Seakeeping in the Ship Design Process. NATO, Military Agency for Standardization, Brussels, 2000.

NORDFORSK 1987: Assessment of ship performance in a seaway. NORDFORSK, Copenhagen, 1987.

Norwegian Deepwater Programme 2014: http://www.epim.no/norwegian-deepwater-programme/projects/metocean/main-activities.

Stevens, S.C. and Parsons, M.C. 2002: Effects of Motion at Sea on Crew Performance: A Survey. Marine Technology, Vol. 39, No. 1, January 2002, pp. 29–47.

*Maritime-Port Technology and Development – Ehlers et al. (Eds)*
*© 2015 Taylor & Francis Group, London, ISBN 978-1-138-02726-8*

# Real time ship exhaust gas monitoring for compliance to $SO_x$ and $NO_x$ regulation and $CO_2$ footprint

T. Wijaya
*Energy Research Institute at NTU (ERI@N), Singapore*

T. Tjahjowidodo
*School of Mechanical and Aerospace Engineering, Nanyang Technological University, Singapore*

P. Thepsithar
*Energy Research Institute at NTU (ERI@N), Singapore*

ABSTRACT: IMO MARPOL Annex VI regulation has been set for global application to reduce the negative consequences of ship gas emissions $SO_x$, $NO_x$, and $CO_2$. To enforce the regulation, real time gas emission monitoring system is in need. The system in this paper consists of a microcontroller, gas analyzers, sensors and Xbee radio module. The data acquired from the sensors and gas analyzer is transmitted using a microcontroller over the Xbee module to the receiver Xbee module on the main microcontroller for real-time compliance check. Real time monitoring system expedites the checks by officials in enforcing the regulation. Ultimately, this application enables engine manufacturers, ship-owners and Administrations to ensure in real time that all applicable marine diesel engines are in compliance with the limiting values determined by IMO MARPOL Annex VI.

## 1 INTRODUCTION

The increasing number of ships operating on the sea has increased the awareness of IMO (International Maritime Organization), the primary regulatory agency made of 170 Member States tasked with developing regulations for the control of pollution from international shipping activities. IMO created the IMO MARPOL Annex VI regulation to limit the emission of $SO_x$, $NO_x$, and $CO_2$ from the ships. Tier II of the regulation has been enforced since 1st January 2011. Since the regulation requires the ship owner to keep complying throughout the time, there is a need for IMO to have a tool to enforce the regulation. The tool must have the sensors performance compliance as have been detailed by IMO. The tool of such has been developed by many to cater the need of ship owners, but not for the enforcement by IMO. Therefore, IMO lacks of a system that enables it to monitor the ship gas emission in real time.

In this paper, the monitoring of $SO_x$, $NO_x$, and $CO_2$ follows the 2009 Guidelines for Exhaust Gas Cleaning Systems, and Chapter 6.4 Dircct Measurement and Monitoring Method of the $NO_x$ Technical Code (2008). Meanwhile, the limit value stated at IMO MARPOL Annex VI regulation 13 and regulation 14 are described in Figure 1 and Figure 2.

This paper offers the solution of the lack of technology mentioned above. Important parameters for compliance are measured by sensors that give electrical signal output of various types. After that, the data value from the sensors is collected by the microcontroller (data acquisition system). The microcontroller processes the data calculation and output the value useful for end-user such as $SO_x$, $NO_x$ and $CO_2$ emission. This end-user data is stored in the local memory of the microcontroller. Thereafter, the end-user data is transmitted through the Xbee to the receiver Xbee on the main microcontroller. This end-user data obtained by the main microcontroller is then stored and used

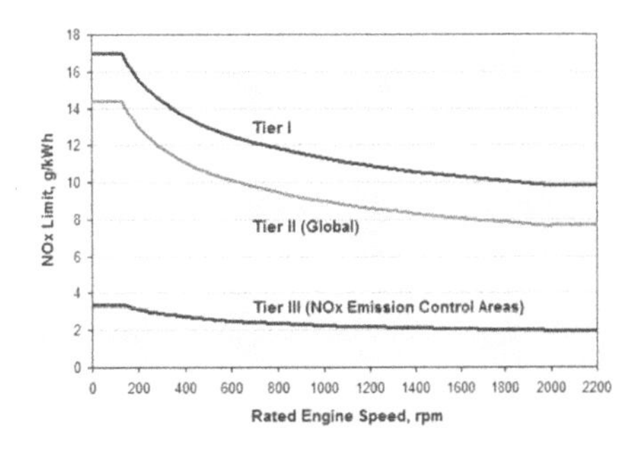


Figure 1. $NO_x$ emission limit.

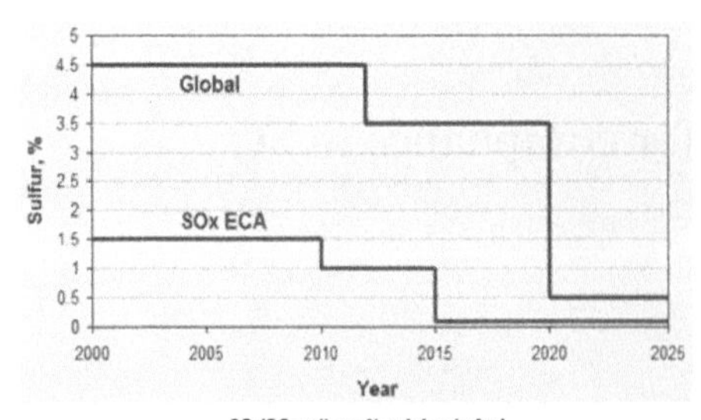


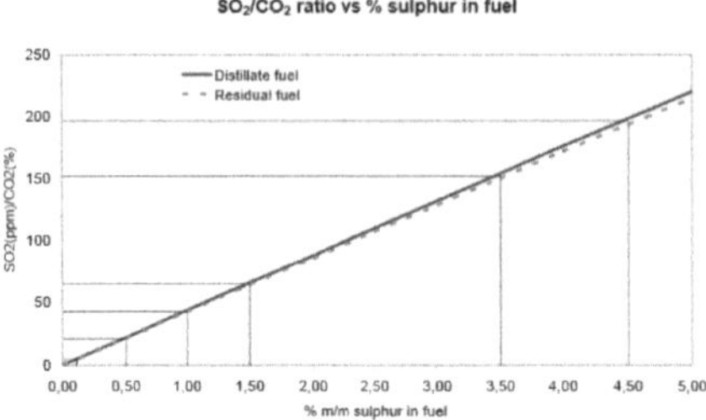


Figure 2. $SO_x$ and $CO_2$ emission limits.

accordingly for monitoring purposes. Ultimately, the aim of this solution is to be the standard for IMO to enforce the IMO MARPOL Annex VI.

## 2 PARAMETERS MEASUREMENT

The parameters required for compliance refer to the independent variables identified in the 2009 Guidelines for Exhaust Gas Cleaning Systems, and Chapter 6.4 Direct Measurement and Monitoring Method of the $NO_x$ Technical Code (2008). These independent parameters are listed in Table 1.

The parameters above will be used for calculation using the following formulas below which are adjusted specifically for incomplete combustion and compression ignition engines with intermediate air cooler. Calculating $gas_x$ in equation (1) is our goal.

$$gas_x = \frac{\sum_{i=1}^{i=n}(q_{mgasi} \cdot W_{Fi})}{\sum_{i=1}^{i=n}(P_i \cdot W_{Fi})} \tag{1}$$

$$P = P_m + P_{aux} \tag{2}$$

$$q_{mgas} = u_{gas} \cdot c_{gas} \cdot q_{mew} \cdot K_{hd} (\text{for } NO_x) \tag{3}$$

$$q_{mgas} = u_{gas} \cdot c_{gas} \cdot q_{mew} (\text{for others}) \tag{4}$$

$$q_{mew} = q_{maw} + q_{mf} \tag{5}$$

$$c_{gas} = c_w = k_{wr2} \cdot c_d \tag{6}$$

$$k_{wr2} = \frac{1}{1 + \alpha \cdot 0.005 \cdot (C_{co2d} + c_{cod}) - 0.01 \cdot c_{H2d} + k_{w2} - \frac{P_r}{P_b}} \tag{7}$$

$$\alpha = 11.9614 \cdot \frac{W_{ALF}}{W_{BET}} \tag{8}$$

$$c_{H2d} = \frac{0.5 \cdot \alpha \cdot c_{COd} \cdot (c_{COd} + c_{CO2d})}{c_{COd} + 3 \cdot c_{CO2d}} \tag{9}$$

$$k_{w2} = \frac{1.608 \cdot H_a}{1000 + (1.608 \cdot H_a)} \tag{10}$$

$$H_a = \frac{6.22 \cdot p_a \cdot R_a}{p_b - 0.01 \cdot R_a \cdot p_a} \tag{11}$$

$$p_a = (4.856884 + 0.2660089 \cdot t_a + 0.01688919 \cdot t_a^2) \cdot \frac{101.32}{760} \tag{12}$$

$$K_{hd} = \frac{1}{1 - 0.012 \cdot (H_a - 10.71) - 0.00275 \cdot (T_a - 298) + 0.00285(T_{sc} - T_{SCRef})} \tag{13}$$

Table 2 lists the information for parameters in the equations above that is not found in Table 1.

## 3 GAS ANALYZER AND SENSORS SELECTION

The gas analyzer is prescribed in Appendix 3 of $NO_x$ Technical Code (2008) and also point 6 Emission Testing of 2009 Guidelines for Exhaust Gas Cleaning Systems. CO, $CO_2$, and $SO_2$, should be measured by Non-Dispersive Infrared (NDIR) absorption type. NO, $NO_2$ should be measured by Chemiluminescent Detector (CLD) or Heated Chemiluminescent Detector (HCLD). Appendix 8 point 2.1.1 of the $NO_x$ Technical Code (2008) however, mentions that other systems or analyzers can be used as long as it is approved by the Administration and yield equivalent results to the advised equipment. The specification and arrangement of the gas analyzers shall follow the Appendix 3 and Appendix 4 of the $NO_x$ Technical Code (2008).

In measuring exhaust gas concentration; there are two types of gas analyzer, dry basis and wet basis. The dry basis gas analyzer requires the removal of water molecule $H_2O$ from the sample gas before entering the gas analyzer. The removal or gas conditioning cools the sample gas temperature to less than 4°C. Depending on the gas conditioning method, the dry basis gas analyzer often removes small part of $NO_2$, $SO_2$, and some hydrocarbon. Thus the reading will be less accurate.

The wet basis gas analyzer on the other hand, measures the sample gas in high temperature of 180–200°C and does not require removal of $H_2O$. Thus it has higher accuracy compared to the dry basis side.

Table 1. Independent parameters for compliance calculation.

| Symbol | Term | Unit |
|---|---|---|
| $q_{maw}$ | Intake air mass flow rate on wet basis^ | kg/h |
| $q_{mad}$ | Intake air mass flow rate on dry basis^ | kg/h |
| $q_{mf}$ | Fuel mass flow rate^ | kg/h |
| $c_{SO2d}$ | Concentration of $SO_2$ in the exhaust in dry basis^ | ppm |
| $c_{NOd}$ | Concentration of NO in the exhaust gas in dry basis^ | ppm |
| $c_{NO2d}$ | Concentration of $NO_2$ in the exhaust gas in dry basis^ | ppm |
| $c_{CO2d}$ | Concentration of $CO_2$ in the exhaust gas in dry basis^ | ppm and (%V/V) |
| $c_{COd}$ | Concentration of CO in the exhaust gas in dry basis^ | ppm |
| $p_b$ | Total barometric pressure^ | kPa |
| $R_a$ | Relative humidity of the intake air^ | % |
| $T_a$ | Intake air temperature determined at the engine intake^ | K |
| $T_{SC}$ | Charge air temperature^ | K |
| $PAH_w$ | Polycyclic Aromatic Hydrocarbon concentration of wash water^ | μg/L $PAH_{phe}$ |
| $Turbidity_w$ | Turbidity of the wash water^ | FNU or NTU |
| $pH_w$ | pH of the wash water^ | 1 |
| $T_w$ | Temperature of the wash water^ | K |
| $P_{aux}$ | Declared total power absorbed by auxiliaries fitted for the test and not required by ISO14396* | kW |
| $P_m$ | Maximum measured or declared power at the test engine speed under test condition* | kW |
| $W_{ALF}$ | H content of fuel* | % m/m |
| $W_{BET}$ | C content of fuel* | % m/m |
| $T_{SCRef}$ | Charge air reference temperature* | K |
| $W_F$ | Weighting factor** | 1 |
| u | Ratio of exhaust component and exhaust gas densities** | 1 |
| $T_b$ | Temperature of water vapor after cooling bath of the analysis system** | °C |

^Require sensor or gas analyzer to measure; *Obtained from the manufacturers Technical File; **Obtained from NOx Technical Code (2008).

Table 2. Parameters information.

| Symbol | Term | Unit |
|---|---|---|
| $gas_x$ | Specific emission of gas | g/kWh |
| $P_i$ | Uncorrected brake power on individual mode | kW |
| $q_{mgas}$ | Emission mass flow rate of individual gas | g/h |
| $c_x$ | Concentration in the exhaust (with suffix of the component nominating, d = dry or w = wet) | ppm/% (V/V) |
| $K_{hd}$ | Humidity correction factor for $NO_x$ for diesel engines | 1 |
| $q_{mew}$ | Exhaust gas mass flow rate on wet basis | kg/h |
| $k_{wr2}$ | Dry to wet correction factor for the raw exhaust gas | 1 |
| $c_{H2d}$ | Concentration of H2 in the exhaust gas in dry basis | ppm |
| $p_r$ | Water vapour pressure after cooling bath of the analysis system | kPa |
| $p_a$ | Saturation vapour pressure of the engine intake air determined using a temperature value for the intake air measured at the same physical location as the measurements for $p_b$ and $R_a$ | kPa |
| $H_a$ | Absolute humidity of the intake air (g water/kg dry air) | g/kg |

The suffix w in $c_w$ and $c_d$ in the equation (6) and other equations for gas concentration denote this difference. The suffix w is meant for wet basis gas analyzer while the suffix d is meant for dry basis gas analyzer. Since the concentration that we need is in wet basis, therefore there is a correction factor that we shall calculate for dry basis gas analyzer.

In order to maximize the cost efficiency for end-users while maintaining the performance of the gas analyzer, this paper uses NDIR dry basis gas analyzer to measure $SO_2$, $CO_2$ and including the NO, and $NO_2$.

The sensors selection refers to the parameters listed in Table 1. The sensors specification shall follow the Appendix 4 of the $NO_x$ Technical Code (2008) for compliance.

The differences that come in the sensor selection are from the marine diesel engine performance to be measured. Different marine diesel engine gives different requirement for the sensors. This paper refers to Daihatsu 6DK-26 for the sensors selection. This Daihatsu 6DK-26 has 6 cylinders, 1570 kW with exhaust turbo charger and intermediate air cooler. Therefore, readers are advised to adjust the sensors according to the marine diesel engine that is used.

## 4 EQUIPMENT SETUP

This setup considers vast variety of engine, where different engine requires different kind of sensors to serve the engine performance. Hence, the setup is so much flexible such that when the sensors are replaced to adjust to the engine, the same data acquisition and transmission system can still be used.

The schematics to describe the equipment setup can be found in Figure 3. The equipment is arranged utilizing several slave microcontrollers (slave MC) and one master microcontroller (master MC). The slaves collect the data from various sensors and a gas analyzer and store it in its local microSD memory card. Thereafter, this data is sent for calculation to the master by using XBee radio communication module. After the calculation is done, the data is stored in the master microSD memory card. The data is also displayed on the screen attached to the slave and master microcontrollers.

Programming of the software is done on the Arduino platform, the Arduino Web IDE. The language for Arduino Web IDE programming is C++.

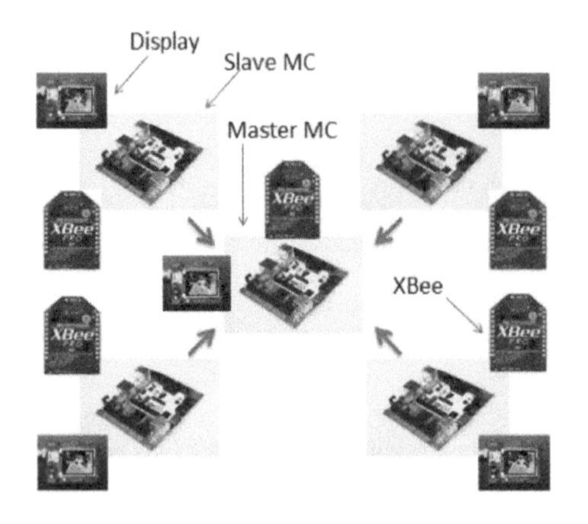


Figure 3. Data acquisition and transmission arrangement.

## 5 CONCLUSIONS AND DISCUSSION

The development of the system has gone through detailed analysis of the parameters required for measurement in order to fulfill the compliance. It is to be noted that for engine owner or engineers responsible for the engine compliance shall understand each of the parameters mentioned in Table 1 and 2. The selection of equipment such as sensors and gas analyzer requires detailed analysis of their performance as well.

The system developed in this paper provides the flexibility for IMO to install to ships for law enforcement. On the other side, this flexibility also gives the ship owner the advantage to monitor the exhaust gas on their ships. Through the future application of Tier III and also awareness on the global warming effect, this system is expected to come in place to make everything easier for many parties.

## ACKNOWLEDGEMENT

The authors wish to acknowledge the funding support for this project from Marine Port Authority Singapore and the collaboration partners from VTT Finland in giving very precious insight to the project. The authors also wish to acknowledge ClassNK for the support as the industrial partner in this project. Furthermore, the authors also wish to thank Daihatsu in engine consultation matters. Lastly, the authors also wish to thank vendors of various sensors and gas analyzer in providing valuable insight to the selection of the equipment.

## REFERENCES

Corbett, J.J., Winebrake, J.J., Green, EH., Kasibhatla, P., Eyring, V., & Lauer, A. (2007). Mortality from ship emissions: a global assessment. *Environmental Science & Technology*, *41*(24), 8512–8518.

International Maritime Organisation (2008), Revised MARPOL Annex VI, Resolution MEPC.176(58), Report of the Marine Environment Protection Committee on Its Fifty-eight Session.

International Maritime Organisation (2008), NOx Technical Code—Resolution MEPC.177(58), Amendments to the Technical Code on Control of Emission of Nitrogen Oxides from Marine Diesel Engine.

International Maritime Organisation (2009), Guidelines for Exhaust Gas Cleaning System—Resolution MEPC.184(59).

Liu, X., Cheng, S., Liu, H., Hu, S., Zhang, D. and Ning, H. (2002), A Survey on Gas Sensing Technology, *Sensor*, V. 12, pp. 9635–9665.

Thepsithar, P. and Chong, W.S. (2011), Emissions from Ships—Upcoming Regulations and Their Implications, Technology Review 2011—Keppel Marine & Offshore Technology Centre, pp. 105–114.

*Maritime-Port Technology and Development – Ehlers et al. (Eds)*
*© 2015 Taylor & Francis Group, London, ISBN 978-1-138-02726-8*

# Modelling of LNG fuel systems for simulations of transient operations

E.L. Grotle & V. Æsøy
*Aalesund University College, Aalesund, Norway*

E. Pedersen
*Norwegian University of Science and Technology, Trondheim, Norway*

ABSTRACT: The use of Liquefied Natural Gas (LNG) as a fuel is motivated by reduced emissions, availability and cost, compared to conventional fuels like Heavy Fuel Oil (HFO) and Marine Diesel Oil (MDO). Development of Natural Gas (NG) fuelled engines and fuel technology has been an ongoing research activity for more than 25 years. In Norway, more than 40 ships are now in operation on LNG either using dual fuel engines or pure gas engines. LNG is a low temperature, volatile fuel with very low flash point, and the main challenges are related to fuel storing and handling. The main components in the LNG fuel systems are the tanks, evaporators/heaters, Pressure Build-up Units (PBU), and the gas regulating units. Control of the overall system performance during transient operations and ship motions is vital. For optimal design through better understanding of the behaviour of the fuel system, simulation models are being developed and simulations performed. Operational experiences and full scale measurements are adapted to effectively contribute to more accurate models. This paper discusses the challenges of modelling such a system and presents relevant component models, performance simulation methods and operational experience. In particular, the liquid/gas phase transition dynamics in the LNG tank as well as simulations of the tank are addressed.

## 1 INTRODUCTION

More than 40 ships worldwide use LNG fuelled propulsion systems, and it has already proven to be a promising fuel for marine applications. The use of LNG instead of other conventional fuels, like MDO or HFO, is mainly motivated by the incentive of reduced emissions. Expectation of lower gas prices, compared to oil, is also a driving factor for ship owners choosing LNG as a fuel.

The future demand of energy is analysed and described in BP Group, 2014. The $CO_2$ emissions from shipping were estimated to be around 3.3% in 2007, which corresponds to 1,046 tons (Second IMO GHG Study 2009). Since LNG normally consists of 95–100% methane, the $CO_2$ emissions are reduced by more than 25% compared to traditional marine fuels. However, the total greenhouse house gas emission effect is slightly lower considering the methane slip from incomplete combustion. By using Lean Burn engines the $NO_x$ emissions are reduced by up to 90% (Æsøy *et al.* 2011). The IMO Tier 3 $NO_x$ regulations can thereby be fulfilled by using LNG as a fuel. No sulphur is contained in the natural gas composition, so $SO_x$ emissions are therefore reduced to zero and emissions of Particle Matter (PM) are close to zero. The most critical challenges using LNG as a marine fuel are related to infrastructure, safety issues and investment costs (Stenersen & Æsøy, 2013).

LNG composition varies from different locations and production facilities. Methane is the major component with typically 95–99% of the total composition. The boiling point of LNG varies slightly with its composition but is typically −162°C at atmospheric pressure. The density of LNG is typically 450 kg/m$^3$ (Mokhatab *et al.* 2014). With operational temperatures below −150°C it is therefore a cryogenic liquid, and special care must be taken in its handling and use.

This paper discusses some of the operational challenges in LNG fuel systems and LNG handling on-board. A simulation environment is in progress to investigate the dynamic response in the fuel system during load variations and environmental disturbances, as well as better to understand the behaviour of fuel system and propulsion machinery system as a whole.

## 2 LNG FUEL SYSTEMS FOR MARINE APPLICATIONS

LNG fuel systems supply natural gas to the engines and consist generally of storage tanks, pipes, valves, vaporizers, gas heater, filters and other components. The liquid is pressurized, vaporized and heated before entering the regulating engine, ensuring correct pressure and mass flow to the engines (see Fig. 1 for a simple schematic overview).

### 2.1 *LNG fuelled engines*

Different types of marine engines have been designed for natural gas as the main fuel, or as dual fuel engines. LNG fuel systems are designed to serve two main types of engine installations (Einang *et al.* 1998):

- Pure gas engines—LBSI (Lean Burn Spark Ignition)
- Dual Fuel engines (DF)
  - Lean burn with diesel pilot (low pressure)
  - High pressure gas injection (high pressure)

Pure gas engines (LBSI) are preferred on smaller vessels like ferries and coastal cargo/passenger vessels. DF (low pressure) engines are installed on ships where fuel flexibility is required due to availability and operations in/out of Emission Control Areas (ECA's). Larger vessels, like tankers, bulk or cargo vessels use slow speed engines with a two-stroke cycle. If gas is used as fuel in these engines, it is combined with a diesel oil pilot and the gas supply is normally at high pressure, approximately 250–350 bars (Einang *et al.* 1998).

### 2.2 *Low pressure LNG fuel gas supply systems*

A typical LNG fuel system is shown in Figure 1. It consists of a pressurized tank, Pressure Build Up unit (PBU) and a gas regulating unit. The process system is built into a "cold box," which is an integrated part of the tank. The room comprising the tank and cold box is called "tank connection space" (see Fig. 2). The pressure in the tank can be controlled by supplying heat in a closed loop with LNG tapped off the tank and supplied back as vapour. This is a very simple and robust solution, avoiding pumps as critical components for wear and maintenance. Further the low temperatures and temperature variations may cause thermal stress on metal structures, and this is one of the issues concerning piping and valves.

Most low pressure gas fuel systems are designed with a Pressure Build Up (PBU) vaporizer, a main supply vaporizer and a gas heater as separate units (Fig. 1). In the PBU liquefied natural gas is vaporized using heat from water/glycol. The static pressure increases after temperature increase and phase transition as the driving force. The enthalpy supplied by the heating agent should be sufficient to keep the tank pressure at the necessary level. If sudden pressure changes are situated in the tank, it can be challenging to regulate this system, and the result is vaporizers that are oversized. This will of course increase the costs for the ship owner and/or contractors.

All in all the design of low pressure LNG systems seems to follow the same philosophies, with some minor differences. Cryo AB (part of Linde Engineering Division) use a somewhat different concept with the PBU vaporizer, main vaporizer and gas heater integrated as one unit.

### 2.3 *High pressure LNG fuel gas supply systems*

The high pressure gas systems are very similar to low pressure systems, but a liquid pump or high pressure compressor arrangement is required. The necessary requirements for high pressure systems are naturally more stringent compared to low pressure systems. It requires high pressure components and higher level safety requirements. High

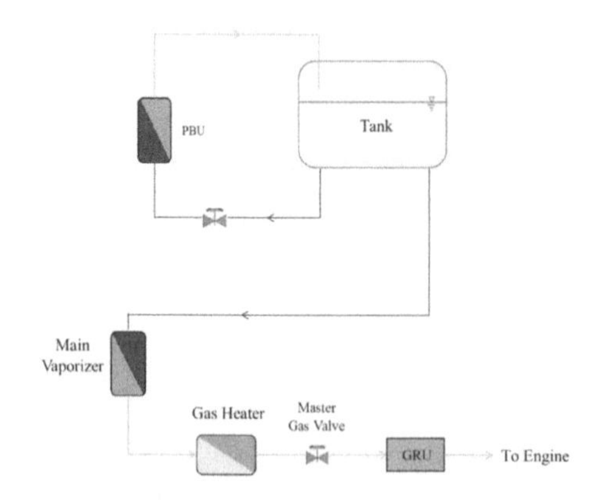


Figure 1. Low pressure fuel supply system with Pressure Build up Unit (PBU).

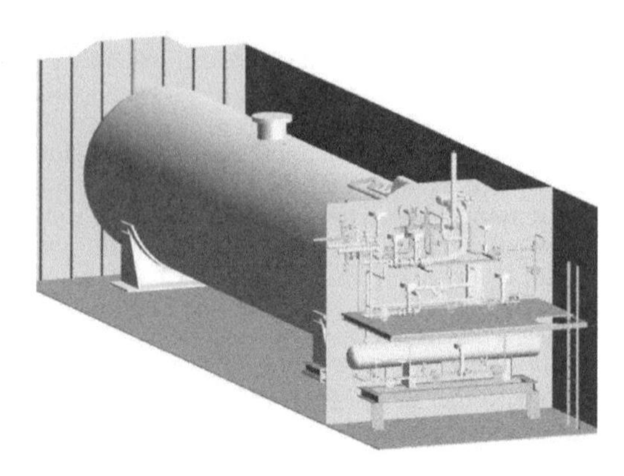

Figure 2. Process system as an integrated part of the tank.

pressure systems will not be discussed further in this paper.

### 2.4 *LNG storage tanks*

LNG storage tanks can be divided into two main types:

- Vacuum insulated pressure tanks (C-tanks)
- Atmospheric pressure membrane tanks (A-tanks)

The most common fuel tank solution now in use is the C-type tanks where pressure is kept in the range of 4–8 bars. Vacuum insulation provides excellent thermal insulation and a second barrier in case of leakage. The main disadvantages for C-tanks are limited unit size, installation requirements and cost. A-type tanks are still mainly used for cargo transport, and will most likely be preferred for larger fuel capacities in the future. However, these tanks require a special care taken to boil off gas, since they only allow a certain small pressure build up in gas phase. For A-tanks liquid pumps are required to supply engines with needed gas pressure.

### 2.5 *Safety requirements in design of LNG fuel supply systems*

Existing guidelines and codes regarding safe design, the International Code for the Construction and Equipment of Ships Carrying Liquefied Gases in Bulk (IMO, 1993), are issued by committees under the International Maritime Organization (IMO). The Interim Guidelines for LNG fuelled ships (IMO, 2009a, Resolution MSC.285(86)) and the IGC code serve as a standard for how equipment is constructed and how each system is designed.

With LNG fuelled ships, the Interim Guidelines require area classifications on-board. This is typically for hazardous or non-hazardous areas, since LNG vapour is considered explosive and flammable. Hazardous areas are divided into three zones. Zone 0 is in the interiors of equipment carrying LNG in liquid or vapour form, for example inside the tank or pipes. Zone 1 is inside the tank room or compressor rooms. Zone 2 includes areas within 1.5 meters surrounding distance to open or semi-closed spaces of zone 1 (IMO, 2009b). Emergency Shutdown (ESD) is required for spaces containing ignition sources. As it is a potentially hazardous area, a specified ventilation capacity must be provided together with gas detection. On the latest vessels gas safe machinery spaces are used instead of ESD-protected. This philosophy requires that all gas-containing elements are enclosed by double barrier systems like double wall piping.

## 3 OPERATIONAL CHALLENGES WITH LNG FUEL SYSTEMS

This section discusses issues related to LNG fuel systems with emphasis on the dynamic behaviour of the overall system. The discussion is based primarily on experience from existing LNG fuelled ships. The critical operations involving dynamics are mainly:

- Bunkering and other fuel transfer operations
- Transient load conditions such as manoeuvring
- Operations in heavy sea states

### 3.1 *LNG bunkering operations*

An important part of the operation of LNG fuelled ships is the bunkering process. The use of potentially hazardous fuel like LNG brings up certain questions about requirements that are relevant for the handling of low flashpoint liquids and gases in general. With LNG handled as cargo on-board LNG carriers there are more than 50 years of experience, and the rules and regulations for LNG fuelled ships have more or less adopted the already existing rules comprising transport of LNG. The new challenges are related to LNG as fuel handled in a micro-scale.

Bunkering facilities are still under development as more or less prototype installations, and may be very different depending on their location. Most commonly, LNG is supplied from trucks (Fig. 3) or onshore terminals (Stenersen *et al.* 2012).

Specially constructed ships are also used for ship to ship bunkering. Important safety issues are related to bunkering and LNG quality such as composition and temperature. The temperature on the LNG supplied may have important implications

Figure 3. Simultaneous loading of "Pioneer Knutsen" and a road trailer at Kollsnes production plant (Source: Gasnor).

for the operation of the ship, where both high and low temperatures cause different challenges during bunkering and operation.

Bunkering processes involves purging, checking for leakages and cooling down the piping system before starting, as well as gas freeing/purging with $N_2$ after the LNG transfer is finished. During bunkering the pressure in the tank is controlled by either spray filling from the tank top or bottom fill. The spraying will cause condensation of the gas in the tank and thereby reduce the pressure. Bottom filling will compress the top gas volume and increase the pressure. An example showing measured pressure and tank level is shown in Figure 4, where both spray and bottom filling are used. Experienced from bunkering is that gas pressure control is very sensitive to difference in LNG supply temperature. In some cases where the LNG temperature is too high, the tank pressure will increase even during spraying.

### 3.2 *Dynamic effects during operation*

Dynamic effects for LNG fuelled ships are quite different from a land based process facilities. First of all, the ratio of fuel storage capacity to fuel consumption is quite small. Secondly, the fuel consumption may vary significantly in time, as in a manoeuvring situation. Thirdly, a ship is moving which results in motions of the liquid contained in the tanks and vaporizers. The engines require a steady supply pressure and temperature which can be challenging to achieve under certain conditions depending on:

- Load transients
- Tank level
- Ship motions
- LNG qualities (temperature and composition)

In Figure 5 a situation is given where the gas was fed directly from the tank top. As the ship

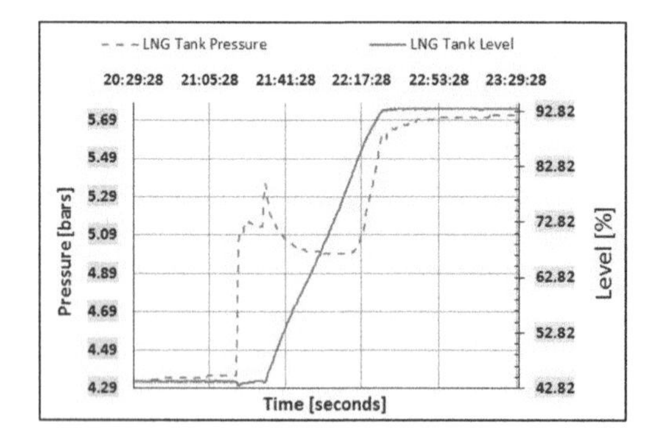


Figure 4. Tank pressure and level measurements from bunkering (data from a Norwegian gas operated ferry).

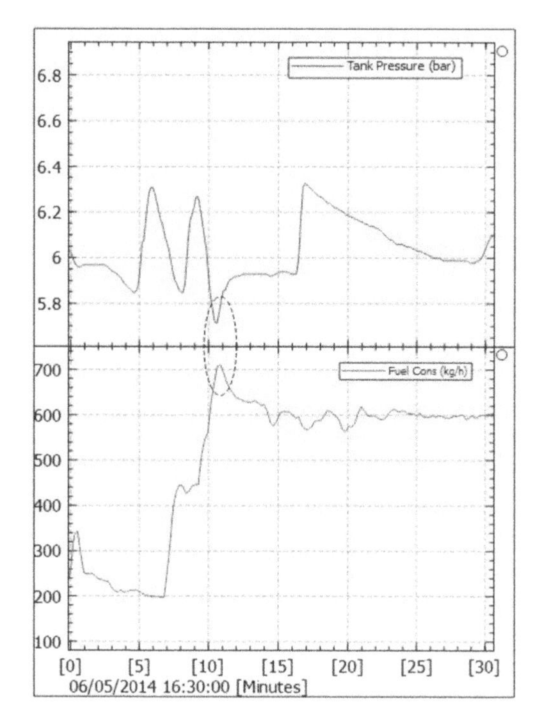


Figure 5. LNG tank pressure dynamics during load variations (data from a Norwegian gas ferry).

accelerated, and the fuel consumption reaches its maximum, and the pressure in the tank drops below the minimum GRU regulating pressure. After some time, the pressure is back to normal after pressure build-up by both main evaporator and PBU.

### 3.3 *Sloshing*

One of the most complex challenges in ensuring reliable performance of LNG fuelled ships is the liquid motions inside the large fuel tanks. Sloshing has traditionally been a concern in transportation of liquids, both as cargo and ballast water as well as with stabilizer tanks, where structural damages have resulted from sloshing. Several studies have therefore been done, particularly investigation of the liquid motions inside the tanks (Faltinsen & Timokha, 2009). However, structural impact is not the major challenge for LNG fuel tanks. Sudden pressure drop in the LNG storage tanks has been experienced for LNG fuelled ships in heavy sea state. The pressure drop is certainly influenced by the ship motions. However the sloshing mechanisms influencing phase transition between liquid and vapour are not completely understood yet (ref. section 4.3.4).

Tank geometry, location relative to ship motions and internal bulkheads are the most important factors, which must be further investigated. Later investigations have shown that the density ratio of liquid and gas has significant influence on the impact pressures (Maillard & Brosset, 2009).

## 4 MODELLING DESCRIPTION OF THE LNG FUEL SYSTEM

The system consists of several different components which more or less are challenging as comprised by complicated physical laws. A thermo-fluid system is a branch of science and engineering divided into four sections including heat and mass transfer, thermodynamics, fluid mechanics and combustion. The most important considerations here, excluding the engine, are the three first sections mentioned. Natural gas consists of several components but mainly methane and therefore it can be modelled as a single component fluid. Phase transition is important considering the fuel system, and we are dealing with two-phase flow and boiling in forced flow regimes.

Models of the different components within the system are presented and discussed in the following sections. An energy balance considering the tank as a control volume is given, and a modelling approach for phase transition inside the tank is presented. This is highly relevant for the bunkering process. Since it deals with system modelling, the method using a lumped capacity approach (Pedersen & Engja, 2010) will form the basis in the modelling. One-dimensional energy flows are summed up and integrated and the states must be defined by proper physical laws.

Literature related to modelling of thermo-fluid systems is among others Thoma and Bouamama, 2000 and Bouamama, 2003. The latter paper presents a model library relevant for this work.

Since we have a two-phase system some basic assumptions are made for the models:

1. Two phase flow only occurs in vaporizers
2. Flow in pipes and valves are either in liquid or gas phase
3. Compressible flow only occurs for the gas phase

The gas fuel system consists of the following main components:

- Pipes
- Valves
- Tank
- Vaporizers and heat exchangers

Pipes and valves are previously modelled by Pedersen & Æsøy 2011. In this paper, the focus has been on the tank. A short description of how to model different components in the system follows.

### 4.1 *Modelling of vaporizers*

Vaporizers in the fuel system are simple shell-and-tube heat exchangers, and may consist of U-tube or straight tube bundles. We consider boiling in forced flow, and the different flow regimes in a horizontal pipe are shown in Figure 6, which is the most relevant. Gravity pushes the liquid towards the bottom and the heat transfer is changed significantly.

The pressure drop calculation can be done by recognizing the different flow regimes inside the pipes. Flow of liquid drops in a gas can be treated as homogenous, but if stratified flow occurs, it might be calculated with the separated flow model (section 4.2).

If the necessary conditions for boiling are fulfilled near the wall of the tube, bubbles will form even though the mean temperature in the liquid is below the saturation temperature. The necessary requirement for sub-cooled boiling to start is that the bubbles needs to be slightly over-pressured, as the surface tension forces on the bubble surface are preventing it from growing. The necessary bubble pressure is given as $P_b = P_L + 2\sigma/r$ (Stephan, 1992), where $\sigma$ is the surface tension and $P_L$ is the pressure of the surrounding liquid. Compared to a plane interphase between liquid and vapour ($r \rightarrow \infty$) the surrounding liquid needs to be superheated for a bubble to grow. In sub-cooled boiling the thermodynamic quality is less than zero. Bubbles that depart from the wall are condensed again while transported towards the liquid core. Heat is transferred more effectively and the heat transfer coefficient is dominated by the heat flux.

When the thermodynamic quality is zero, the mean temperature in the pipe cross-sectional area is equal to the saturation temperature and saturated boiling occur. Bubbles no longer condense, but may join with other bubbles and form larger plugs. The temperature in the liquid and wall is nearly constant throughout the nucleate boiling section. It varies slightly with the pressure drop.

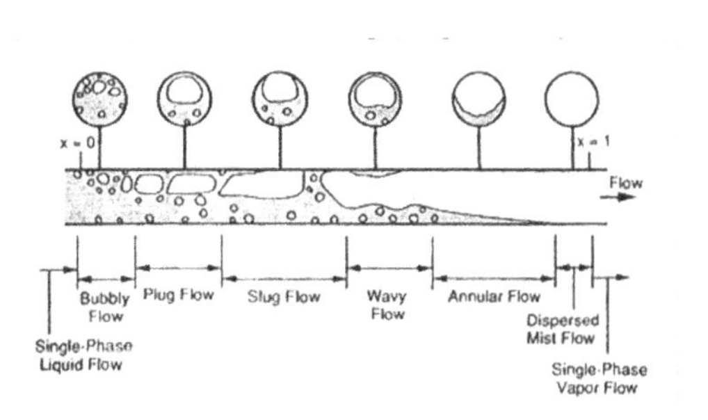


Figure 6. Flow patterns in a horizontal vaporizer tube (ref. Two Phase Flow Laboratory).

The difference between wall temperature and saturation temperature is decreasing as the thermal resistance is reduced along the tube.

The fluid is accelerated with increasing vapour quality, and the liquid film along the wall is decreasing. The convective heat transfer gets more pronounced and less influenced by the bubble formation (heat flux). With less thermal resistance between wall and vapour, the temperature at the wall is decreased, and bubble formation is suppressed (Stephan, 1992).

As the film is completely evaporated and the wall becomes dry, the heat transfer coefficient is dropping significantly and the wall temperature rises rapidly. The necessary heat flux for dry-out to occur is termed the critical heat flux. At this point, the thermodynamic quality is equal to the real quality ($x_{real} = m_G/m_L$) and is lower than unity. The thermodynamic quality is equal to $(h - h_L)/\Delta h_{fg}$.

### 4.2 *Two-phase flow in evaporator pipes*

Determination of the pressure gradient in two-phase flow is depending on the type of flow regime. In general, the force balance takes the same form as with single-phase, but one would need to take into account the phase densities, expressed with the void fraction. The pressure gradient in one-dimensional steady pipe flow takes the form (Whalley, 1987)

$$-\frac{dp}{dx} = \frac{4\tau_w}{d_p} + [\alpha\rho_g + (1-\alpha)\rho_l]g\sin(\theta) + \frac{d}{dx}[\alpha\rho_g u_g^2 + (1-\alpha)\rho_l u_l^2] \qquad (1)$$

where $\alpha$ is the void fraction, given as

$$\alpha = \frac{1}{1 + \left(\frac{u_g}{u_l}\frac{1-x}{x}\frac{\rho_g}{\rho_l}\right)} \qquad (2)$$

$u_g/u_l$ is the slip ratio, $S$, and $x$ is the vapour quality. In homogenous flow, the slip ratio is equal to 1, and if the quality varies linearly through the pipe, the expression for the homogenous density together with the pressure gradient can be integrated over the pipe section. For determination of the void fraction, different correlations can be used. Commonly used is the CISE correlation (Whalley, 1987).

Regarding the frictional pressure gradient, an empirical expression must be used, and except for homogenous flow, integration of the pressure gradient must be done numerically. For homogenous flow we get:

$$\left(-\frac{dp}{dz}\right) = \frac{4\pi}{d} = \frac{4}{d}C_{fh}\frac{1}{2}\frac{G^2}{\rho_h^2} \qquad (3)$$

This is analogous to single-phase flow. $G$ is the mass flux of the total flow. $\rho_h$ is the homogenous density, given as

$$\frac{1}{\rho_h} = \frac{x}{\rho_g} + \frac{(1-x)}{\rho_l} \qquad (4)$$

$C_{fh}$ is the homogenous friction factor coefficient, and can be found as with single-phase flow by calculating the Reynolds number for the homogenous flow. The homogenous dynamic viscosity is expressed the same way as the density.

For separated flow, the two-phase frictional pressure gradient is found with two-phase multipliers. An accurate correlation is the Friedel correlation. The pressure gradient then takes the form

$$\left(-\frac{dp}{dz}\right)_F = \left(-\frac{dp}{dz}\right)_{lo}\phi_{lo}^2 \qquad (5)$$

where the subscript $_{lo}$ means the single-phase pressure gradient of liquid calculated with the total mass flux of vapour and liquid. The two-phase multiplier, $\phi^2$, can be found with the Friedel correlation (Whalley, 1987).

The pressure drop in annular flow can be found by solving the triangular relationship and the interfacial roughness correlation, which connects the average film thickness and the shear stress. Turbulent velocity distribution across the film is assumed and integrated to give the mass flow. The liquid film is evaporating at its surface, and heat is transferred mainly by convective vaporization.

Calculations of heat and mass transfer in forced flow rely on correlations that are valid for a specified regime. Even though calculations are performed by assuming steady-state conditions, this is hardly ever achieved inside a vaporizing tube. In order to make a dynamic model, one can divide the tubes into several elements and solve one-dimensional spatial differences numerically.

### 4.3 *Tank model*

The discussion here starts with defining a control volume around the tank, and by looking at the different situations as shown in Figure 7. The heat ingress is only considered when the tank is stored. Operation of a ship is a combination of all three cases shown, but the importance differs mainly by the time dependency of each in the overall process.

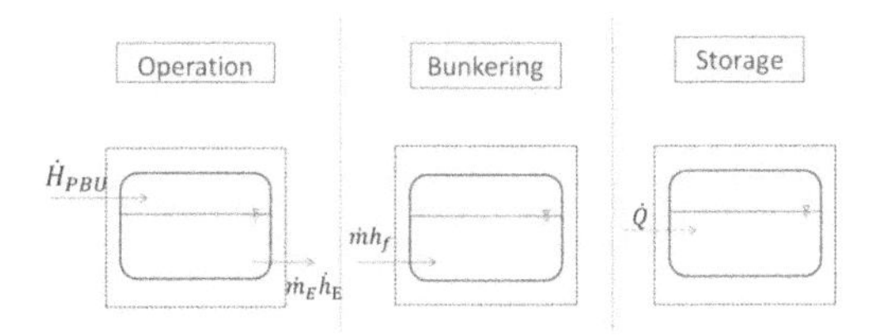


Figure 7. Control volume around tank in operation and in bunkering.

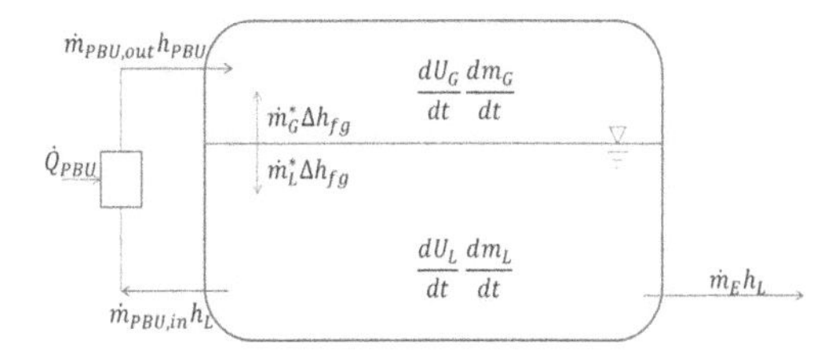


Figure 8. Tank with PBU—mass and energy balance.

While in operation, the liquid mass is reduced with a varying rate, depending on the fuel consumption. The process is transient in this respect, but also because the regulation of pressure is not ideal and happens with varying rate. The tank will operate within a certain pressure range, where the set pressure is the minimum, and the maximum is somewhat above. This study is also highly relevant for the bunkering process, where there are issues with pressure control. In heavy sea states, sloshing in the tanks occur and may cause sudden changes in the bulk pressure.

Later on, it may also be relevant to include substances other than methane, at least a binary mixture that illustrates the difference from modelling a pure substance.

#### 4.3.1 *Energy balance considering the tank as one control volume*

During operation the tank is considered to be adiabatic, since the ratio of energy consumption to the heat ingress is assumed to be much larger than unity. Assuming that the cross sectional area of the outlet and inlet on the tank is small relative to the total area of the control volume border, the kinetic and potential energy accompanied by mass is negligible. With no external forces acting on the control volume the energy balance is (Warberg, 2006)

$$\left(\frac{dU_T}{dt}\right)_{cv} = \sum \dot{H}_{in} - \sum \dot{H}_{out} \tag{6}$$

This assumption does not hold in sloshing situations, as the work done on the tank can no longer be disregarded. The energy balance during normal operation is found by considering the input and output of the control volume border, as shown in Figure 8.

$$\frac{dU_T}{dt} = \left(\dot{m}_{out}h\right)_{PBU} - \left(\dot{m}_{in}h\right)PBU - \dot{m}_E h_L \tag{7}$$

$$\frac{dm_T}{dt} = \left(\dot{m}_{out} - \dot{m}_{in}\right)_{PBU} - \dot{m}_E \tag{8}$$

The energy and mass balance for the PBU is

$$\frac{dU_{PBU}}{dt} = \left(\dot{m}_{in}h_L\right)_{PBU} + \dot{Q}_{PBU} - \left(\dot{m}_{out}h\right)_{PBU} \tag{9}$$

$$\frac{dm_{PBU}}{dt} = \left(\dot{m}_{in} - \dot{m}_{out}\right)_{PBU} \tag{10}$$

If a steady-state condition is assumed for the PBU vaporizer, then the mass and energy balance is reduced to

$$\frac{dU_T}{dt} = \dot{Q}_{PBU} - \dot{m}_E h_L \tag{11}$$

$$\frac{dm_T}{dt} = -\dot{m}_E \tag{12}$$

The assumption of steady-state is too scarce for a realistic operation. But it simplifies the balance and provides better understanding at this point.

#### 4.3.2 *Energy balance for liquid and gas with separate control volumes*

It might be relevant to divide the total volume into separate gas and liquid volumes. If the kinetic energy of the moving free surface is neglected, then the energy and mass balance reads

$$\frac{dU_G}{dt} = \left(\dot{m}_{in}h_L\right)_{PBU} + \dot{Q}_{PBU} + \dot{m}^*_{evap}\Delta h_{fg} \tag{13}$$

$$\frac{dU_L}{dt} = -\left(\dot{m}_{in}h_L\right)_{PBU} - \dot{m}_E h_L + \dot{m}^*_{cond}\Delta h_{fg} \tag{14}$$

$$\frac{dm_G}{dt} = \dot{m}^*_G \tag{15}$$

$$\frac{dm_L}{dt} = \dot{m}^*_{cond} - \dot{m}_E \tag{16}$$

$$\dot{m}^*_{evap} = -\dot{m}^*_{cond} \tag{17}$$

#### 4.3.3 *Equilibrium conditions in the tank*

The normal procedure when performing calculations of similar problems is to assume phase equilibrium. With equal temperature and pressure it gives the requirement of equal chemical potential in each phase (Callen, 1985) where differential form is given as

$$-s_L dT_L + v_L dp_L = -s_V dT_V + v_V dp_V \tag{18}$$

As $dh = Tds$, combining this with the above equation gives us the Clausius-Clapeyron equation:

$$\frac{dp}{dT} = \left(\frac{dp}{dT}\right)_{sat} = \frac{\Delta h_{fg}}{T_{sat}(v_g - v_f)} \tag{19}$$

This equation can be used together with the ideal gas law or other equations of state to determine the saturation pressure for given temperature or opposite.

Equilibrium at the interphase means that the thermal resistance is neglected. Depending on the case, it may be too scarce assuming that the interphase is always at the saturation point. On a molecular scale, the interphase between liquid and gas is always in violent agitation (Ghiaasiaan, 2008). Equilibrium is a dynamic phenomenon.

The maximum heat flux can be found by assuming that all molecules that hit the liquid surface join the liquid. The net mass flux rate of evaporation can be determined from the equation (Bond & Struchtrup, 2004)

$$m''_{evap} = \frac{2\psi_e}{2-\psi_e}\left[\frac{M_v}{2\pi R_u}\right]^{1/2}\left[\frac{P_{sat}(T_l)}{\sqrt{T_l}} - \frac{P_v}{\sqrt{T_v}}\right] \tag{20}$$

where $M_v$ is mole weight, $R_u$ the universal gas constant and $T_l$ and $T_v$ are the interphase temperatures on the liquid and vapour side respectively. $\Psi_e$ is a evaporation coefficient. $P_v$ is the vapour pressure which can be assumed uniform, although not entirely correctly.

Mass and heat will flow either when a temperature gradient is imposed across the system, so that $T_L \neq T_V$, or by perturbing the vapour pressure away from the saturation pressure. Allowing for temperature variations from the interphase to the bulk gives the total heat flux towards the interphase:

$$q''_s = m''_s h_l - k_L \frac{dT}{dz} = m''_s h_v - k_v \frac{dT}{dz} \tag{21}$$

where $k$ is the thermal conductivity of liquid and vapour and $h_l$ and $h_v$ are the enthalpies at the interphase on liquid and vapour side respectively. In Bond and Struchtrup 2004, a simplified approach is presented, where the liquid temperature at the interphase is close to the saturation temperature so that $T_l \approx T_{sat}(p_v)$. Further it can be assumed that $h_v - h_l = \Delta h_{fg}$. If the temperature difference at the interphase is ignored, which is more relevant than ignoring variations towards the interphase on each side, the mass flux can be approximated as

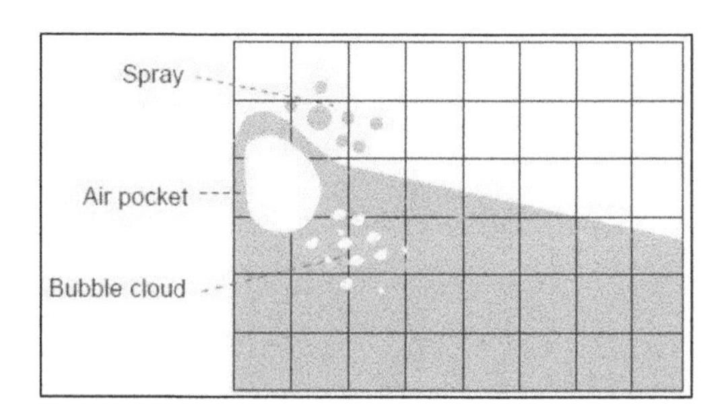


Figure 9. Two-phase phenomenon at a breaking wave; spatial grid resolution.

$$m''_s \approx \left[\frac{k_V}{L_v} + \frac{k_L}{L_l}\right]\frac{T_{bl} - T_{sat}(p_V)}{\Delta h_{fg}} \tag{22}$$

or

$$m''_s \approx \frac{k_L}{L_l}\frac{T_{bl} - T_{sat}(p_V)}{\Delta h_{fg}} \tag{23}$$

where the heat flux is then given as

$$q''_{evap} = \overset{*}{m}_{evap}\Delta h_{fg} \tag{24}$$

This means that equilibrium exist at the interphase. The temperature in the liquid is varying linearly from the interphase to the bottom, $T_{bl}$.

### 4.3.4 *Sloshing*

Investigating details in a sloshing tank requires a very high resolution grid to capture (see Fig. 9) spatial and/or time differences. The computational effort can be significant in these cases. Such models could be used in parts of the system, where spatial resolution and three-dimensional state variations are important.

Even though the use of CFD models would capture details in a sloshing tank fairly well, it is still difficult to be sure if this represents the actual physics, and such methods still rely on proper validation of results. The methods could however provide better insight into the mechanisms situated in the tank.

## 5 SIMULATION OF PHASE TRANSITION IN THE TANK

A simple simulation using the assumptions given at the end of the last chapter is presented. The case is an initial liquid temperature at $t = 0$ that is approximately 3 degrees lower than the gas. The temperature is assumed to vary linearly with the liquid depth (see equation 24), and the energy in the incompressible liquid is calculated according to the mean temperature. The case is illustrated in Figure 10.

The gas temperature is assumed to always be equal to saturated conditions given by the respective pressure. The gas is assumed to behave as a perfect gas, so that the relation $pV = mRT$ is used. Further, this means that the internal energy can be evaluated with $E = mC_vT$, where zero absolute temperature is used as the reference condition.

Using one single control volume in such a problem is a bit controversial, since the above relation between energy and temperature is best suited when the temperature is constant. As mentioned in Bond and Struchtrup, 2004, the assumption of linear variation over the total depth of the liquid is valid for small mass fluxes.

The discussion of the temperature profile for specific non-equilibrium conditions in the tank is quite essential in how accurate our model is.

The energy in both liquid and gas is integrated at each time step, and the temperature is evaluated. The density and specific heat capacity, found in the NIST library (NIST, 2014), are updated at each time step. The specific heat capacity is found from a fourth order polynomial function of temperature, assumed to vary only as in ideal gas conditions. The saturation temperature is found from the pressure using the same library.

The mean temperature in the liquid bulk can be used to find the mass flux across the interphase by using

$$T_L = \frac{T_{bl} + T_{sat}}{2} \tag{25}$$

which gives

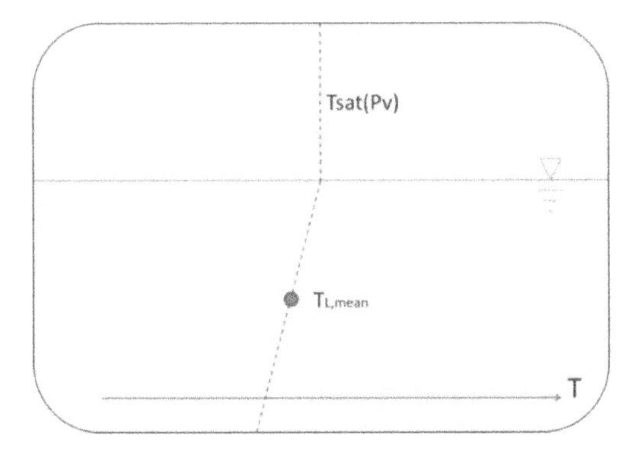


Figure 10. Temperature profile in the tank.

$$m_s'' = 2\frac{k_L}{L_l}\frac{T_{L,mean} - T_{sat}(p_V)}{\Delta h_{fg}} \tag{26}$$

In order to implement this in a simple and effective way, 20-sim was used. The bond graph model is shown below. The method of bond graphs is not described in detail here. As seen from Figure 11, the model can easily be extended to include discharge or filling of mass and energy, which are the two types of state variable used in this model.

Information of effort (temperature and pressure) and flow (energy and mass) is exchanged between the C- and R-elements. The states are defined inside the two C-elements, where each consists of single phase methane.

The starting value of the mean liquid bulk temperature is 130 Kelvin, and the gas temperature at t = 0 is equal to the saturation temperature at $P_v = 4.5$ bars, which is approximately 133.47 Kelvin.

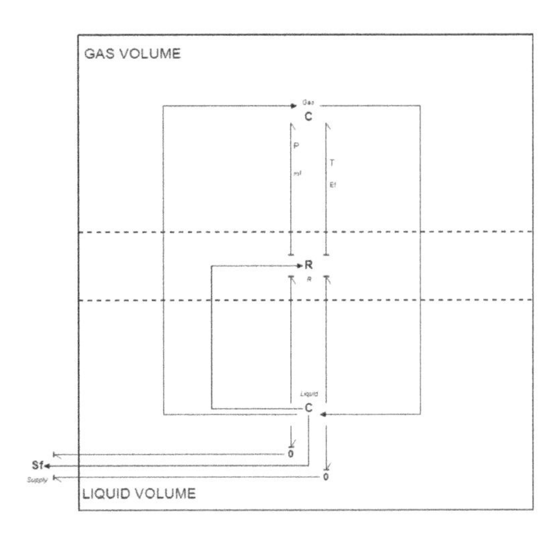


Figure 11. Bond graph model.

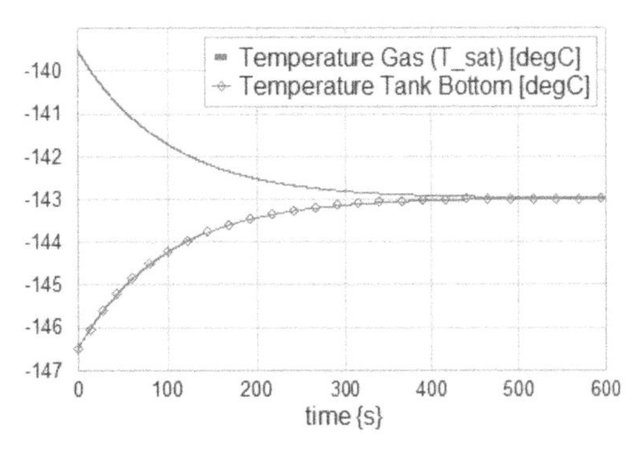


Figure 12. Liquid and gas temperature.

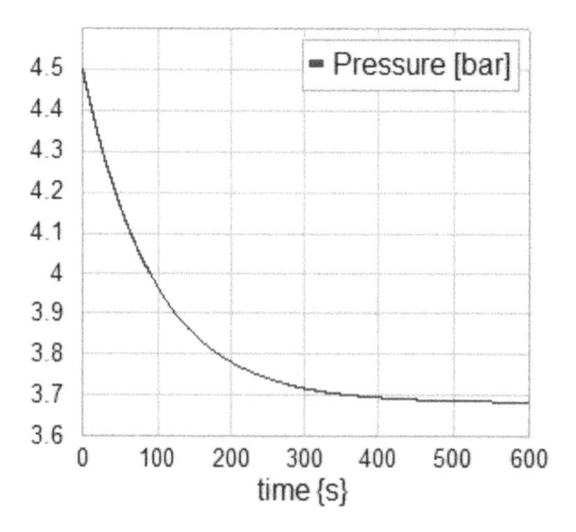


Figure 13. Pressure in the tank.

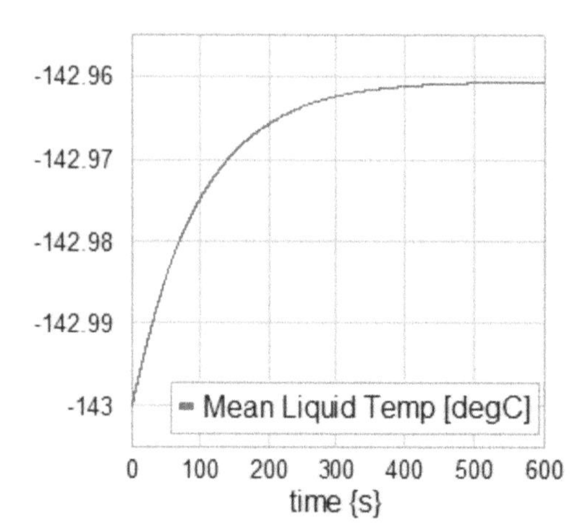


Figure 14. Mean liquid temperature.

The thermal conductivity coefficient $k_L$ is adjusted to a value 1000 times higher to speed up the results. The initial volume of liquid is half of the total volume.

The resulting gas/liquid temperatures, pressure and mean liquid temperature together with the mass flow across the interphase are shown in Figures 12, 13, 14 and 15 respectively. The mass flow across the interphase (Fig. 15) is unreasonably high as is expected since the thermal conductivity was adjusted to a greater value. With realistic values this would be in the order of $\sim 1e^{-6}$.

Incorporating non-equilibrium conditions must be done with care in using a lumped capacity model, as we do not know the temperature distribution. Most likely it will be closer to exponential,

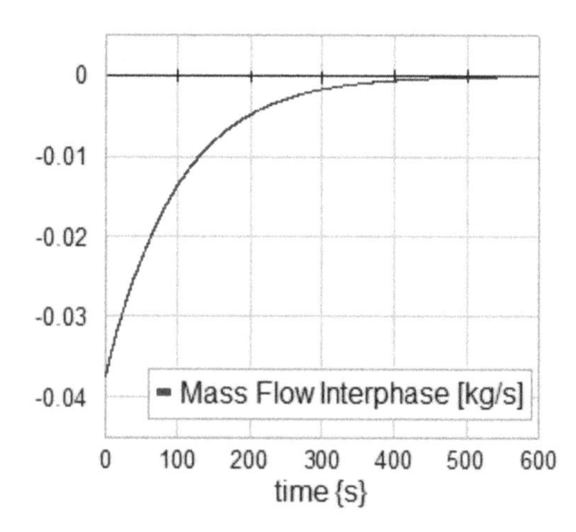


Figure 15. Mass flow across the interphase (condensation).

where the sharpest gradient is close to the interphase. A distinct temperature gradient in the gas is also likely to exist, which is different from the liquid temperature distribution. The assumptions used in this simulation would therefore be too scarce in cases where a sudden drop in pressure occurs.

## 6 CONCLUSION AND FURTHER WORK

This paper presents a pre-study of dynamics in LNG fuel system for ships. Operational experiences and some main challenges are presented and discussed. Some theory background in modelling thermo-fluid systems is presented with a main focus on the LNG tank. Further, a modelling approach to the vaporizers is initiated. The theory and modelling procedure presented was tested with a simple tank model that can easily be extended to include filling or discharge, or even sudden change in pressure or temperature. Investigation on how to treat the temperature distribution in each phase must be done.

The next step will be to model the total fuel system. Sub-models of vaporizers and heat exchangers will be developed, and their influence on the total system will be investigated. Particular care must be taken in modelling the PBU and its vaporizer, as pressure control is important to ensure reliable operation.

Further work is planned to study dynamics during critical operations such as bunkering, manoeuvring and ship in heavy sea conditions. Different scenarios involving non-equilibrium conditions should be investigated in this respect.

Not all of the theory presented in chapter 3 is ready to be used in dynamic situations. Investigation on how to incorporate some of the mentioned theory is left to be determined.

## REFERENCES


Æsøy, V., Stenersen, D, Einang, P.M., Hennie, E., Valberg, I., 2011. LNG-Fuelled Engines and Fuel Systems for Medium-Speed Engines in Maritime Applications. *Society of Automotive Engineers of Japan (SAE).*

Æsøy, V., Stenersen, D., 2013. Low Emission LNG Fuelled Ships for Environmental Friendly Operations in Arctic Areas. *Proceedings of the 32th International Conference on Ocean, Offshore and Arctic Engineering. OMAE2013.*

Bouamama, B.O. 2003. Bond Graph Approach as Analysis Tool in Thermofluid Model Library Conception. *Journal of the Franklin Institute* 340(2003):1–23.

BP Group. 2014. Energy Outlook. [ONLINE], Available: http://www.bp.com/en/global/corporate/about-bp/energy-economics/energy-outlook/outlook-to-2035.html. [Accessed 07. August 2014].

Braeunig, J.P., Brosset, L., Dias, F., Ghidaglia, J.M. 2010. On the Effect of Phase Transition on Impact Pressures due to Sloshing. *The International Society of Offshore and Polar Engineers (ISOPE).*

Callen, H.B. (2nd edition). 1985. *Thermodynamics and an Introduction to Thermostatics.* John Wiley & Sons.

Einang, P.M. 2007. Gas Fuelled Ships. *CIMAC Congress.*

Faltinsen, O.M., Timokha, A.N. (1st edition) 2009. *Sloshing*, Cambridge University Press.

Ghiaasiaan, S.M. (1st edition). 2008. *Two-Phase Flow, Boiling and Condensation.* Cambridge.

GL Group. 2008. Rules for Classification and Construction—Part 1 Ship Technology. [ONLINE] Available: http://www.gl-group.com/infoServices/rules/pdfs/gl_i-1-6_e.pdf. [Accessed 07. August 2014].

International Maritime Organization (IMO). 1993. International Code for the Construction and Equipment of Ships Carrying Liquefied Gases in Bulk (IGC Code). 1993 Edition (with supplements of the resolutions of 1994 and 2004).

International Maritime Organization (IMO). 2009a. Second IMO GHG Study 2009. London, UK.International Maritime Organization (IMO). 2009b. RESOLUTION MSC.285(86)—Interim Guide-lines on Safety for Natural Gas-fuelled Engine Installations in Ships.

ISO. 2013. ISO/DTS 18683—Guidelines for Systems and Installations for Supply of LNG as Fuel to Ships.

Juliussen, L.R., Kryger, M.J., Andreasen, A. 2011. MAN B&W ME-GI Engines. Recent Research and Results. *International Symposium on Marine Engineering (ISME).* Kobe, October 17–21.

MAN & DNV. 2014. Quantum 9000-Two-stroke LNG. [ONLINE] Available at: http://www.mandieselturbo.com/files/news/filesof16391/5510-0108-00ppr_low.pdf. [Accessed 07 August 2014].

Maillard, S., Brosset, L. 2009. Influence of Density Ratio between Liquid and Gas on Sloshing Model Test Results. *Proceedings of the Nineteenth (2009) International Offshore and Polar Engineering Conference (ISOPE).* Osaka, June 21–26.

McGuire, G., White, B. (3rd edition). 2000. SIGGTO. *Liquefied Gas Handling Principles On Ships and in Terminals.* London: Witherby & Company Limited.

Mokhatab, S., Mak, J.Y., Valappil, J.V., Wood, D.A. (1st edition). 2014. *Handbook of Liquefied Natural Gas.* Oxford: Elsevier.

National Institute of Standards and Technology (NIST). 2014. NIST Standard Reference Database 23. [ONLINE] Available: http://www.nist.gov/srd/nist23.cfm. [Accessed 07 August 2014].

Pedersen, E., Engja, H. 2010. Mathematical Modelling and Simulation of Physical Systems. *Lecture Notes in TMR4275.*

Pedersen, E., Æsøy, V. 2011. Modeling and Simulation for Design and Testing of Direct Injection Gaseous Fuel Systems for Medium-Speed Engines. *SAE International.*

Stenersen, D., Æsøy, V., Einang, P.M., 2012. LNG Fuel Infrastructure for Low Emission Ships Operating on Natural Gas. *World Maritime Technology Conference, St. Petersburg.*

Stephan, K. 1992. *Heat Transfer in Condensation and Boiling.* Springer.

TGE Marine. Design Concepts for LNG FSRU/FPSO with IMO C Cargo Tanks. *Presentation by Klaus Gerdsmeyer.*

Thoma, J., Bouamama, B.O. (1st edition). 2000. *Modelling and Simulation in Thermal and Chemical Engineering.* Springer.

Warberg, T.H. (1st edition). 2006. *Den termodynamiske arbeidsboken.* Kolofon forlag.

*Maritime-Port Technology and Development – Ehlers et al. (Eds)*
*© 2015 Taylor & Francis Group, London, ISBN 978-1-138-02726-8*

# Future Internet enabled ship-port coordination

A. Rialland & Å. Tjora
*Norwegian Marine Technology Research Institute (MARINTEK), Norway*

ABSTRACT: This paper presents a Future Internet based concept for integrated planning of port call, meaning the coordination of resource planning among all actors involved: port, terminal, ship, agents, and other stakeholders. In the project FInest, ICT tools based on Software-as-a-Service principles were sketched and a scenario-based demonstrator was built in order to showcase the possibilities offered by Future Internet technologies. Data from the Port of Aalesund were used to illustrate how the planning challenge at port and terminal can be tackled by getting the right information at the right time, from and to the right actor. The business model supported by the FInest platform is the semi-automatic handling of port call based on real-time information enabled by synchronization of resource planning among the port and terminal service suppliers. The ship can register its port call and book services through a platform displaying real-time information about resource availabilities, while the port- and terminal service providers can coordinate their services and resources also based on real-time information provided by the platform. The main expected benefit is the coordination of information among multiple actors, enabling more visibility, more efficient planning of services at port and terminals, easier booking of services, better capacity utilization, as well as support for optimization of ship voyage.

## 1 INTRODUCTION

In order to exploit the potential offered by Future Internet services in supporting business interoperability, the EU funded project FInest (FInest) has designed a cloud-based Collaboration Platform for the Transport and Logistics sector. Five demonstrators have been mocked-up, each exploring the Collaboration Platform in a particular case of the sector, tackling specific challenges with tailor-made solution. The Resource Hub is one of these demonstrators: with the goal of improving interaction among business actors and efficiency in operation, the Resource Hub is a particular instance of the Finest Collaboration Platform offering functionalities for resource management, operation planning, and information exchange to support the operations linked to the process of vessel calling at port.

### 1.1 *The role of the port in the maritime value chain*

Ports are central actors in the maritime transport value chain, not only as a transfer point for cargo between maritime and inland transport, but more importantly given their role in the coordination of material and information flow (Carbone and De Martino, 2003). A port has a large and varied community of stakeholders to serve and several roles to fulfill in different domains (Froese and Zuesongdham, 2008). Navigation and ship traffic monitoring, ship services and supply, inspections, safety and security, logistics and cargo handling are among the main activities performed by ports.

A *port call*, which is the visit of a vessel to a port, can be seen as an entire project, with a multiple processes to follow and services to be planned and delivered by several actors. These are summarized in Figure 1.

The port often stands out as a big bottleneck in maritime and intermodal transport, causing congestion costs, mainly in form of time loss (Meersman et al., 2012): a *bottleneck for cargo* in transit between seaborne and land transport, and a *bottleneck for ships* queuing at port.

The overall goal of a port being to create port throughput by attracting more cargo, it is vital to ensure operational efficiency and service quality. As an example, the Port of Aalesund defines five main priorities to achieve this goal (Rialland and Tjora, 2013).

- Set infrastructure and services at port users' disposal
- Increase efficiency through better use of IT and internet
- Keep up dialog with decision-makers and stakeholders
- Improve information exchange and dialogue with port users

Cooperation between port actors is necessary to ensure overall service performance and value-

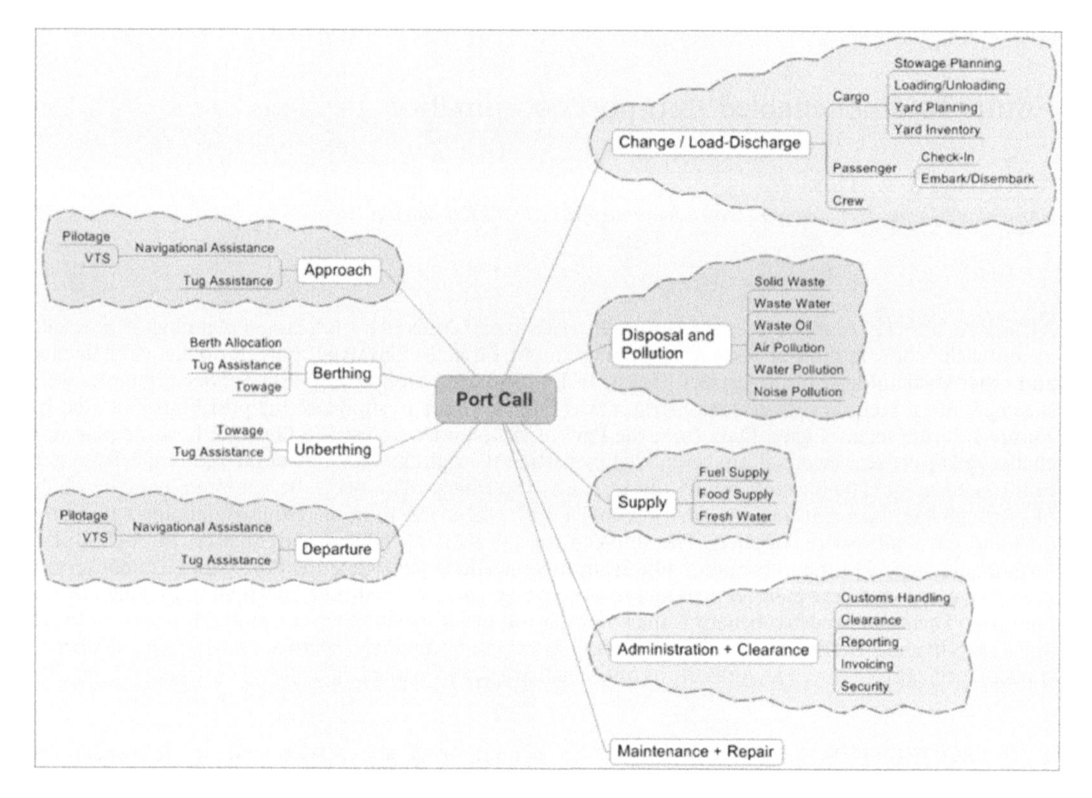


Figure 1. Port call: Main processes and ports services (Source: Mathes, 2007).

creation to port users (Vitsounis and Pallis, 2012). To support business networks at port, partnerships and common goal setting are necessary, but also a well-functioning communication framework.

### 1.2 *New technologies for business cooperation/business interoperability*

The Resource Hub is a concept utilizing the capabilities offered by Future Internet services to support information exchange between the different actors, and thus improving information quality, reliability and timeliness.

Future Internet is a general term for research activities on new architectures for the Internet for improving and extending Internet usability (Domingue et al., 2011). One initiative is the FI-PPP program hosted by EU to accelerate the development and adoption of Future Internet technologies in Europe, advance the European market for smart infrastructures, and increase the effectiveness of business processes through the Internet. Future Internet covers the concepts of Service oriented Architecture (SoA) and also Internet of Things (IoT). SoA is a software architecture where the actors within a community can send/receive and publish/subscribe to information by invoking services. These services are loosely coupled and data is transported in messages. IoT denotes the advanced connectivity of devices, systems and services that goes beyond machine-to-machine communications and which covers a variety of protocols.

Future Internet technologies and also the related concepts of cloud computing offer a unique opportunity to improve collaboration in the supply chain by removing barriers to system deployment, increasing transparency, encouraging new on-demand software applications, integrating Internet of Things with users in real time and encouraging novel business models that capture value in new and innovative ways (Arendt et al., 2012). This is done by providing availability of operations and data at anytime, anywhere, according to the goal of Future Internet.

For the supply chain, the technologies can enable and encourage novel business model that capture value in new and innovative ways, by providing availability of operation and data anytime, anywhere.

The rest of the paper is structured as follows: Section 2 gives a summary of the methodology used for development of the FInest cases, section 3 describes current practices and challenges for resource coordination in a port, section 4 gives a description of the Port Resource Hub concept, including a demonstration scenario showing the handling of deviations to the original port call plan, and section 5 describes some of the expected impacts and thoughts on future development and possibilities of the technologies discussed in the paper.

## 2 METHODOLOGY

The Resource Hub is a product of several research projects. The concept was first created in the VITSAR project (VITSAR), and further developed in the MIS project (MIS) which goal was to design a maritime information centre for increasing efficiency in information exchange. The concept served also as a pilot for the SiSaS project (SiSaS) which worked on the development of methods and tools for Software as a Service. Finally, the Resource Hub was redesigned, finetuned and put to life in a scenario in the EU-funded project FInest. As one of the pilot project for the Future Internet Puplic-Private-Partnership (FIPPP) program, the FInest project designed a Collaboration Platform based on Future Internet technologies, and a set of use case scenarios from the logistics sector to demonstrate the capabilities and potential offered by the platform.

The methodology used in the FInest project aimed specifically at developing use case scenarios for designing, experimenting and evaluating Future Internet-based services.

Although neither a business process reengineering nor a business model exercise, the work performed in FInest has enabled designing a concept for Resource Hub that is highly relevant for small and medium sized ports and their stakeholders, confirming the work previously done in the MIS project.

The utility of the Use Cases in the FInest project has been to define relevant and realistic scenarios illustrating how transport business operations could be conducted and facilitated through the help of a Future-Internet based collaboration platform. The use case team has been concerned in designing scenarios which took into account current business and technical challenges, and showed improvements in business operations compared to current practice. This resulted in five use case scenarios illustrating the interplay between IT support and business practice, one of which is the Resource Hub.

The method followed consisted of a qualitative process, with in-depth interviews and regular workshop with industrial actors, constant interaction between industrials representatives and IT developers, analysis of business processes and challenges, use of scenarios and mock-ups to simulate the concept in use.

The working process is summarized below step by step, and schematized in Figure 2.

1. *High Level Use Case Description*
   Description of business actors, business processes (vessel, terminal and port perspective) and information exchange (as-is), to establish insight into the case and input to the system design.
2. *Main Challenges*
   Description of main challenges related to collaboration and integration, experienced by the use case actors, and potential for improvement.
3. *Root-Causes*
   Root-cause of each challenge (human, technical, organizational) and main problems encountered, in order to identify areas for improvement and potential for Future Internet based support.
4. *As-Is Scenarios*
   Sketch of real-life business operation scenarios describing how the challenges are experienced and how events (critical to operations) are handled with today's practice and technology.
5. *Search for Solution*
   Formulation of specific needs for improvement, accompanied with requirements for IT support (to serve as input to the functionalities of the FInest platform).
6. *To-Be Scenarios*
   Based on solutions envisioned and the list of capabilities offered by the FInest platform, new business operation scenarios are drawn, describing how actors interact and conduct business by using the FInest Collaboration Platform.

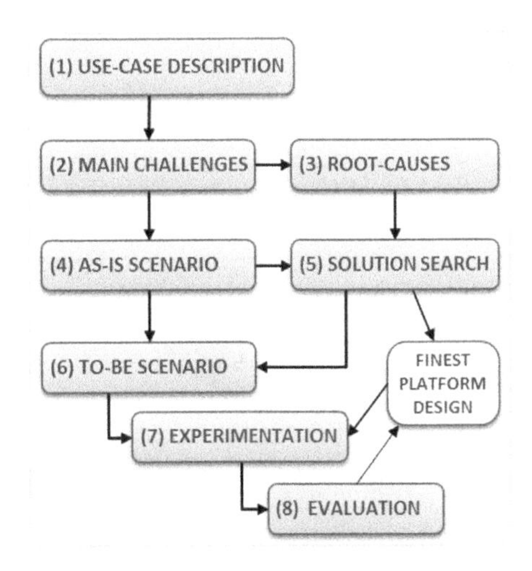


Figure 2. Working process for use cases in FInest.

Projecting business users into a virtual world is useful to ensure that the system under development is adapted to business practices and targets the relevant challenges.

7. *Experimentation Specification*
   Used for testing the technical capabilities of the system, test protocols are defined as a stepwise description of actions conducted by business users. The test aims at verifying whether the system works as promised and provides the expected support to operations.
8. *Evaluation*
   A set of KPIs measuring operational performance and based on the business requirements earlier defined, is developed to enable the assessment of the FInest platform and its contribution to operational performance.

## 3 RESOURCE COORDINATION AT PORT

Planning and execution of port call require the coordination of multiple resource and activity plans. A port has a large community of suppliers providing many different kinds of services during a port call, as well as many customers and stakeholders requiring services and information. Figure 3 is a rough illustration of the main groups of actors around a port call.

Value co-creation among all actors in the port community is complex, due to the degree and types of interdependencies (Vitsounis and Pallis, 2012), but also the degree of interaction and exchange of information, the format and timing of this information, and a multitude of IT systems which are not always communicating well (Baltzersen et al., 2009). In order to best deliver services involving several actors at the same time, seamless coordination of resources and operation planning among actors is crucial, yet challenging.

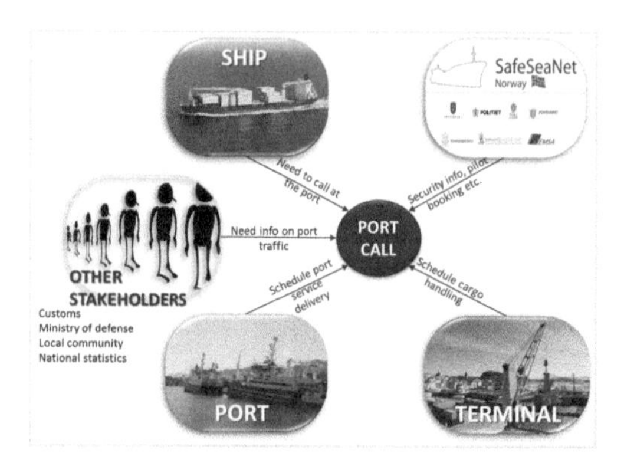


Figure 3. Port call: Actors, roles and requirements.

### 3.1 *Current practices*

By resource coordination we mean the synchronization and alignment of operational planning and re-planning activities, in order to make best use of resources available and enable efficient and effective service delivery. This operational integration requires collaboration and effective interaction among service suppliers in order to serve the ship and handle the cargo simultaneously, as well as between the port community and the ship in order to schedule the port call in a way that best fits the ship's voyage plan and doesn't disturb ongoing ship traffic.

To monitor operational planning at ports and terminals, there are already many different IT tools in use. Taking the example of the Port of Aalesund and business partners involved in the FInest project, the following support systems can be mentioned (source: FInest):

- Softship: operational planning system for cargo shipping
- ShipLog: AIS vessel tracking system
- PortIT: Port Community System enabling more data exchange via XML
- PortWin: Port Community System
- Greenwave: terminal transaction tool, supporting data exchange via XML.
- SafeSeaNet: Centralised European platform for vessel traffic monitoring and information exchange related to ships, ship movements, and hazardous cargoes.

Resource coordination remains a complex exercise, particularly at small and medium sized ports, where activity is still mainly conducted through manual work and human intervention. When ICT systems are in place, the problem often lies in the lack of interoperability (Baltzersen et al., 2009).

### 3.2 *Challenges*

Despite the huge progress in technology to support ship-port communication, ship tracking, resource planning at port, Single-Window concept etc. (Rødseth, 2011), the process of port call still represents an important area of improvement for both ship and port.

Many small and medium-sized ports, like the Port of Aalesund, do not have a strict slot-time regime for port calls and berth allocation. The most current practice for both large and small ports is that of First Come, First Served and queuing at port. In addition, little communication ship-port during voyage and no incentive for agents to talk with the port in order to adjust arrival add to the problem of high congestion at port (Rosaeg, 2010). A part from some vessel priority rules intending to regulated ship arrival (see as example the ports of Newcastle[1] and Napier[2]), no real slot allocation system has been implemented to reduce the unnecessary queuing at ports (Alvarez et al., 2010). This issue was raised by the project Synchroport (2007–2010), which designed a framework and IT system for improving cooperation among actors and synchronization of planning activity (Haga, 2010; Rotty, 2010), and designed a queuing system (and associated draft charterparty clause) to enable adjustment of ship arrival. A queue number is issued based on a Standardized Estimated Time of Arrival communicated when the ship leaves the previous port (Rosaeg, 2010).

This system, tested successfully in the Synchroport project, enables vessel to adapt speed and save fuel, but also contributes to higher port productivity through better resource management.

The project FInest share a lot of the principles raised in Synchroport, treating the issue mainly from the perspective of the port. While Synchroport focused primarily on the scheduling of ship arrival, the business case in FInest has explored the possibility to integrate internal resource coordination at port and external communication ship/port for scheduling vessel call. Furthermore, the Synchroport's main purpose was to reduce the unnecessary heavily queueing at large ports, whereas the focus in the FInest business case has been on smaller ports, not systematically struggling with queue, but definitely seeking for higher efficiency of resource utilization.

In the study of the port of Aalesund conducted in the FInest project, high number of daily calls, delays or earlier arrivals, unannounced calls, queue at port, last minute changes in ETA or ETD or booking of resources, unavailability of resources at port and terminal, delays in cargo handling, etc. were identified the main problems related to port and terminal activity planning (Rialland et al., 2012).

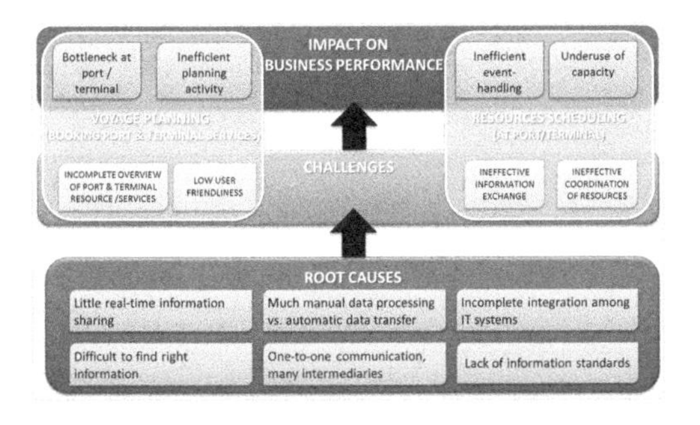


Figure 4. Challenges and root-causes related to operational planning of port call.

These problems can be mostly explained by inefficiency and unreliability in information exchange among business actors, which in turns affects both the planning of voyage for the ship, and the scheduling of resources at the port and terminal. These challenges and root-causes are summarized in Figure 4.

### 3.3 *Requirements*

In the search for potential support to be provided by Future Internet technologies, the following business and functional requirements have been described by industry representatives in the project FInest.

The goal was to explore the ways in which Future Internet technologies could support integrated planning (Ramstad et al., 2012) of activity among port & terminal service users and providers.

- Better information exchange
  - Correct/updated information, in real-time, at right time
  - Easy access to information, single information source
  - Standard format, harmonization of info
  - Higher coordination among all involved actors, less one-to-one communication
  - Automatic information exchange and processing, less manual work
- Support for resource coordination at port and terminal
  - Better capacity overview (one virtual meeting place for all actors)
  - Facilitate port call and useful use of slot-time
  - Better resource coordination and capacity utilization
- Support for event-handling
  - Immediate notification about planned port call, delays, changed bookings etc.
  - Early detection of deviations
  - All information are distributed through the portal in real-time.

## 4 THE PORT RESOURCE HUB

The port Resource Hub (Tjora et al., 2012) is a proposed tool for integrated planning for the port. It combines functionality for resource management, service generation based on multiple resources and a marketplace for service search and booking. It also supports real-time information

[1]http://www.newportcorp.com.au/site/index.cfm?display=111679.

[2]http://www.portofnapier.biz/shipping-info/berth-crane-allocation/.

sharing on resource availability among the business partners.

The main purpose for the system is allowing the users to view resource availability in real time and enable them to book services through an online interface, without the need for contacting agents and service providers by the traditional means of e-mail, fax or telephone. Bookings can also be connected to other information like the port arrival ID so relevant information from e.g. SafeSeaNet, AIS and port community systems can be fetched automatically when the booking is created.

Central to the resource hub is the resource management functionalities. The resources in this context include equipment (e.g. trucks and tools), personnel (e.g. mooring crew) and space (e.g. berths and storage). The functionality is to keep track of resource availability as well as information on dependencies between resources. The resource management must also be able to handle deviations with the resources, such as unexpected downtime for equipment or quay unavailability.

The resources are the basic "building blocks" for the port services, and it is these services that are actually offered to customers through the system. In its simplest form, a port service may be the use of a single resource, i.e. the resource itself is offered as a service. Complex port services may include resources from different stakeholders and a timing plan for the use of the resources.

Typically the providers will predefine the services they offer, typically with some room for customization by the customer. The offered services may also include cooperation with other service providers or resource owners. The services may again be combined with other services, adding timing plans for execution and further dependencies between services, in order to generate plans for a port call involving several service providers at the port.

Booking of the services is similar to booking processes in other online system. The booking process must take care of the dependencies between the resources, e.g. by a two-phase booking process. The system must also have functionality for booking updates, including adding or cancelling services or changing the timing of the booked services.

The concept and its functionalities are summarized in Figure 5.

While the described functionalities must be available through the system, the system will in most cases function as a frontend for the same functionality in other systems. Many service providers have their own systems for resource management and booking specialized for their business, the resource hub is not meant to replace these systems, but add value by information sharing and coordination

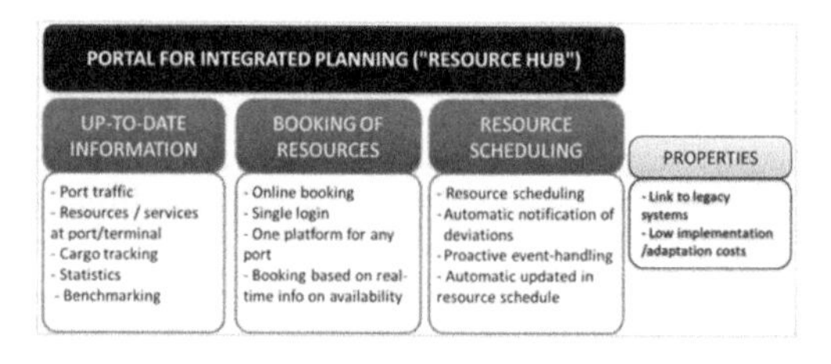


Figure 5. Functionalities and properties of the Resource Hub.

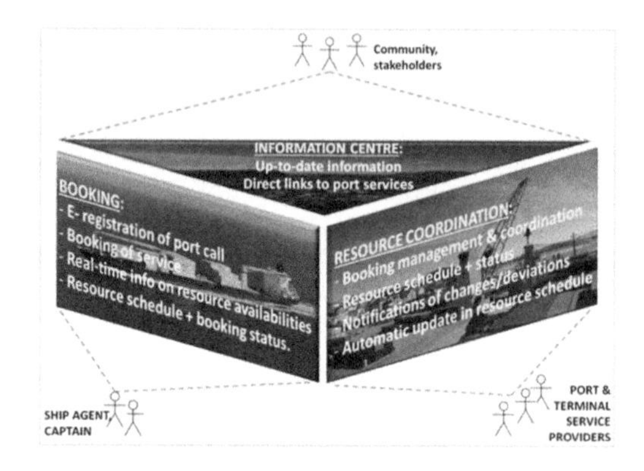


Figure 6. Main users of the Resource Hub.

between actors, as well as giving a common interface for the customer to the systems of the various service providers at the port.

The main users of the Resource Hub are the port and terminal service providers, the ship and ship agents, while the remaining port stakeholders include senders and receivers of information related to ship traffic. The value added of the Resource Hub is the possibility given to all actors to interact simultaneously through the same platform, as schematized in Figure 6.

This concept of common platform for operational planning and information sharing does not pretend to compete with advance Port Community Systems. It is rather directly for small and medium sized ports seeking to increase their productivity and resource exploitation by enabling better business collaboration and synchronization of planning activities inside the port and with the vessel.

### 4.1 *How does the Resource Hub work*

As schematized in Figure 7, the Resource Hub allows the business interaction and exchange of information through a common cloud-based platform. The platform integrates the information from existing legacy systems and other informa-

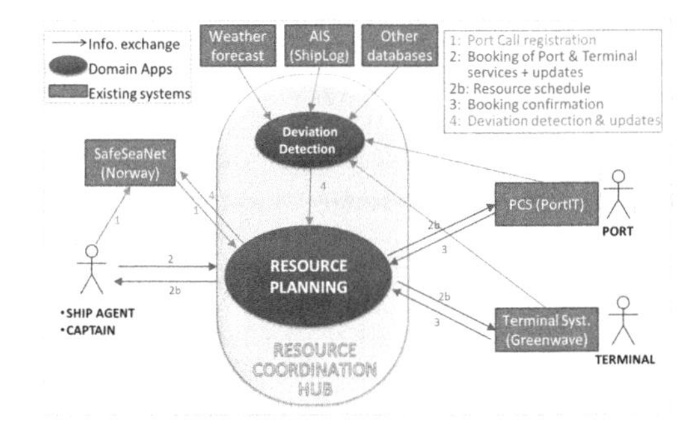


Figure 7. To-be business interaction using the Resource Hub (Rialland and Tjora, 2013).

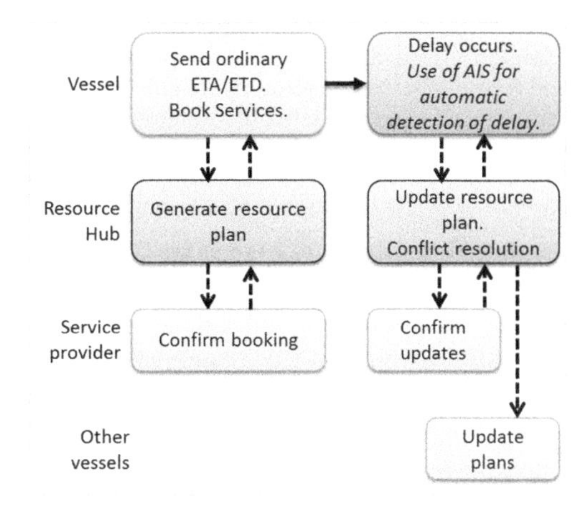


Figure 8. Resource Hub's handling of information and deviations.

tion sources. This information is analyzed by the platform's Applications forming the core of the Resource Hub, and used to support planning of resources as well as detecting potential deviations from plan and make necessary adjustments. Each user has its own interface, and uses the Resource Hub in individual manner.

### 4.2 *Scenario*

To demonstrate the concept in use, the FInest project has set up business scenarios (Rialland and Tjora, 2013) featuring, on one side, vessel representatives (vessel captain or ship agent) responsible for planning the call at each port along the route, and on the other side, a port, terminal and other services providers offering ship and cargo handling services.

The scenario describes how business operations are achieved by using the Resource Hub (FInest platform), and how typical events and deviations from plan are handled in a more effective and efficient way compare to current practices. The story, generalized and schematized is Figure 8, is the following.

Four container ships, sailing on regular routes along the Norwegian coast, will call at the Port of Aalesund on Friday. The port calls are registered in SafeSeaNet, the online portal for ship reporting, two days before arrival. This information (ETA, POD, POA, vessel name etc.) is automatically fetched by the Resource Hub and communicated to Port and Terminal operators, so that berth and time slot can be allocated.

One day before the planned arrival, the ship agent confirms ETA and ETD, reserve a quay and book services from the relevant port service providers. This is done directly through the online interface of the Resource hub, displaying resource availability in real time.

Based on information about the ships' ETA, ETD and the services required, the Resource Hub establishes a draft resource plan, taking into account the resources available at the port and terminal. This plan, which is a collective plan for port and terminal operators, is transferred to all service providers simultaneously, who can then confirm the bookings directly via the platform.

On the day of arrival, two deviations occur, which will be handled by the Resource Hub.

Deviation 1: delay in arrival causing a need for update of slot time. A vessel (ship 1) is behind schedule on its route towards Aalesund. This delay is detected 6 hours before arrival by the Resource Hub through cross-check of AIS data with the ship's schedule. A warning is immediately sent by the system to the port and terminal, together with an indication of potential resource conflicts due to likely overlapping slot times at the quay.

Furthermore, this delay triggers a re-planning process in the Resource Hub, and a new draft of collective resource plan is created and sent to the port and terminal service providers for control. Once this new resource plan is checked and confirmed, the delayed ship receives an updated slot time, as well as the next vessel in line for that quay (ship 2), so that it can adapt its speed and avoid queuing at port.

Deviation 2: change of quay for access to special ship services. 4 hours before arrival, a vessel (ship 3) notifies the ports that it needs ship repair services, which are not possible at the quay currently allocated. The ship needs to berth at another quay, already allocated to another ship at the given slot time. This conflict of triggers a re-planning process, which this time needs to be solved in an ad-hoc manner. The Resource Hub sends a warning to the service providers concerned, together with an overview of the current resource plan. Thanks to this common visualization of resource plan, the port and terminal services operators are able to re-plan the vessel calls in simultaneously. The ship (ship 3) is able to switch quay, while the other ship (ship 4), originally planned at the given quay, is assigned to another quay. Notification is sent to both vessels.

### 4.3 *Benefits*

Existing solutions already offer some of the benefits of this cloud-based concept, such as automation of business processes like online notification of call, online booking of ship and cargo handling services—like it is the case at the Port of Aalesund, automatic transfer of reports etc. Nevertheless, the main value-added of the Resource Hub is the seamless integration of all information systems, and the provision of up-to-date information which enable better informed decisions, higher operational efficiency and quick reactions in case of deviations (Rialland et al., 2012).

### 4.4 *Status of development*

The research has focused on modelling the functionalities of a Future Internet based Collaboration Platform (FInest), including a detailed mock-up of the Resource Hub. Although no prototype has been built during the FInest project, the ongoing project FIspace is working on building modules that can be used for several of the functionalities envisioned. Meanwhile, the Port of Aalesund has launched an online system for booking of services to supplement the traditional e-mail/phone/fax/radio type of port call notification.

## 5 IMPACT & FURTHER DEVELOPMENT

The research gives an example on how the Future Internet technologies may support new solutions for better interaction in the transport and logistics sector.

It is believed that the possibilities that Future Internet technologies offer may solve many of the information exchange and cooperation challenges experienced today. Especially in communities with many actors and diverse software solutions, where system integration, although technically feasible, is not always achievable, the use of Future Internet technologies may reduce the problem, allowing cross-company cooperation through cloud-based systems.

In order to make the most out of the technology available and future internet capabilities offered, a careful study of the business processes and challenges is necessary, so that the solution designed for increasing collaboration and integration among actors is well-adapted to the existing business practices and with minimal barriers to uptake.

The collaboration platform envisioned as the Port Resource Hub is a demonstration of how cloud technology can support the coordination of operations around a port call, in order to achieve higher operational efficiency and better use of resources.

In addition, as a part of the Internet of Services, the services of the Resource Hub may be used and combined with other functionalities in order to create new services, thus serving a larger part of the maritime transport value-chain.

Furthermore, while the port and its network are the main users of the Resource Hub, these kinds of solutions may be useful for other sectors with similar challenges and processes. The main results from the FInest project were proposed solutions for a wide range of transport and logistics businesses, centered around common Future Internet software components, such as transport planning, open electronic market-places, shipment tracking, proactive event handling. Concepts and technologies from FInest are further developed in the follow-up project FIspace, working on the creation of a Collaboration Platform and associated Applications.

Although the expected benefits from the resource Hub are already identified, quantifying the potential in performance improvement for each group of users is a cumbersome task. It requires a careful study of planning and operational performance along with identification of performance measurements, and a correct estimation of potential future performance given distinct contexts and scenarios.

## ACKNOWLEDGEMENTS

The research leading to these results has received funding from the European Community's Seventh Framework Programme FP7/2007–2013 under grant agreement 285598 (FInest). The authors thank the port of Aalesund for their collaboration in in setting up the resource hub demonstrator.

## REFERENCES

Alvarez J.F., Longva, T. and Engebrethsen, E.S. 2010. A methodology to assess vessel berthing and speed optimization policies, *Maritime Economics & Logistics*, (2010)12: 327–346.

Arendt, F., Meyer–Larsen, N., Müller, R. & Veenstra, W. 2012. Practical approaches towards enhanced security and visibility in international intermodal container supply chains, *International Journal of Shipping and Transport Logistics*, 4(2/2012): 182–196.

Baltzersen, P., Hagaseth, M., Kvamstad, B. & Mathes, S. 2009. *EFFORTS: Service Oriented Architecture for Interoperable Port Systems*, WCTRS—Special Interest Group 2, CRITICAL ISSUES IN THE PORT AND MARITIME SECTOR, 7–8 May 2009.
Carbone, V. & De Martino, M. 2003. The changing role of ports in supply-chain management: an empirical analysis, *Maritime Policy Management* 30(4): 305–320.
Domingue, J., Galis, A., Gavras, A., Zahariadis, T., Lambert, D., Cleary, F., Daras, P., Krco, S., Müller, H., Li, M.S., Schaffers, H., Lotz, V., Alvarez, F., Stiller, B., Karnouskos, S., Avessta, S. & Nilsson, M. (Eds.). 2011. *The Future Internet—Future Internet Assembly 2011: Achievements and Technological Promises*, Series: Lecture Notes in Computer Science, Vol. 6656, Subseries: Computer Communication Networks and Telecommunications, 1st Edition., 2011.
FInest. FInest project website, http://www.finest-ppp.eu.
FIspace. FIspace project website, http://www.fispace.eu/.
Froese, J. & Zuesongdham, P. 2008. Port Process Map, *Project EFFORTS*, Deliverable 3.1.3.
Haga, A. 2010. Kommuniserer for miljøets skyld, BA *Bergens Avisen*, January 19th 2010.
Mathes, S. 2007. Process Ontology and Process Modelling Platform, Project EFFORTS, Deliverable 3.1.2.
Meersman, H. van de Voorde, E. & Vanelslander, T. 2012. Port congestion and implications to maritime logistics. In D.-W. Song, P. Panayides (Eds.), *Maritime logistics: contemporary issues*: 49–68. Emerald Group Publishing Limited, 2012.
MIS. MIS project website, http://www.sintef.no/Projectweb/MIS/.
Port of Aalesund. *Port of Ålesund Website*: http://www.alesund.havn.no/?sc_lang=en.
Ramstad, L.S., Halvorsen, K. & Holte, E.A. 2012, Implementing Integrated Planning—Organizational enablers and capabilities. In: Rosendahl, T., & Hepsø, V. (Eds.) (2012). *Integrated Operations in the Oil and Gas Industry: Sustainability and Capability Development*. IGI-Global.
Rialland, A. & Tjora, Å. (ed.). 2013. Final Use Case Specification and Phase 2 Experimentation Plan, *Project FInest*, Deliverable D2.5.
Rialland, A., Ramstad, L., Koc, H. & van Harten, E.J. 2012. *Detailed Specification of Use Case Scenarios*, Project FInest, Deliverable D2.3.
Rosaeg, E. 2010. A System for Queuing in Ports (October 25, 2010). Available at SSRN: http://ssrn.com/abstract=1697404.
Rotty, S. 2010. *A Vision of Greener Port Operations*, APSN Meeting, Shanghai, 16 september 2010. Det Norske Veritas (DNV).
Rødseth, Ø.J. 2011. A Maritime ITS Architecture for e-Navigation and e-Maritime: Supporting Environment Friendly Ship Transport, IEEE ITSC 2011, Washington USA, 5th to 7th October 2011.
SiSaS. SiSaS Methodology Website, http://sisas.model-based.net/.
Synchroport. Synchroport project, https://grieglogistics.no/rd/synchro-port.
Tjora, Å., Solberg, A., Steinebach, C., Fjørtoft, K.E. & Hagaseth, M. 2012. A SaaS-Based Port Resource Hub, *e-Freight Conference 2012*, Delft, the Netherlands, 2012.
VITSAR. *VITSAR project website*: http://prosjekt.marintek.sintef.no/vitsar/.
Vitsounis, T.K. & Pallis, A.A. 2012. Creating Value in Seaports: Port Value Chains and the role of Interdependencies. In Song D.W. & P. Panayides (eds), *Maritime Logistics: Contemporary Issues*: 155–174. Emerald Group Publishing Limited, 2012.

*Maritime-Port Technology and Development – Ehlers et al. (Eds)*



# Ship port pre-arrival reporting and ship survey status as eMar services

E. Vanem
*DNV-GL, Strategic Research and Innovation, Høvik, Norway*

D. Yarmolenka
*DNV-GL, Ship Management Operations, Høvik, Norway*

G. Korody
*DNV-GL, Digital Channels and Collaboration, Høvik, Norway*

ABSTRACT: This paper presents two pilot services that have been developed based on the Common Reporting Schema (CRS) for the maritime industries. Both pilot services demonstrate how existing services may be enhanced if they can be offered as XML Web services according to a standardized schema. The information provided by the web services can then be consumed by other services and also interpreted and processed programmatically in order to facilitate improved information exchange and communication between ship and shore and between different stakeholders within maritime transportation. The two basic pilot services presented in this paper are simplified ship to shore pre-arrival reporting and provision of ship certificate and survey status. It is noted that for both pilot services, further work may extend the functionalities of the pilot services, but it is demonstrated that the definitions in the CRS are sufficient for exchanging the necessary information.

## 1 INTRODUCTION AND BACKGROUND

The e-maritime initiative can be considered as an extension of the e-Navigation concept currently being discussed at the International Maritime Organization (IMO) that promotes the use of all maritime data and information, distributed by way of information and communication technologies, to facilitate maritime transportation and provide value added services.

eMar is an EU-funded research project supporting the e-Maritime initiative with the aim of empowering the European maritime sector with an eMar ecosystem of fully integrated services over an upgraded information management infrastructure. A recent survey of services supporting interaction with class, Safety, Security and Environmental risk management systems identified two potential eMar services; pre-arrival reporting from ship to shore and ship survey status. These services have been implemented as pilot eMar services and will be outlined in this paper. Both these pilot services are developed as eMar XML web services and are implemented by adopting the Common Reporting Schema (CRS).

The pre-arrival reporting service aims at further simplifying port clearance procedures for arriving and departing ships and to alleviate the administrative burden on the captain on the bridge, and may communicate directly with Authorities through national single windows. Using this service, the approaching ship may receive immediate confirmation that the arrival report is successfully received on shore or otherwise be notified. This may result in smoother port clearance procedures, reduce delays and possible penalties during port calls and ensure overall safer navigation.

The ship survey status service will provide basic ship information together with status of class and statutory certificates and class and statutory surveys as XML web services that may be interpreted programmatically or consumed by other services.

### 1.1 *e-Navigation*

The concept of e-Navigation was initially introduced by the International Maritime Organization (IMO) to increase safety and security in commercial shipping by way of improved organization of data and improved data exchange and communication between ship and shore. The scope of e-Navigation as defined by the IMO and formulated by IALA (International Association of Lighthouse Authorities) is *the harmonized collection, integration, exchange, presentation and analysis of marine information onboard and ashore by electronic means to enhance berth to berth navigation and related services for safety and security at sea and for protection of the marine environment*. Hence, the strategic vision of e-Navigation is related to the utilization of existing and new navigational tools, i.e. electronic ones such as ECDIS (Vanem et al. 2008), in a holistic and systematic way.

One main motivation for the initiative was various studies that indicated that the combination of navigational errors and human failures was increasingly contributing to ship accidents. A number of ship- and shore-based technologies have been developed and are currently available to support e-Navigation, including AIS (Automatic Identification System), ECDIS (Electronic Chart Display and Information system), IBS/INS (Integrated Bridge Systems/Integrated Navigation System), ARPA (Automatic Radar Plotting Aids), radio navigation, LRIT (Long Range Identification and Tracking) systems, VTS (Vessel Traffic Services) and the GMDSS (Global Maritime Distress Safety System). It is believed that such technologies may contribute to significantly reduce navigational errors and failures, and simultaneously reduce the burden on the navigator, but that they need to be co-ordinated and integrated in order to reap the full benefits of such technologies. Hence, e-Navigation aims at integrating existing and new navigational tools, in particular electric tools in an all-embracing system that will contribute to enhanced navigational safety (Kystverket 2012).

Even though e-Navigation is essentially meant to improve safety and security aspects of maritime transportation, it also has the potential to increase efficiency of ship operations, a main concern for ship-owners, operators and their service providers. The Marine Electronic Highway system in the straits of Malacca and Singapore is a trial project where possible e-Navigation solutions can be tried out (Marlow & Gardner 2006; Dahalan et al. 2013).

An extensive list of possible e-Navigation solutions was proposed in IMO (2012), and it was found necessary to focus on a limited number of generalized and prioritized solutions. Hence, the correspondence group on e-Navigation to NAV 59 prioritized the following five main potential e-navigation solutions (IMO 2013):

- Improved, harmonized and user-friendly bridge design
- Means for standardized and automated reporting
- Improved reliability, resilience and integrity of bridge equipment and navigation information
- Integration and presentation of available information in graphical displays received vie communication equipment
- Improved communication of VTS service portfolio

These solutions have been subject to risk and cost-effectiveness assessment according to the Formal Safety Assessment (FSA) which has been used as basis for proposing various risk control options for implementation at IMO.

Central in the e-Navigation concept is reliable and automated communication and information exchange, which rely on a Common Maritime Data Structure (CMDS). When relevant, it will make use of the International Hydrographic Organization (IHO) standard for hydrographic data (IHO 2010, Ward and Greenslade 2011).

### 1.2 *The e-Maritime strategic framework*

e-Navigation was not intended to cover commercial interests, but commercial users may expect capacity and information services to improve operational efficiency of shipping beyond what is needed for improving the safety and security of navigation. Such a system for adding maritime commercial interests to e-Navigation is often referred to as e-Maritime. e-Maritime can thus be construed to include e-Navigation and promotes the use of all available maritime data and information to facilitate maritime transport and provide value added services. Hence, e-Maritime aims at reducing the cost for shipping and coastal states and at the same time providing benefits to the commercial shipping industry.

The e-Maritime Strategic Framework (EMSF) is a target model for maritime transport pertaining to common industry interests and business benefits to be realized in short or long term. It includes a description of processes, actors, rules, information flows and other domain entities. In particular, it will describe information exchange requirements for different user communities and explain how these can be achieved through appropriate processes, standards and policies. The specification of the e-Maritime Strategic Framework is addressed in the eMar deliverable D1.3 (Cane et al. 2013). The EU-funded research project eMar was initiated to support the development and specification of the e-Maritime Strategic Framework.

### 1.3 *The eMar ecosystem*

The eMar ecosystem is a platform that utilizes the e-Maritime Strategic and supports the execution of various e-Maritime applications built up from elementary e-Maritime services. It should support the implementation of the e-Maritime Strategic Framework, utilizing existing technologies and components. It should also provide a repository for e-Maritime application and services, a runtime environment that supports the operation and interaction with the e-Maritime applications and a software development environment for producing additional applications and to integrate with existing ones. The eMar architecture and base software platform are outlined in the eMar deliverable D2.1 (Katsoulakos et al. 2013).

A Service-Oriented Architecture (SOA) will be adopted to support functional reusability and independence of technology. An e-Maritime service is an elementary piece of software that may be used as building blocks for e-Maritime applications. It is interoperable software which translates directly from the e-Maritime Framework.

The eMar ecosystem should also provide connectivity and security services to support different categories of users and application developers. Furthermore, shared information and knowledge will be provided and semantic services will be developed within the eMar project to give all actors a common understanding of relevant concepts, processes and object within the e-Maritime system. This includes an e-Maritime ontology with the formal terminology of all information exchange between all stakeholders as specified by the e-Maritime Strategic Framework, support for semantic annotation of e-Maritime web services, support for run-time interoperability among different platform users and maritime transport specific interface-based mechanisms for automated discovery and integration of suitable services.

The e-Maritime applications consist of software components and data feeds that perform meaningful functions for the actors in maritime transportation. Reference e-Maritime applications will be developed for improved business, supporting port operations, ship operations and transport logistics, and for interfacing with administrations and regulatory systems.

### 1.4 *The Common Reporting Schema*

In order to support semantic interoperability in information exchange, there is a need for agreeing on data types and code lists. Hence, the development of a Common Reporting Schema (CRS) for the maritime industries was initiated in the EU-project e-Freight and is being continued and extended in the eMar project. Initially, the focus was on promoting a standard framework for freight information exchange, and important components are a single European transport document for carriage of goods, a single window concept for all forms of reporting to all authorities and the e-Freight connectivity infrastructure. A Common Reporting Schema (CRS) was developed, and it was strongly recommended that this CRS should be accepted as standard interface for the national single windows. Standardization is deemed important for industry take-up of the framework (Vayou et al. 2012). However, even though e-Freight is Eurocentric, the inherent international nature of shipping makes global standardization an important success factor and it is recommended to aim for a common standard in the international communities.

The common reporting schema is a single, standardized, electronic reporting document that aims at including all information fields that are necessary and sufficient for reporting to Authorities in all member states. It defines the structure and content of information that must be reported. It has been further extended to include elements other than freight-related information, something which is currently being continuously investigated within eMar.

## 2 e-Maritime SOLUTIONS FOR COOPERATION IN SAFETY, SECURITY AND ENVIRONMENTAL PROTECTION

The SafeSeaNet (SSN) and the National Single Windows (NSW) are important initiatives for the development of the e-Maritime strategic framework. These concepts will be briefly outlined in the following, together with a recent survey of other potential e-Maritime services related to safety, security and environmental protection.

### 2.1 *SafeSeaNet*

The SafeSeaNet (SSN) is important for the development of the e-Maritime vision. SSN is a vessel traffic monitoring and information system established as a centralized European platform for maritime data exchange between maritime authorities across Europe. Maritime authorities may provide and receive information of ships, ship movements and hazardous cargo through the SSN, where information are collected from AIS-based position reports and notification messages sent to designated authorities in participating countries. The overall objective is to enhance maritime safety, port and maritime security, marine environmental protection and the efficiency of maritime traffic and maritime transport.

The SSN architecture includes a European Index Server (EIS) which acts as a secure and reliable index system within a network where users can provide and/or request data. The EIS is able to locate and retrieve information on vessels related to one member state upon a request made by another. The main notification reports submitted to SSN are Ship Notification (ships' voyage and cargo information), Port Notification (ships bound for a specific port), Hazmat Notification (ships carrying hazardous materials) and Incident Report (information on a specific incident). Users may exchange messages through two different interfaces, i.e. an XML-based interface allowing applications of member states to communicate programmatically with SSN (automatically between systems) and a browser-based web interface that enables users to visualize the information stored in EIS.

### 2.2 *National Single Windows*

The National Single Window (NSW) is a means to provide operators with a single point of contact for all reporting requirements relating to vessel movement and cargo. It should ensure that the relevant information is transmitted automatically to the various national authorities such as the SSN, e-Customs etc. National Single Windows by different nations are being developed with different national approaches and is influenced by developments in SSN, e-Customs and output from research projects. For example, e-Freight is developing a National Single Window for Latvia as well as a common reporting interface to link businesses to NSWs and this will serve as a baseline for the work in eMar.

A Maritime Single Window (MSW) may not be restricted to be national, and it is foreseen that such single windows may be implemented at EU or international level in order to increase the efficiency of maritime transportation. A guide and checklist for the development of EU Maritime Single Windows can be found in Katsoulakos (2013). According to the EU directive 2010/65/EU—the Reporting Formalities Directive (EU 2010)—Eu Member States are required to accept reporting formalities in electronic format, transmitted via a single window no later than June 1 2015.

### 2.3 *Other potential services*

A recent survey of potential e-Maritime services related to classification and to support interactions with class, safety, security and environmental protection systems identified a number of potential services that could be enhanced by being offered as eMar services within the eMar ecosystem. Some of the services that were identified are related to emergency response services, various environmental monitoring services, voyage optimization services, security services related to the observation and protection of critical maritime infrastructure, navigational risk management services, e-Class services, condition monitoring services and education and training services (Vanem 2014). All these types of services have in common a substantial amount of information exchange between different actors within maritime transportation, which could be optimized by the e-Maritime Strategic Framework.

Among the identified services, two particular services were selected for pilot implementation within the eMar project. These are services for simplified ship-to-shore reporting and for providing up-to-date information about the status of statutory and class certificates and surveys. These services as well as the implementation of them as eMar pilot services will be outlined in the following.

## 3 THE SHIP TO PORT PRE-ARRIVAL REPORTING SERVICE

Every ship arriving at a European port is required to submit information in various forms to the relevant authorities in the member state, according to international and national legislations (see e.g. IMO 2011, EU 2010). Some required forms are related to general declarations, cargo declaration, crew and passenger lists, dangerous goods and declaration of health. Security information with ship particulars and list of previous calls at port facilities should also be submitted. It is out of scope to list all reporting requirements when arriving at different ports, but it is emphasized that all this reporting can be a laborious task which represents a distraction for the mariners on the bridge. Hence, services that simplify ship pre-arrival reporting and alleviate the reporting burden can give more time for the mariners to navigate the vessel and have the potential to improve navigational safety. The Navigator Port service is such a service which is currently being offered to the maritime industries and which can be further improved by being offered as an eMar web service. The current service will be briefly outlined in the following, and a pilot implementation of this service as an eMar XML web service will then be presented.

### 3.1 *The Navigator Port service*

The Navigator Port service is a software service aiding the mariners with information exchange between ship and shore. It simplifies the port clearance procedures and helps ensure that the crew on board a ship has the right information whenever it is needed.

The port manager is the core of the service, providing vessels with timely and correct information and forms according to the Port State Authorities' requirements. The main aim is to reduce paperwork on the bridge, simplify and speed up the ship to shore reporting, ensure smoother port clearance procedures, reduce delays and possible penalties during port calls and ultimately to ensure safer navigation by allowing the master to focus on his main navigational tasks. The information is structured and combined in a way to make it easily accessible through port entry and departure checklists.

More than 1600 port clearance forms are currently available, covering 11 000 ports, terminals and pilot stations in more than 180 countries. These are programmatically linked to the logs and the databases of the vessel. In this way, pre-filled forms with the relevant data automatically appear as soon as they are opened, reducing the time needed for preparation of such reports to a minimum. Included are inter alia eNOAD reporting to the US Coast Guard as well as electronic reporting to many other countries.

In addition to the port state reports, the Navigator Port report generator may generate company specific forms, environmental reporting forms and daily reports meeting requirements from the charterers. Some examples of generated reports are shown in Figure 1. Standard xml-based data interfaces are used for data sharing and integration with other software systems.

The core service is supplemented with different optional modules. The passage planner module is a tool designed to assist the navigation officer in the voyage-planning process. The work and rest hours module is a tool for registering planned and executed work hours and provides a cost-effective way of demonstrating compliance with relevant regulations on seafarers' rest hours. The fleet manager module facilitates data exchange between vessels via the office and distribution of information from the office to the vessels. It consists of two parts, one for the vessel and one for the office. Data is easily exported to Excel for transfer to the office and all ship data is stored and maintained in the office for use in planning for individual ships or the fleet as a whole.

Depending on requirements defined by the authorities or other parties, completed forms and reports may be submitted electronically in different

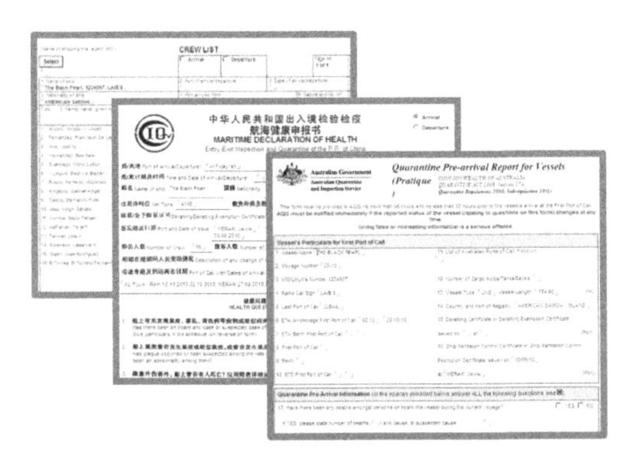
CREW LIST

中华人民共和国出入境检验检疫
航海健康申报书
MARITIME DECLARATION OF HEALTH

Quarantine Pre-arrival Report for Vessels

Figure 1. Some reports generated by the Navigator Port service.

formats. If Internet is available, secure online data submission is possible. Otherwise the electronic files may be submitted by e-mail. Typically, the required forms are submitted by e-mail, and confirmation on successful reception of the forms is obtained by calling the relevant authorities on the phone after some time. However, by implementing this service as an eMar web service, utilizing the CRS, the pre-arrival reporting can be simplified even further and the ship may get immediate confirmation about the success of the submission.

### 3.2 *Navigator Port as an eMar XML web service*

In order to allow the ship to report pre-arrival information to the relevant Port Authorities, the data must first be entered into the Navigator Port. Then, when the user is initiating a pre-arrival report, the data is automatically formatted according to the CRS specification and submitted via a web service to a common reporting gateway that is hosted remotely. A confirmation message will immediately be received and shown to the user indicating that the ship notification has been successfully received and processed at the reporting gateway. If, on the other hand, the submission is incomplete or is not successful for any reason, an error message should be received indicating what went wrong. This way the Navigator Port eMar pilot acts as a web service consumer, submitting onboard data to the onshore server hosting the web service.

In the initial pilot version of the service, only information regarding the vessel ID, i.e. the IMO number, the vessel name, the port of departure, the port of arrival and the estimated time of arrival is submitted. This information is already available in the system and it has been demonstrated that it can successfully be transferred from the Navigator Port client to a remotely hosted testbed, with an immediate confirmation of success. Furthermore, useful error messages are received if the forms are incorrectly filled in or some necessary information is missing. For example, omitting the vessel name gives the error message "Error: Datafield (operand 1) "VesselName" cannot be recognized, or is empty".

To submit an eMar message to the reporting gateway from the Navigator Port eMar pilot, simply update the logs with the vessel data and the voyage history and press the "submit to eMAR" button from the voyage history view. The user interface for doing this is shown in Figure 2, with a confirmation of successful submission. Verification about the successful submission of the information can be obtained by logging into the common reporting gateway at the testbed, where information about recent ship notifications can be seen, as shown in Figure 3.

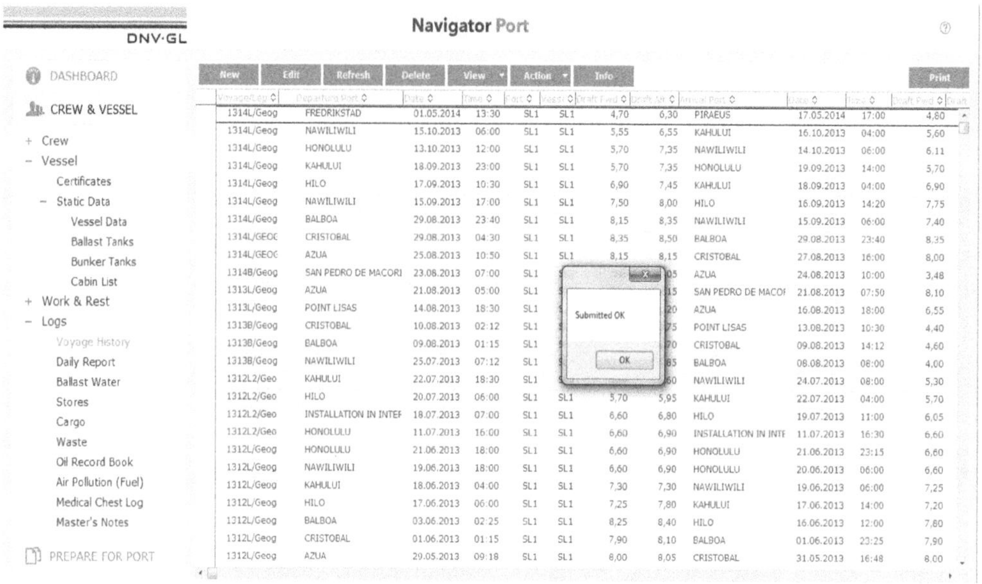

DNV·GL

**Navigator Port**

DASHBOARD

CREW & VESSEL

+ Crew
- Vessel
  - Certificates
  - Static Data
    - Vessel Data
    - Ballast Tanks
    - Bunker Tanks
    - Cabin List
+ Work & Rest
- Logs
  - Voyage History
  - Daily Report
  - Ballast Water
  - Stores
  - Cargo
  - Waste
  - Oil Record Book
  - Air Pollution (Fuel)
  - Medical Chest Log
  - Master's Notes

PREPARE FOR PORT

New | Edit | Refresh | Delete | View | Action | Info | Print

| Voyage/Leg | Departure Port | Date | Time | Port | Vessel | Draft Fwd | Draft Aft | Arrival Port | Date | Time | Draft Fwd |
|---|---|---|---|---|---|---|---|---|---|---|---|
| 1314L/Geog | FREDRIKSTAD | 01.05.2014 | 13:30 | SL1 | SL1 | 4,70 | 6,30 | PIRAEUS | 17.05.2014 | 17:00 | 4,80 |
| 1314L/Geog | NAWILIWILI | 15.10.2013 | 06:00 | SL1 | SL1 | 5,55 | 6,55 | KAHULUI | 16.10.2013 | 04:00 | 5,60 |
| 1314L/Geog | HONOLULU | 13.10.2013 | 12:00 | SL1 | SL1 | 5,70 | 7,35 | NAWILIWILI | 14.10.2013 | 06:00 | 6.11 |
| 1314L/Geog | KAHULUI | 18.09.2013 | 23:00 | SL1 | SL1 | 5,70 | 7,35 | HONOLULU | 19.09.2013 | 14:00 | 5,70 |
| 1314L/Geog | HILO | 17.09.2013 | 10:30 | SL1 | SL1 | 6,90 | 7,45 | KAHULUI | 18.09.2013 | 04:00 | 6,90 |
| 1314L/Geog | NAWILIWILI | 15.09.2013 | 17:00 | SL1 | SL1 | 7,50 | 8,00 | HILO | 16.09.2013 | 14:20 | 7,75 |
| 1314L/Geog | BALBOA | 29.08.2013 | 23:40 | SL1 | SL1 | 8,15 | 8,35 | NAWILIWILI | 15.09.2013 | 06:00 | 7,40 |
| 1314L/GEOC | CRISTOBAL | 29.08.2013 | 04:30 | SL1 | SL1 | 8,35 | 8,50 | BALBOA | 29.08.2013 | 23:40 | 8,35 |
| 1314L/GEOC | AZUA | 25.08.2013 | 10:50 | SL1 | SL1 | 8,15 | 8,15 | CRISTOBAL | 27.08.2013 | 16:00 | 8,00 |
| 1314B/Geog | SAN PEDRO DE MACORI | 23.08.2013 | 07:00 | SL1 | [illegible] | [illegible] | [illegible]05 | AZUA | 24.08.2013 | 10:00 | 3,48 |
| 1313L/Geog | AZUA | 21.08.2013 | 05:00 | SL1 | [illegible] | [illegible] | [illegible]15 | SAN PEDRO DE MACOI | 21.08.2013 | 07:50 | 8,10 |
| 1313L/Geog | POINT LISAS | 14.08.2013 | 18:30 | SL1 | [illegible] | [illegible] | [illegible]20 | AZUA | 16.08.2013 | 18:00 | 6,55 |
| 1313B/Geog | CRISTOBAL | 10.08.2013 | 02:12 | SL1 | [illegible] | [illegible] | [illegible]75 | POINT LISAS | 13.08.2013 | 10:30 | 4,40 |
| 1313B/Geog | BALBOA | 09.08.2013 | 01:15 | SL1 | [illegible] | [illegible] | [illegible]70 | CRISTOBAL | 09.08.2013 | 14:12 | 4,60 |
| 1313B/Geog | NAWILIWILI | 25.07.2013 | 07:12 | SL1 | [illegible] | [illegible] | [illegible]85 | BALBOA | 08.08.2013 | 08:00 | 4,00 |
| 1312L2/Geo | KAHULUI | 22.07.2013 | 18:30 | SL1 | [illegible] | [illegible] | [illegible]60 | NAWILIWILI | 24.07.2013 | 08:00 | 5,30 |
| 1312L2/Geo | HILO | 20.07.2013 | 06:00 | SL1 | SL1 | 5,70 | 5,95 | KAHULUI | 22.07.2013 | 04:00 | 5,70 |
| 1312L2/Geo | INSTALLATION IN INTEF | 18.07.2013 | 07:00 | SL1 | SL1 | 6,60 | 6,80 | HILO | 19.07.2013 | 11:00 | 6,05 |
| 1312L2/Geo | HONOLULU | 11.07.2013 | 16:00 | SL1 | SL1 | 6,60 | 6,90 | INSTALLATION IN INTE | 11.07.2013 | 16:30 | 6,60 |
| 1312L/Geog | HONOLULU | 21.06.2013 | 18:00 | SL1 | SL1 | 6,60 | 6,90 | HONOLULU | 21.06.2013 | 23:15 | 6,60 |
| 1312L/Geog | NAWILIWILI | 19.06.2013 | 18:00 | SL1 | SL1 | 6,60 | 6,90 | HONOLULU | 20.06.2013 | 06:00 | 6,60 |
| 1312L/Geog | KAHULUI | 18.06.2013 | 04:00 | SL1 | SL1 | 7,30 | 7,30 | NAWILIWILI | 19.06.2013 | 06:00 | 7,25 |
| 1312L/Geog | HILO | 17.06.2013 | 06:00 | SL1 | SL1 | 7,25 | 7,80 | KAHULUI | 17.06.2013 | 14:00 | 7,20 |
| 1312L/Geog | BALBOA | 03.06.2013 | 02:25 | SL1 | SL1 | 8,25 | 8,40 | HILO | 16.06.2013 | 12:00 | 7,80 |
| 1312L/Geog | CRISTOBAL | 01.06.2013 | 01:15 | SL1 | SL1 | 7,90 | 8,10 | BALBOA | 01.06.2013 | 23:25 | 7,90 |
| 1312L/Geog | AZUA | 29.05.2013 | 09:18 | SL1 | SL1 | 8,00 | 8,05 | CRISTOBAL | 31.05.2013 | 16:48 | 8.00 |

Submitted OK

OK

Figure 2. User interface for submitting a pre-arrival report as an eMar XML web service according to the CRS specifications.

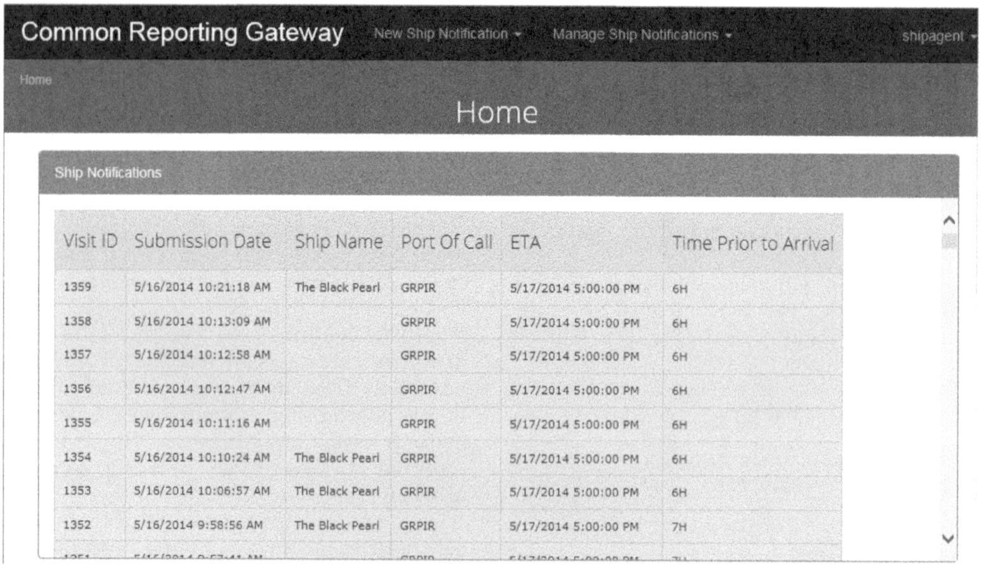

**Common Reporting Gateway** New Ship Notification ▾ Manage Ship Notifications ▾ shipagent ▾

Home

## Home

Ship Notifications

| Visit ID | Submission Date | Ship Name | Port Of Call | ETA | Time Prior to Arrival |
|---|---|---|---|---|---|
| 1359 | 5/16/2014 10:21:18 AM | The Black Pearl | GRPIR | 5/17/2014 5:00:00 PM | 6H |
| 1358 | 5/16/2014 10:13:09 AM | | GRPIR | 5/17/2014 5:00:00 PM | 6H |
| 1357 | 5/16/2014 10:12:58 AM | | GRPIR | 5/17/2014 5:00:00 PM | 6H |
| 1356 | 5/16/2014 10:12:47 AM | | GRPIR | 5/17/2014 5:00:00 PM | 6H |
| 1355 | 5/16/2014 10:11:16 AM | | GRPIR | 5/17/2014 5:00:00 PM | 6H |
| 1354 | 5/16/2014 10:10:24 AM | The Black Pearl | GRPIR | 5/17/2014 5:00:00 PM | 6H |
| 1353 | 5/16/2014 10:06:57 AM | The Black Pearl | GRPIR | 5/17/2014 5:00:00 PM | 6H |
| 1352 | 5/16/2014 9:58:56 AM | The Black Pearl | GRPIR | 5/17/2014 5:00:00 PM | 7H |

Figure 3. Logging into the Common Reporting Gateway verifies that the submission has been successfully received.

The ship pre-arrival eMar service requires a broadband connection between the ship and the shore, and can simplify further the ship to shore reporting procedures on board compared to the current service.

### 3.3 *Potential for further development*

Currently, only a limited set of functionalities are included in the eMar Navigator Port pilot service and only a limited subset of the available information is submitted. This provides a proof of the concept and demonstrates that the ship pre-arrival reports can successfully be submitted as an eMar web service adopting the CRS, provided an internet connection is available onboard the ship. However, the pilot service can easily be updated to handle more information and more functionality.

Some plans for further work are to extend the service to include other information that is readily available on the ship's logs, such as crew information, cargo information and recent port calls. Furthermore, the service can be extended with other functionalities such as daily reports, environmental reporting, etc. Furthermore, it should be integrated with actual systems of coastal administrations that could process the information received and provide further feedback and confirmations if all necessary information is successfully received by all relevant authorities.

## 4 THE SHIP SURVEY STATUS SERVICE

The status of statutory and class certificates and surveys for a particular vessel can be of interest to different stakeholders for many reasons. Currently, such and other information are available through the web-based tool DNV Exchange. However, if the information can be made available via an eMar XML web service, the information can be consumed by other web-services and interpreted programmatically to yield value added services to the maritime community. In the following, the web-based tool for ship survey and certificate status will be briefly outlined and the pilot implementation of the eMar web service will be presented.

### 4.1 *DNV Exchange and ship survey status information*

The DNV Exchange web tool provides an interface for clients and other parties to obtain information about their business relationship with class as well as other general information from, from anywhere and at any time. Different subscription types yield different service levels, contents and features.

The basic service gives secure, web-based access to class survey certificate and general class reference documentation. It features access to reports detailing certificates, periodic surveys, any conditions of class, memos to owner, retroactive requirements, continuous machinery and continuous hull surveys at any time. It also provides online survey requests and graphical survey scheduling tools as well as updated rules and standards and general casualty information.

Additional services may be included depending on subscription level, such as electronic scanned vessel certificates, additional vessel documentation, port state control checklists and guidance material. The amount of information that is available is different for owners and managers, flag and port state authorities, ship yards, etc. The owner may allow other parties to access the information pertaining to their vessels, as required. However, some of the information is publically available in the register of vessels. This is the vessel particulars and general information as well as certificates and survey lists and possible overdue certificates and surveys.

The vessel info in the DNV Exchange web tool is searchable by vessel name, class ID, IMO number, Builder's number and flag and contains the following information: Vessel summary, dimensions, classification, registry, hull summary, machinery summary and yard owner. A screenshot of this information for an arbitrary anonymised ship is shown in Figure 4.

The class and statutory certificates status in the DNV Exchange web tool lists all relevant certificates and green, yellow and red flags, respectively, indicates whether the certificates are valid (green flag), due within the next few months (yellow flag) or overdue (red flags). Similar views for the class and statutory survey status are publically available in the DNV Exchange web-tool. An example of a screenshot showing the class and statutory survey status is given in Figure 5. For this arbitrary vessel, there are no red flags indicating that there are no overdue surveys, but green and yellow flags indicate which surveys that will be due shortly.

### 4.2 *Ship survey status as an eMar XML web service*

The information currently available through the DNV Exchange web-tool is displayed on the screen and needs to be interpreted by a human being. However, if the information can be made available in XML format as a web-service, it can be used programmatically by other web services in order to provide value added services, possibly in combination with other information. With this in mind, the vessel info and the ship certificate and survey status have been implemented as a pilot eMar web service using the specifications of the Common Reporting Schema (CRS). Currently, a first version of this service is available, containing the public information from DNV Exchange, i.e. the vessel info and certificate and survey status for class and statutory certificates and surveys. Information is searchable by, inter alia, vessel name class ID and IMO number and the first version of the pilot service can be accessed at https://exchange.dnv.com/exchange/VesselInfoWS.asmx?WSDL.

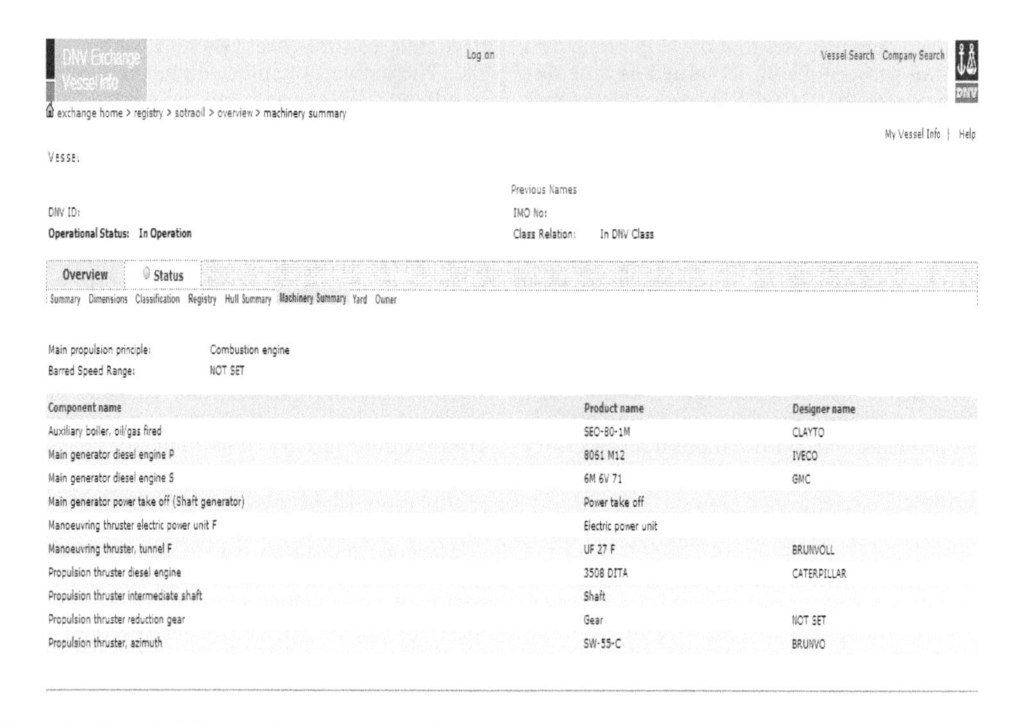


Figure 4. Vessel info in the web-based DNV Exchange tool.

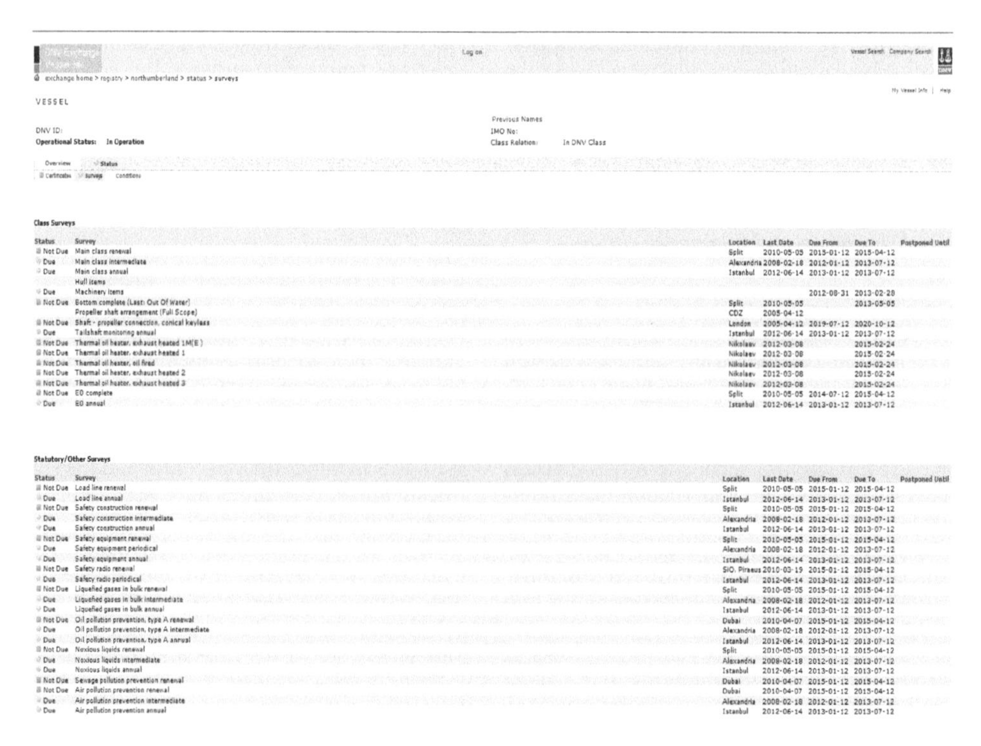


Figure 5. Class and statutory survey status in the web-based DNV Exchange tool.

### 4.3 *Possible further extensions*

One possible extension of the ship survey status as an eMar XML web service is to include other, non-public information. This would require additional security services related to the protection and authentication of the information and identification management for authorization of access to the information. Furthermore, it is possible to extend the search criteria options to be used and to allow for aggregated information at fleet or company level. Finally, the service could be updated with more robust error handling and more informative error messages, for example if some of the search criteria are not found or is inconsistent. For now, however, this is left for possible future work and further development of the service.

## 5 DISCUSSION AND RECOMMENDATIONS

The two eMar pilot services that have been implemented demonstrates that the current version of the Common Reporting Schema is quite adequate for facilitating improved information exchange between different stakeholders within the maritime industries. Indeed, the XML specifications were sufficient for exchanging all the information in the two implemented services, containing information needed for ship pre-arrival reporting and the status of certificates and class, without the need for any amendments.

Whenever the services will be extended to include even more functionality and information, it needs to be investigated whether there is a need to extend the specifications in the CRS further. Even if this should be needed, it should be rather straightforward to do so, and it is believed that the CRS is a promising candidate for international standardization within the maritime industries.

## 6 SUMMARY AND CONCLUSIONS

This paper has presented two pilot eMar services that have recently been developed to facilitate easier information exchange within the maritime industries. These are a ship to shore pre-arrival reporting service and a ship survey status service. Both services utilize the Common Reporting Schema, which is found to serve its purpose in both the applications, to offer the services as XML web services.

For the pre-arrival reporting service, it is believed that the service may significantly simplify the port clearance procedures on the bridge of a ship, allowing the mariners more time to focus on the safe navigation of the vessel and hence to improved navigational safety. However, a prerequisite for using this service is reliable broadband connections between the ship and shore, which is currently of limited availability. However, it is foreseen that this is something that will improve significantly in the years to come.

The ship survey status service provides information that can be interpreted programmatically and that might be consumed by other web services in order to provide other value-added services, possibly in combination with other information from other web services, to the maritime industries.


## ACKNOWLEDGEMENTS

The work presented in this paper has been carried out with support from the EU-funded FP7 DG MOVE research project eMar: e-Maritime Strategic Framework and Simulation based Validation; grant agreement no 265851. The project website is available at http://www.emarproject.eu/.


## REFERENCES


Cane, T., Katsoulakos, T., Lambrou, M. & Pedersen, J.T. 2013. e-Maritime Strategic Framework Specification Version 1. eMar report D1.3.

Dahalan, W.S.A.W., Zainol, Z.A., Hassim, J. & Ting, C.H. 2013. e-Navigation in the Straits of Malacca and Singapore. International Journal of Computer Theory and Engineering 5(3): 388–390.

EU 2010. Directive 2010/65/EU of the European Parliament and of the Council. Official Journal of the European Union.

IHO 2010. S-100—Universal Hydrographic Data Model, Edition 1.0.0—January 2010. International Hydrographic Bureau, Monaco.

IMO 2011. FAL convention 2011 edition. International Maritime Organization, London.

IMO 2012. Development of en e-Navigation strategy implementation plan. NAV 58/6, submitted by Norway.

IMO 2013. Development of an e-Navigation strategy implementation plan. NAV 59/6, submitted by Norway.

Katsoulakos, T. 2013. EU Maritime Single Window Development Guide and Check-list. eMAR White Paper MSW 1, ver. 1.1.

Katsoulakos, T., Lambrou, M., Zorgis, Y., Cane, T., Theodossiou, D. & Christofi, S. 2013. eMar Architecture and Base Software Platform. eMar report D2.1.

Kystverker 2012. e-navigation—enhanced safety of navigation and efficiency of shipping. Norwegian Coastal Administration (Kystverket). Online (accessed May 15 2014): http://www.kystverket.no/Documents/e-navigation/e-nav%20folder%202012.pdf.

Marlov, P.B. & Gardner, B.M. 2006. The marine electronic highway in the Straits of Malacca and Singapore—an assessment of costs and key benefits. Marine Policy & Management 33(2): 187–202.
Vanem, E. 2014. Possible e-Maritime applications for improved safety, security and environmental protection in maritime transport. In Proc. 6th International Conference on Maritime Transport, Barcelona, Spain, 25–27 June 2014.
Vanem, E., Eide, M.S., Lepsøe, A., Gravir, G. & Skjong, R. 2008. Electronic Chart Display and Information Systems—navigational safety in maritime transportation. European Journal of Navigation 6(1): 2–10.
Vayou, M., Pedersen, J.T. & Katsoulakos, T. 2012. e-Freight Policy Recommendations. e-Freight report D6.5.
Ward, R. & Greendale, B. 2011. IHO S-100 The Universal Hydrographic Data Model. IHO information paper, January 2011.

*Maritime-Port Technology and Development – Ehlers et al. (Eds)*
*© 2015 Taylor & Francis Group, London, ISBN 978-1-138-02726-8*

# Author index